Systems Practice:
How to Act in a Climate-Change World

T0206960

Ray Ison

Systems Practice:
How to Act in a
Climate-Change World

Ray Ison
The Open University
Milton Keynes
United Kingdom

First published in 2010 by
Springer London

in association with
The Open University
Walton Hall, Milton Keynes
MK7 6AA
United Kingdom

This book forms part of the Open University course TU812 *Managing systemic change: inquiry, action and interaction*. Details of this and other Open University courses can be obtained from the Student Registration and Enquiry Service, The Open University, P.O. Box 197, Milton Keynes MK7 6BJ, United Kingdom (tel.: +44 (0)845 300 60 90, email: general-enquiries@open.ac.uk).

www.open.ac.uk

Whilst we have made every effort to obtain permission from copyright holders to use the material contained in this book, there have been occasions where we have been unable to locate those concerned. Should copyright holder wish to contact the Publisher, we will be happy to come to an arrangement at the first opportunity.

ISBN 978-1-84996-124-0   e-ISBN 978-1-84996-125-7
DOI 10.1007/978-1-84996-125-7
Springer London Dordrecht Heidelberg New York

Library of Congress Control Number: 2010926040

Springer is part of Springer Science+Business Media (www.springer.com)

Acknowledgements

Completion of this book would not have been possible without the support, intellectual contributions, encouragement and hard work of Pille Bunnell. I am immensely grateful.

A book is but a reification at one moment in time of the product of much joint activity (participation) and the realisation of this makes a mockery, somewhat, of the concept of 'author'. That said, a book is also an act of responsibility – in this case for what I say and do.

I acknowledge the seminal influences Humberto Maturana and David Russell have had on my thinking and doing in relation to the focus of this book by dedicating it to them.

My understandings and practices are immensely richer for the cooperative scholarship and joint projects I have enjoyed with Rosalind Armson (with whom I first developed the juggler isophor), Peter Ampt, Chris Blackmore, Kevin Collins, John Colvin, Alexandra di Stefano, Marion Helme, Chris High, Stephany Kersten, Bernard Hubert, Janice Jiggins, David McClintock, Francis Meynell, Dick Morris, Martin Reynolds, Pier Paolo Roggero, Niels Röling, Sandro Schlindwein, Patrick Steyaert, Drennan Watson and Bob Zimmer.

Some of the book is built around and extends material written for The Open University course 'Managing complexity: A systems approach'; I am grateful to members of that course team who provided a creative milieu for the development of my thinking. My membership of other Open University course teams, especially 'Environmental Decision Making: a systems approach', brought forth many new insights which I draw upon, including collaborative writing within those course teams. My fellow codirectors of the Systemic Development Institute, Richard Bawden, Bruce McKenzie and Roger Packham, have been a source of support and inspiration. I have gained much from an ongoing conversation with Peter Checkland and his writings that has now extended over 20 years.

The final form of the book has been helped by the creative development of figures by Pille Bunnell, Simon Kneebone, Phil Wallis and staff of the design studio at The Open University. Simon's cartoons (described as illustrations in the chapters) are a pleasure to include. Rosalind Armson, Simon Bell and Tim Haslett kindly helped by supplying copies of Figures. At The Open University, Robin Asby, Marilyn Ridsdale, Pat Shah, Monica Shelley and Gemma Byrne helped bring the task to completion. I am also grateful to Ben Iaquinto for his engagement with the text and work with the references. As ever Cathy and Nicky have, through their presence, helped to give the task meaning and relevance.

Preface

I invite the reader to engage with this book as a work in progress. My invitation arises because I have come to understand that systems practitioners benefit from appreciating that their practice is dynamic and in transition. It is never complete. Though this book will be printed, thus reifying my understandings at a moment in time, my own systems practice continues to develop.

The writing of this book has been carried out at a time of significant personal, organisational and societal upheaval and transformation – a situation that has been an emotional roller-coaster and a test of my own juggling praxis.[1] Included in these changes has been the reorganisation of academic groupings at The Open University which now means that Systems is part of a much larger grouping within a new faculty structure. Despite these changes a positive development has been the significant investment that The Open University (UK), or more specifically the new Faculty of Maths, Computing and Technology, has made in a new MSc programme in Systems Thinking in Practice. This book is a product, in part, of that investment. It is also this initiative that, to a large extent, has determined the conceptual boundaries within which I have written and composed.

As a result of the investment made by The Open University in the new MSc, four new books including this one have been produced. One is devoted to the individuals who are generally recognised as systems thinkers [2]. This work presents a biographical history of the field of systems thinking by examining the life and work of 30 of its major thinkers. It discusses each thinker's key contributions, the way this contribution was expressed in practice and the relationship between their life and ideas. This discussion is supported by an extract from the thinker's own writing, to give a flavour of their work and to give readers a sense of which thinkers are most relevant to their own interests.

---

[1]What I mean by juggling praxis will be explained in later chapters.

Another book is devoted to the main methodologies that have been developed by Systems scholars and are often deployed as part of systems practice [3]. In this book the five methodological approaches covered are:

1. System dynamics (SD) developed originally in the late 1950s by Jay Wright Forrester
2. Viable systems method (VSM) developed originally in the late 1960s by Stafford Beer
3. Strategic options development and analysis (SODA: with cognitive mapping) developed originally in the 1960s by Colin Eden
4. Soft systems methodology (SSM) developed originally in the 1970s by Peter Checkland
5. Critical systems heuristics (CSH) developed originally in the early 1980s by Werner Ulrich

These two books establish two of the boundary conditions for my own book. I do not dwell greatly on the pantheon of individuals who are recognised as some of the main systems thinkers, though I do ask the question: how did these people come to do what they did? Secondly, I do not spend time explicating some of the main systems methodologies though, through an explication of 'systemic intervention' (Chapter 12), I attempt to create a bridge to the material in Reynolds and Holwell [3].

This book will deal with a simple logic:

1. What are the situations where systems thinking helps?
2. What does it entail to think and act systemically?
3. How can practices be built that move from systemic understanding to action that is systemically desirable and culturally feasible?
4. How can situations be transformed for the better through systems practice?

The book is introduced in Part I against the backdrop of human induced climate change. This and other factors create a societal need, I argue, to move towards more systemic and adaptive governance regimes which incorporate systems practice. I consider how systems thinkers have chosen to characterise situations that they have encountered and where systems thinking offers both useful understandings and opportunities for situation improving change in Part II. I will argue that how we see a situation, and thus how it is engaged with, are critical first steps and that this is a dynamic heavily dependent on the history, and thus understandings, of the practitioner. For this reason I explore the history of the practitioner (as a biological and social person) and their unfolding relationship with complex situations. My inclusion of the practitioner as a human being means that, unlike many former Systems texts, this book is not methodology led. Particular understandings coming from second-order cybernetics, one of the many systems traditions, will inform aspects of the book.

The systems practitioner referred to in this book is anyone managing in situations of complexity and uncertainty – it is not a specialist role or that of a consultant or hired 'intervener'. Thus the book is structured so as to build a general model of systems practice. Because thinking and acting systemically is a form of practice, just as say playing a piano or being a nurse (nursing) are forms of practice, and because there are many ways of thinking and acting systemically, I first provide a conceptual model of *practice* in general. I follow this by focusing on systemic practice in particular and then invite you to use these conceptual models to look at your own practice as well as the practice of others.

In Part III, some of the main factors that constrain the uptake of systems practice are considered. Three very accessible approaches to systems practice, viz., systemic inquiry, systemic action research and systemic intervention are then introduced. I conclude in Part IV by looking critically at how systems practice is, or might be, valued at levels ranging from the personal to the societal. I do this in part by looking at the claims we can make about effectiveness and by exploring different forms of evaluation.

A third boundary condition for this book is established by the fourth book, a reader that Blackmore [1] has created concerned with social learning systems and communities of practice. Her focus is on practice in multi-stakeholder situations that call for collaborative or concerted action within groups. My primary focus is on the "practitioner... in context" as well as systemic inquiry, systemic action research and constraints to institutionalising systems practice. Whilst the boundary between the individual in a context and a group as a whole is harder to demarcate and articulate than the boundaries that have been created with the other books, I am fortunate to have collaborated with Chris Blackmore for many years. This has enabled us to negotiate what we believe to be an appropriate boundary. Further aspects of my own research appear in Blackmore's [1] reader.

When planning this book, I decided that I wished to expose the reader to a range of ways of doing systems thinking and practice, not just my own. As my preference was to enable others to 'speak' with their own voice, I began to imagine this book as a type of hybrid. Throughout the text, I will be introducing 'readings' of published material by authors I regard as systems practitioners. For this reason when you first pick up and scan the book you may find it somewhat different in structure to others. In organisation, the book is a hybrid between a book and a reader. The latter is usually a collection of previously published work, brought together for some purpose, such as presenting the seminal papers in a particular subject area or the collected works of a particular author (e.g. [1, 3]). Unlike a reader, this book also draws upon material from my own research, teaching and life experience as well as work of colleagues and other scholars.

Thus the book is a combination of new work and readings from published works. The criteria for selection of these readings were:

1. The reading gives a good systemic analysis of a complex situation – and in the process enables a systemic understanding to be gained (this does not have to be a case where the author claims to be providing a systemic analysis)
2. They exemplify a range of types of practice in general and systems practice in particular – which move from understanding a complex situation, to systemic analysis, intervention, inquiry and process management
3. The examples are not gender biased in selection and reflect different systems traditions on the part of the authors

4. The situations in which practice is conducted vary and thus exemplify different contexts in which Systems can be used
5. They describe a complex situation where it seems obvious that some form of systemic action could have made a difference
6. The papers are written in a style that is accessible to the reader

As you read the book you will come to appreciate, I hope, that this additional material is designed to introduce different perspectives on what systems practice is, or might be. My own metaphor for the way I see and use this material comes from my daughter's practices of braiding and adding coloured 'extensions' to her hair. The material, which I will call 'readings', although they are actually 'doings', is designed to develop your practice in 'braiding'. The text in parts II–IV will create opportunities for you to braid this additional material with my arguments and concepts as well as your own experience.

You will find that, in my writing, I have a habit of moving between different conceptual levels. One of the skills essential to systems thinking and practice is being able to move up and down levels of abstraction.[2] In writing this book I have forsaken the use of in-text Harvard style referencing to aid the flow of the narrative; all chapters conclude with a set of references. I also provide extensive footnotes which point to sources and evidence for my claims. These also provide pathways for further exploration of the points being made. I draw on a range of systems concepts in the book and attempt to explain them when they are introduced.

The book has at least one major weakness for which I feel the need to offer an apology. The absence of a critical engagement with the French, Italian and German, possibly Spanish and Brazilian literatures on systems thinking and practice is an indictment of the Anglo-Saxon systems community in general and me in particular. Missing from these pages are reflections on the contributions of the likes of Edgar Morin, Jean-Louis Le Moigne and Frederic Vester. Edgar Morin, who was very familiar with the revolution in genetics initiated by the discovery of DNA, contributed to cybernetics, information theory and a theory of systems.[3] Le Moigne is known for his encyclopaedic work on constructivist epistemology and his contribution to a General Systems theory.[4] Frederic Vester was known as a 'pioneer of networked thinking, a combination of cybernetic and systemic ideas with complexity issues.'[5] There are also many others.

The gulf between systems practitioners in the different linguistic and cultural communities is unfortunately mirrored in the systems and cybernetics field itself. Whilst many are aware of this and working to rebuild relational capital between

---

[2]Key systems concepts that depict this are sub-system, system, supra-system or 'how', 'what', 'why' – the latter three are observer dependent and your 'what' could be my 'how' – which is one reason why it is so easy to talk across each other in meetings unless the level at which something is being discussed is clarified.

[3]See http://en.wikipedia.org/wiki/Edgar_Morin; Accessed 1 October 2009.

[4]See http://en.wikipedia.org/wiki/Jean-Louis_Le_Moigne; Accessed 1 October 2009.

[5]See http://en.wikipedia.org/wiki/Frederic_Vester; Accessed 1 October 2009.

disparate groups there has not been, to my knowledge, a recent synthesis that sets out intellectually and practically how a reinvigorated cyber-systemic praxis field might emerge at this time of significant societal need. I hope this work will in some way contribute to facilitating such a transition.

References

1. Blackmore, C.P. (Ed.). (2010) Social Learning Systems and Communities of Practice. Springer: London.
2. Ramage, M. and Shipp, K. (2009) Systems Thinkers. Springer: London.
3. Reynolds, M. and Holwell, S eds (2010) Systems Approaches to Managing Change. Springer: London.

Contents

Part I
Thinking and Acting Differently

Chapter 1
Introduction and Rationale

1.1 Managing in a Climate That We Are Changing

This book is about how we can take responsibility for the world we are creating by paying much more attention to how we think and act. If we look around us, it is easy to see that we are not making a very good job of it at the moment. When atomic bombs were invented, human beings, for the first time, had to face the prospect of producing the circumstances for their own destruction. So far we have survived the atomic threat! Now we have human-induced climate change with challenges that, for many, are still beyond imagination. In the face of such complexity and uncertainty many will be tempted to give up or to feel that nothing can be done. I admit to not being overly optimistic myself. I certainly do not have a magic wand to wave. What I do have, however, is a strong conviction that thinking and acting differently will have to be at the core of our strategies of action.

The acceptance that humans are changing the climate of the earth is the most compelling, amongst a long litany of reasons, as to why we have to change our ways of thinking and acting.[1] Few now question that we have to be capable of adapting quickly as new and uncertain circumstances emerge and that this capability will need to exist at the personal, group, community, regional, national and international levels all at the same time. The phenomenon of human-induced climate change is new to human history and it is accompanied by 'peak oil', rising population and consumerism, changing demographics and over exploitation of the natural world. In the face of such complexity and uncertainty it is tempting to say it is all too hard! It certainly won't be easy.

At this important historical moment what can we learn from our past? When we look around us what different ways of thinking and acting could be helpful? This book argues that development of our capabilities to think and act systemically is an

---

[1] I will make the case as the book develops that changing our thinking and acting is not just what we as individuals need to do – it is also what our ancestors have done which shape our current institutions and thus so much of what we take for granted in going about our daily lives.

R. Ison, *Systems Practice: How to Act in a Climate-Change World*,
DOI 10.1007/978-1-84996-125-7_1, © The Open University 2010.
Published in Association with Springer-Verlag London Limited

urgent priority.[2] Systems thinking and practice are not new but individually and socially our capability to do it is very limited. Unfortunately these are not abilities developed universally through schooling or at University. In the latter, the rise of specialised subject matter disciplines, the focus on science and technology at the expense of praxis (theory informed practical action) and reductionist research approaches have driven the intellectual and practical field of Systems,[3] a form of trans-disciplinary or 'meta' thinking, from the curriculum.

In calling for a new politics to begin to deal with climate change, Anthony Giddens [6] argues that 'as far as possible we have to prepare beforehand – adaptation must be proactive' [p. 13]. As compelling as his arguments are, he has very little to say about the forms of practice or praxis that will be required. I will argue that re-engaging with and revitalising systems thinking and practice is one of the most significant opportunities we have.[4] One of many reasons for this is because systems thinkers in the past have recognised that there are particular situations that we confront that only appear amenable to change and improvement through systems thinking and practice – these situations have been described as 'messes', 'wicked problems' or issues of the real-life swamp. I will argue that it makes sense to see climate change as part of a lineage of understanding these types of situations in particular ways. The good news is that we have some experience of how to use systems thinking and practice to engage with and change such situations for the better. The bad news is that these capabilities are not widespread, often they are not done very well and many organisations set up rules and practices that get in the way of thinking and acting systemically.

Change of course starts at home, with each of us but only if the circumstances are conducive, amongst which includes knowing what change for the better might look like. In my experience it is not easy to think and act differently. How we think and act is patterned into the very fabric of our existence from birth. It is affected by and sustained by our physiology, particularly our underlying emotions, by the structures of our language, by our practice of reifying explanations (particular ways of thinking) in rules, procedures, techniques and objects, by our culture and our social

---

[2] There are two adjectives derived from the word 'system', i.e. systemic, pertaining to wholes, though not in the sense that wholes are pre-given, but in the sense of a systemic chemical that has the capacity, through a network of interactions, to affect a whole organism and 'systematic', linear or sequential thinking and acting. The Systems approaches I am concerned with encompass both.

[3] Throughout the text I use the capitalised 'Systems' to cover the broad area of scholarship and practice that could be also described as the 'systems field' or the many 'systems approaches'; others have described 'Systems' as a trans-disciplinary meta-subject' but in some contexts it makes more sense to see Systems as a discipline in its own right or as part of interdisciplinary practices [9, 10].

[4] In making this claim I am not Utopian in outlook – there are many other priorities as well – and using systems approaches will not deliver 'utopian solutions' but they can increase our capacity to act effectively.

relations, all of them as they change over time.[5] How technology functions in our society is an important consideration as well. The result is a hugely complex web, a web of existence, in which we are immersed and of which we are only partly aware.

1.2   What Do We Do When We Do What We Do?

On the bright side it is possible to become more aware of the nature of this web. With awareness, new understandings are possible and from these can flow new practices.[6] One way to raise awareness is to ask new or different questions. The first question I invite you to explore with me in this book is:

What is it that we do when we do what we do?[7]

A question like this is not a typical question. Too often we inhabit a taken-for-granted world where our ways of doing things are not questioned. Questions like this that invite critical reflection on our circumstances are not common. Answering this sort of question is also not easy because we are not used to doing the thinking needed to supply an answer. To answer questions like this requires us to take a double look – to look at what we do when we do the original doing and to look at our looking at what we do! By the end of this book I hope you will be much more familiar with what this type of question entails.[8]

Here is an example of what the question means to me. As an academic one of the common practices I have had to learn is how to mark exam papers. This usually involves allocating a mark for an answer, perhaps a mark out of ten, against some criteria that I have specified or have in my head. This practice is widespread not only in schools and universities, but can be used in judging research bids, ranking applicants for a job, ranking achievements or evaluating progress in meeting targets. In fact the practice of quantifying a process is so widespread that we tend to take it for granted. But if I reflect on this particular practice (my doing ... or others who do it) then I can become aware of a range of issues which cause me concern. These include:

- My awareness that practice at the Open University, built around distance teaching, is very different to most other universities because we have to develop marking schemes in advance that can be used by other staff to do the marking. In my

---

[5] Reifying is the process of converting a concept mentally into a thing. The process can have the unintended consequence of giving a concept a seemingly material existence, almost as if it was there all of the time, rather than being 'invented' by someone at some historical moment; I expand on this in Chapter 6.

[6] Language constrains me here – I do not imply a linear sequence – awareness, understanding and practice are all sites for transformation and change. We know this from experience – doing something, like exploring your new mobile phone, a practice, can result in new understandings.

[7] I am grateful to Humberto Maturana for introducing me to this question and for offering the explanation of how human beings live in the braiding of language and emotion.

[8] I will refer to this type of question as a second-order question.

experience most academics at other universities do not develop formal marking schemes but use their own judgement as they mark

- An unintended consequence of not having a marking scheme can be that it becomes easier for students to score high marks in quantitative subjects or where there are clear right and wrong answers than in more qualitative subjects based on essays, mainly because in the latter case academics do not like to award marks over the full range 0–100, i.e. they do not much like giving marks over 80% or 90%
- An unintended consequence of having a marking scheme can be that the creative coupling of the answer to a question in context specific ways may go unrewarded or even unrecognised[9]
- If I think really deeply about marking then I realise that I am giving a quantitative performance measure to someone else's learning... or am I? Perhaps I am giving them a reward for mastering a particular technique, such as answering exam papers in a particular way? And how do I understand learning?[10]
- If I am honest with myself I realise that no matter how hard I try I find it hard to be generous when I find it difficult to understand the handwriting
- If I explore further I might realise that the practice of awarding quantitative marks to student work began in the 1790s – before that it was not imaginable that student learning would be treated in such a way (the 'normal' methods then involved discussion, presentation, discourse and professional judgement). Today quantification seems so much part of our daily life we do not question it. Yet prior to 1792, when it was first carried out at the University of Cambridge, this was an unknown practice. Interestingly it was subsequently fostered mainly by military colleges [7, 12][11,12]

---

[9] At the Open University we attempt to address this by developing marking schemes that operate at several conceptual levels and leave space for context sensitive judgement, but experience shows that some tutors are better at this than others.

[10] In April 2008 a group of 34 British Academics under the banner of 'The Weston Manor Group' produced a manifesto calling for major changes in how Universities assess their students. They argued the need to reorientate current assessment fashions characterised by an 'obsession with marks and grades to one which puts more emphasis on developing effectiveness for learning, rather than assessment of what sometimes passes as learning' (see http://www.timeshighereducation.co.uk/story.asp?storyCode=401576&sectioncode=26 accessed 18th June 2008).

[11] Postman [12, p. 13] following Hoskins [5, pp. 135–146], attributes this 'innovation' to William Farish, a professor of Engineering at Cambridge, and claims that this was a major step in 'constructing a mathematical concept of reality'. He makes the further point, valid to my argument here, that 'if a number can be given to the quality of a thought, then a number can be given to the qualities of mercy, love, hate, beauty, creativity, intelligence, even sanity itself'.

[12] It is possible to successfully design and run 'education systems' which do not rely on quantification as part of an 'assessment system' – I have been fortunate to be part of doing this – see Bawden [2]. I would argue that one of the unintended consequences of assessment systems that primarily rely on 'quantification of learning' is that we have collectively become less skilled in processes of deliberation, which are so important to an effective democracy. But this argument is not one I wish to pursue here.

Fig. 1.1 An application form is an example of a widespread social technology – not all are the same but all have several elements in common and the 'forms' mediate similar social practices

I call practices such as grading and examining, which become incorporated into a culture, social technologies. Social technologies are all around us. Sometimes they are beneficial and facilitate effective practices like creating road rules that minimise accidents. Sometimes they incorporate understandings that, experience shows, were inappropriate in the first place or that, on reflection, are no longer valid. So, based on my experience and reflection on 'marking', it is legitimate to ask, or inquire further, as to whether quantification is really in the best interests of student learning?[13]

Writing about UK public sector reform John Seddon gives another example. He describes the 'inspection industry' which 'has become an instrument of the regime [New Labour], a political instrument. Like ministers, it has lost focus on what works. Instead inspection is concerned with compliance. It is now an integral part of dysfunction' [14, p. 56]. If I unpack Seddon's claims I come to see that 'inspection' and the role of 'inspectors' are social technologies and that what is good 'inspection' or a good 'inspector' is open to intellectual and political fashion.[14]

Social technologies are distinct from artefacts such as a hammer or a computer considered in isolation, which is what we usually think about when technology is mentioned. Social technologies are characterised by a set of relationships in which the technology plays a mediating role just as the document template does in Fig. 1.1. In my terms management, or decision making, can be a social technology when it is made up of procedures and rules designed to standardise behaviour – or in other words, sets of techniques used routinely without awareness of the origins and implications of the use of such techniques, the role of the practitioner and the need for contextual understanding about the situation. My examples of 'marking' and 'inspecting' may seem, at first, a far cry from responding to climate change.

---

[13] I return to the role of social technologies in Chapter 6.

[14] In this case the process of 'inspecting' has become reified at some historical moment into a professional role called 'inspector'. The inspector role brings with it historical connotations about 'inspecting' as well as day-to-day political and intellectual considerations that reshape what it is to be an 'inspector'. Etymologically the process of inspection means to 'examine closely' derived from 'en' (in, within, into) and 'spek' (to see or regard) [15].

It is my contention however that the profound and effective responses to major issues will arise when we become more systemically aware of the 'what and why' in the everyday. Marking and inspecting are seemingly benign practices that touch on the lives of a significant proportion of the world's population. But if we have, in some ways, got these 'wrong' think about the possible implications for many of our other practices! I say more about this in Part II.[15]

1.3 Living in Language

A second question I address is:

What are the consequences of living in language?

Neil Postman made the point that a sentence acts very much like a machine and that a language enables or constrains our thinking in particular ways. He points out that neither the form of a question or its content is neutral. The form of a question may ease our way or pose obstacles. Or, when even slightly altered, it may generate antithetical answers, as in the case of the two priests who, being unsure if it was permissible to eat and pray at the same time, wrote to the Pope for a definitive answer. One priest phrased the question: 'Is it permissible to eat while praying?' and was told it was not, since prayer should be the focus of one's whole attention; the other priest asked if it was permissible to pray while eating and was told that it is, since it is always appropriate to pray [12, pp. 125–126]. The form of a question may even block us from seeing opportunities that become visible through a different question.

A consequence of living in language is that the social and political dynamics of explanations becomes very important – as a species we appear to live with a craving for explanations. An explanation does not exist in and of itself – it is part of a social dynamic between an explainer, an explanation (the form of an explanation) and a listener or reader (Fig. 1.2). As I outline in Part II, accepting a new or different explanation changes who we are; the accepting and rejecting of explanations is a key dynamic of being human. My invitation in this book is to explore what it is like to develop systemic explanations and actions in complex and uncertain situations. I will argue that systems thinking and practice are particular ways of living in language – a systems language – that is unfortunately not greatly valued nor well understood or practised.[16]

Of course all explanations have a history and it is possible to explore this history. In my own approach to systems practice I place a lot of emphasis on attempts to

---

[15] In Part II, I will explain how my use of the term 'social technologies' is very close to what some economists, particularly institutional economists, refer to as 'institutions'.

[16] It can be argued that this in part rests on the contemporary focus on efficiency rather than effectiveness – achieving the latter is more difficult.

Fig. 1.2 The dynamic between an explainer, an explanation and a listener (or reader)

become more aware of the traditions of understanding out of which we think and act. In Chapter 7, I will describe how this can be done in a practical way by exploring metaphors and their entailments as part of a process of systemic inquiry. Recent scholarship in the newish academic disciplines in the history and sociology of science and technology demonstrate the importance of understanding the history of ideas, practices and explanations.[17]

1.4 A Failure to Institutionalise[18]

One of my main arguments is that we have failed to institutionalise systems thinking in our society in general and our organisational practices in particular, and that this has been, to a large extent a failure of knowing what systems practice is, valuing what it can deliver and knowing how to do it! So one of the main aims of this book is to give you, the reader, ideas about how to do it, i.e. to think and act systemically. I will also try to make apparent the sorts of benefits doing Systems can provide in a climate-change world. My ambition is that as you read you will engage in an active inquiry into your own ways of thinking and acting, or put another way, that you will transform your situation through changes in understanding and practice, where neither understanding or practice are prime (Fig. 1.3).

---

[17] Fortunately, explanations are open to historical inquiry and reinterpretation but in my view we need to do much more so as to break out of widespread traps in our thinking, traps that make it difficult for us to respond to complex situations such as climate change.

[18] By institutionalise I mean the failure to create systems practice as an 'institution', a norm or 'rule of the game'. I say more about institutions in Chapter 6.

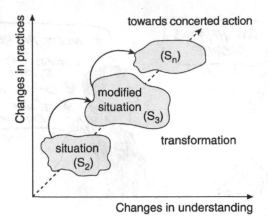

Fig. 1.3 Situations characterised by complexity, uncertainty interdependencies, multiple stake-holders and thus perspectives can be transformed (as indicated by the arrows) through concerted action by stakeholders who build their stakeholding in the process. This leads to changed understandings (knowledge in action) and practices (S = situation; S1 [not shown] is the history of the current situation (S2) which through changes in understanding and practices of stakeholders is transformed to S3 etc. [16])

The transformation I allude to in Fig. 1.3 is towards more effective concerted action by stakeholders in complex and uncertain situations.[19] I could describe this as cooperative, collaborative or collective action, but I prefer 'concerted' (understood as done or performed together or in cooperation) as it evokes for me the idea that, at scales ranging from the local to the global, we, as a species, have to develop more effective performances. My organising metaphor here is that of an orchestra or jazz ensemble. This metaphor enables me to appreciate that what constitutes an effective performance is an emergent property between the actions of an orchestra (i.e. a group of people with different histories, understandings, emotions, instruments who come together and work hard at some common purpose) and an audience (i.e. situation), and that this unfolds in context sensitive ways.[20] A jazz ensemble reveals aspects of improvisation rather than performing a set piece!

---

[19] It is important to my arguments that Fig. 1.3 is read carefully; as a heuristic device I have found it very useful in many face-to-face presentations – but it is useful mainly because of what it enables me to say and the questions it triggers. In Chapter 2, I will say more about how I understand particular situations and use the term 'situation'. Figure 1.3 is also built on a theoretical perspective that sees learning as social rather than individual and learning processes as embedded in the dotted line (which is rarely unidirectional in practice). Another key aspect of the transformation process, but not depicted in Fig. 1.3, is that changes in social relations usually accompany a change in practice or a change in understanding, i.e. in effecting situation improving action one is rarely alone.

[20] All metaphors reveal and conceal – what my use of this metaphor conceals is questions about conductor, score and music composition. I can side-step this partially by deferring to jazz and the improvisation that is part of jazz practice. In my reading of this metaphor I see many different performances operating at different scales, playing different music! I also think that in terms of climate change adaptation we need to write the music as well as perform it!

Illustration 1.1

Of course all metaphors can be interpreted in many ways and this one is no different. If I make clear what I mean by 'action that is systemically desirable', then I can take responsibility for my own normative position, what I would seek in a good performance. My position is that it is not the future of the Earth that is threatened by human actions but our relationship with the Earth, with other species and with other human beings, including future generations. So, for me, an effective performance arises from actions that enhance and sustain the quality of these relationships. At this stage I do not wish to be more specific than this; I will say more about this in Part II.

1.5 Managing in a Co-evolutionary World[21]

Some readers may by this point be struggling to locate themselves in this book. If you are a health professional, a civil servant, an engineer or from a myriad other contexts in which systems thinking and practice can be applied you may not yet have encountered anecdotes or language that resonate with you? I see this as both a challenge and an opportunity... for reader and author alike. This is not a book designed for one specific professional sector – it transcends individual professions. My own experience is that systems thinking and practice can become a skill that is relevant in all aspects of life, personal and professional, individual and group. My ambition and motivation is more than the utilitarian, however. Our circumstances have become such that more of the same, a business as usual approach, even if done better is no longer good enough. We face an unparalleled situation, one which requires responses, small and large, in all aspects of our daily lives.

---

[21] Material in this section comes mainly from Collins & Ison [4].

Illustration 1.2

Atomic bombs and human-induced climate change do not mean the end of the world, but they could ultimately mean the end of the world as we have come to know it, or a world in which we humans are a part. The situation is as serious as that! In the discourses that have built up around the acceptance that humans are actually affecting the climate of the earth, two terms have come to prominence. These are climate change mitigation and climate change adaptation. The former is concerned with acting now to stop climate change, or to minimise it, as it is not really stoppable. The latter concerns how we go on living in a world affected by climate change.

The word 'adaptation' has always been important in scientific fields associated with evolution, ecology and environmental change. The advent of anthropogenic climate change has again positioned 'adaptation' as a key term and concept. Etymologically 'adaptation' means 'fitted or suited' and to adapt is 'to fit' or 'make suitable'. At the level of metaphor two possible conceptions arise from these meanings which have significant practical and policy implications. The first metaphor, and the most widespread understanding, is that of 'adaptation as fitting into'. In this metaphor something (predetermined) is fitted into a situation (also predetermined or knowable in advance) to which it is fit-able or suited, like when doing a jigsaw [2].[22]

The other metaphor is that of 'adaptation as a good pair of shoes'. This metaphor requires a little more explication. What makes a good pair of shoes at a given moment? Well, usually because you have worn them in, they are comfortable, flexible etc. But these same shoes may not be a good pair of shoes if you were to put them in a cupboard for a year before wearing them again. Why? Because your feet will have changed! Within this metaphor a good pair of shoes arises from the recurrent interactions between shoes and feet – this is an example of co-evolution. This has also been described as the structural coupling of a system to its environment over

---

[22] It can be argued that this is a common understanding that informs practices like plant breeding and agronomy.

time [8, 11]. For those who understand the dynamics of co-evolution, and are not so interested in shoes, then the metaphor can become 'adaptation as co-evolution' [4].

Rather than seeing adaptation as one way, co-evolution is different – the idea of a separate environment is set aside in favour of processes of mutual interaction which in human social systems can be seen as processes of learning and development [5, p. 121; 6]. Despite our capabilities we seem to have room for improvement in the realms of learning and development. If we are to manage in a climate-changing world that is essentially unknowable in advance, and where we need to take more responsibility for the systemic effects we as a species have, then adaptation as co-evolution seems to me the only way forward. This requires an effective form of praxis – a systems practice.[23]

An increasing number of policy makers recognise that, in the face of climate change, a global water crisis and the like, a business as usual approach to governance is no longer tenable [1, 3, 6, 17].[24] These same commentators recognise that systems thinking and practice are key to delivering effective policy and practice that address long-term complex and intractable issues. Noting the long-term nature of many of Australia's key policy challenges the Australian Prime Minister, Kevin Rudd, argued the need to "invest in a greater strategic policy capability" by which he meant "a greater capacity to see emerging challenges and opportunities – and to see them not just from the perspective of government, but also from the perspective of all parts of the community" and delivering "genuine joined-up government" [13]. But, as experience in Britain demonstrates, joined-up government is easy to talk about but much harder to enact [14].

The material that follows explicates an 'ideal type' model of systems practice that, with investment, has the potential to deliver the missing praxis elements of joined-up, or systemic and adaptive, governance (Part II). But effective practice is always contextual and at the moment there are various constraints to institutionalising effective systems practice. These constraints as well as opportunities are explored in Part III. Our governance arrangements call out for transformation but such a transformation has to be built on fundamental shifts in thinking and practice as well as what we choose to value (Part IV). In our current context, systems thinking and practice is dangerous: dangerous because it may change who you are and how you act. I can think of no better time than now to live dangerously!

References

1. APSC (Australian Public Service Commission) (2007) Tackling Wicked Problems. A public policy perspective. Australian Government: Canberra.
2. Bawden, R.J (1989) Assessing the Capable Agriculturalist. Assessment & Evaluation in Higher Education 13,151–162.

---

[23] Praxis: the means by which a theory or philosophy becomes a practical social action.

[24] By governance, I mean the ways and means by which social groups 'steer' themselves in relation to feedback processes as they chart an uncertain future (I take this up again in Chapter 9).

3. Chapman, Jake (2003) System Failure: Why governments must learn to think differently. Demos: London.
4. Collins, K.B. and Ison, R.L. (2009) Living with environmental change: adaptation as social learning (Editorial) Environmental Policy & Governance 19, 351–7.
5. Fairtlough, G. (2007) The Three Ways of Getting Things Done. Hierarchy, Heterarchy & Responsible Autonomy in Organizations. Triarchy Press: Axminster.
6. Giddens, A. (2009) The Politics of Climate Change. Polity: London.
7. Hoskins, K. (1979) The examination, disciplinary power and rational schooling. History of Education VIII(2), 135–46.
8. Ison R.L., Röling N., Watson D. (2007) Challenges to science and society in the sustainable management and use of water: investigating the role of social learning. Environmental Science & Policy 10, 499–511.
9. Ison, R.L. (2008) Systems thinking and practice for action research. In P. Reason and H. Bradbury (Eds.). The Sage Handbook of Action Research Participative Inquiry and Practice (2nd edn) (pp. 139–158). Sage: London.
10. Maiteny, P.T. and Ison, R.L. (2000) Appreciating systems: critical reflections on the changing nature of systems as a discipline in a systems learning society. Systems Practice & Action Research 16(4), 559–586.
11. Maturana, H. (2007) Systemic versus genetic determination. Constructivist Foundations 3(1), 21–26.
12. Postman, N. (1993) Technopoly. The Surrender of Culture to Technology. Vintage: New York.
13. Rudd, K. (2008) Prime Minister's Address to Heads of Agencies and Members of Senior Executive Service, Great Hall, Parliament House: Canberra 30 April 2008
14. Seddon, John (2008) Systems Thinking in the Public Sector: the failure of the reform regime and a manifesto for a better way. Triarchy Press: Axminster.
15. Shipley, J.T. (1984) The Origins of English Words: A Discursive Dictionary of Indo European Roots. The John Hopkins University Press: Baltimore/London.
16. SLIM (Social Learning for the Integrated Management of Water). (2004) SLIM Framework: Social Learning as a Policy Approach for Sustainable Use of Water. Available at http://slim. open.ac.uk. Accessed 7 August 2009.
17. Syme, G., Nancarrow, B., Stephens, M., Green, M., and Johnston, C. (2006) Volunteerism, Democracy, Administration and The Evolution of Future Landscapes: A Land & Water Australia Project. CSIRO Land and Water project team.

Part II
Systems Practice as Juggling

Chapter 2
Introducing Systems Practice

2.1 Systems Thinking or Thinking Systemically

I would like to believe that the ability to think and act systemically is more widespread than seems apparent. It is sometimes surprising for students of Open University Systems courses to discover that they are already systems thinkers. Many students describe an "Aha…" moment in the early phases of their study when they realise that their own way of thinking has a name and is a valid way of engaging with the world. Others take longer before the "aha-moment" arrives. For some it never materialises. The limited research on educating the systems thinker suggests that everyone can develop this ability but for some it is a demanding journey. For example, Helm Steirlin [35, p. 164], philosopher, medical practitioner and psycho-analyst, after a lifetime of systems practice said: 'systemic thinking can only be learned through one's work; it cannot be instilled into others; it needs time to gather experience and to make mistakes'. Following Steirlin, I invite you to consider how this book might help you to create the circumstances where it is safe to fail as you attempt to develop your systems thinking and practice. I phrase it like this as it is in circumstances where it is safe to fail that learning is maximised. Engaging with Systems is perhaps like learning a new language – I could refer to it as learning 'systems talking', where 'talking' involves thinking and doing, i.e. practice. It is the sort of learning that can challenge our sense of identity. It is as if 'systems talk' is 'talk that undermines the boundaries between our categories of things in the world, [and thus] undermines 'us', the stability of the kinds of beings we take ourselves to be.'[1] On this note, 'systems talk' can, depending on your perspective, be transforming or dangerous because of its ability to transform who we are.

If systems thinking and practice were more widespread then my task would be simpler; it would require awareness raising rather than the more difficult task of inviting you to consider how you currently think and act in relation to complex situations, something that can be challenging, resisted and in want of justification. Those who do not think systemically usually require explanations of what 'it' is and

---

[1] Here I am following positions espoused by John Shotter and Mary Douglass – see Shotter [32, p. 4].

R. Ison, *Systems Practice: How to Act in a Climate-Change World*,
DOI 10.1007/978-1-84996-125-7_2, © The Open University 2010.
Published in Association with Springer-Verlag London Limited

justification or evidence that 'it works' or that there is a 'value proposition' for engaging with it. There is also a tendency to require explanations of effectiveness in causal terms of the form: 'using systems thinking can cause X to happen' i.e., using a framework of linear causality in which a systemic view is lost because one cannot understand circularity by making it linear: after-all a matrix is a complex network of nested and intersecting circular relationships.[2]

Although a challenge, this book is designed for readers of all backgrounds when it comes to systems thinking and practice. This challenge can be understood as a set of interconnected factors that have to be addressed if one is to provide a reading experience, or build a curriculum capable of 'educating' a systems practitioner. It is even more demanding in a distance teaching setting (as in The Open University UK) because my experience, and that of many of my colleagues, is that systems thinking and practice is best learned experientially.[3]

If systems thinking and practice is best learnt experientially then that creates some design difficulties for the author of a book. The simplest challenge is to create the circumstances whereby those who already think systemically can be affirmed in what it is that they do, whether they are aware of it or not. One way to do this is to develop a language, including conceptual and methodological insights, to better understand the nature of their systems thinking. For those who fall into this category it should also be possible to become better at thinking and acting systemically and to be better able to make choices of when to use or not use this type of thinking. Regardless of where you would position yourself in your systems thinking capabilities it makes sense to invite you to engage actively with the ideas. This is a strategy I pursue. I do so by moving the focus of reading between different authors – my own text and that of other systems thinkers and practitioners. My motivation for doing this is to encourage you to look at what, how and why some systems practitioners do what they do and to compare this to what you currently do, or could do, in similar situations.[4]

If you have made it this far with your reading but do not yet regard yourself as a systems thinker then I would invite you to consider this question:

> What is it that you would have to experience that created the circumstances where you could experiment with thinking and acting systemically?

---

[2] Much can be said on this point: My prejudice is that those who demand answers to these questions often do not make the same demands on what they currently do…or, to rephrase it slightly, the systemic effectiveness of what it is that they do. I would further claim that it is not possible to provide arguments for effectiveness in one way of thinking in terms of another way of thinking. Thus, to impose inappropriate evaluative frameworks is to risk paradigm incommensurability or to conflate explanations across different domains.

[3] Even when one cannot in a course DO the real world application, one can do experiential activities that are isophoric. I will explain later what an isophor is and how it works.

[4] My approach is limited by the format and structure of a book and the act of reading a text – your systems thinking and practice is something you have to 'live' i.e., do. What I write, and my references to other texts, needs to be understood as an invitation to experiment with your ways of doing – it is not a prescription or a demand to do as I say!

You may notice that this is a strange question, a bit like asking "what would you have to experience that created the circumstances in which you could experiment with falling in love?" Thus I cannot answer this question for you so I invite you to return to it at the end of the book by which time the experience that answers the question may have arisen. To begin I offer two pointers based on my own experience: (i) abandon certainty, or to phrase it another way, acknowledge the certainty of uncertainty and (ii) be open to your circumstances.[5] Both of these claims concern attitudes or predispositions, which if adopted realise a particular emotional dynamic in which *another* arises as a legitimate other.[6] By *another* I mean other people with different experiences, cultures, explanations, other species as well as the biophysical and inanimate world.[7]

2.2 Systems Thinking as a Social Dynamic

At this point let me say that I have no intention of defining what systems thinking or practice is, or is not. In my experience definitions are constraining because (i) they are abstractions and thus a limited one dimensional snapshot of a complex dynamic and (ii) we do not appreciate how definitions blind us to what we do when we employ a definition. Instead I claim that systems thinking and practice arise as a particular dynamic in social relations as part of everyday life. So how might someone recognise systems thinking and practice arising as a social dynamic? The following are the most common:

1. When someone experiences what you say or do and claims that this is thinking or acting systemically[8]

---

[5]When one is open to one's circumstances (surrounding conditions) there are generative or innovative possibilities.

[6]Among human beings this is best demonstrated when a conversation starts in which mutual engagement and exploration happens. My perspective is captured in part by Benjamin Whorf who said: 'it is not sufficiently realized that the ideal of world-wide fraternity and cooperation fails if it does not include ability to adjust intellectually as well as emotionally to our brethren in other countries' [6, p. 21]. To this I add future generations, other species and the inanimate, or biophysical, world.

[7]In making this claim it is important to note that I am not, in the process of legitimising others, granting them legitimacy. The "arise as" means that we accept them to be already legitimate at the moment we become aware of them, it or circumstance. Legitimate does not mean you like or condone, its an acceptance of what appears as present as being what it is, without having to account for itself or justify itself to you.

[8]As I will outline later I understand experience as arising in a distinction we are able to make in relation to ourselves – thus without a distinction no experience arises. To experience a systems practitioner I would claim involves being able to distinguish a manner of acting (or living, or being) in a situation that we choose to describe as 'systemic'. This is most apparent when we experience someone making a connection with lineages of ways of thinking and acting systemically in which congruence between what is said and what is done emerges. My claim does not preclude people acting systemically even though they may not distinguish it as such.

2. When you engage in some form of personal reflection and make the claim for yourself that you were thinking and acting systemically
3. When your reflection is more formal as in writing a paper, report or book so that others who read this may agree or disagree to claims about acting systemically

The key aspect of these social dynamics is: Would you, or someone else, agree that you are doing systems thinking or practice?[9] I suggest that the key to agreement in all of these dynamics is the nature and extent of the connection you and others in the social dynamic make with the history of systems thinking and practice, including the main concepts that have been developed and continue to be used (Table 2.1). In practical terms systems practice can arise when we reflect on our own actions and make personal claims in relation to a history of systems thinking (a form of purposeful behaviour) or when others observe actions that they would explain in reference to the history of systems thinking (a form of purposive behaviour).[10,11] From this perspective what is accepted (or not accepted) as systems practice arises in social relations as part of the praxis of daily living.

2.3 Exemplifying Systems Thinking as a Social Dynamic

Some may see my failure to define systems thinking and practice as an abrogation of responsibility or an intellectual 'cop out'. It certainly goes against the grain of most academic practice in which the mainstream perspective seems to be that definitional clarity promotes both operational and conceptual certainty (a position that contradicts my point about the need to acknowledge the certainty of uncertainty!).[12] For this reason I want to introduce my first Reading to, hopefully, make my perspective more apparent. Please take some time now to read 'Dyslexic management

---

[9] Those interested in this question and its framing may find similarities with Wittgenstein when he said: 'It is what human beings say that is true and false; and they agree in the language they use. That is not an agreement in opinions but in form of life' [40].

[10] Two forms of behaviour in relation to purpose have also been distinguished. One is purposeful behaviour, which can be described as behaviour that is willed – there is thus some sense of voluntary action. The other is purposive behaviour – behaviour to which an observer can attribute purpose.

[11] There is another dimension which I will address subsequently – that is the extent to which one experiences in one's own actions, or those of others, a congruence or coherence between what is espoused and what is enacted. I will relate this to the notion of authenticity which as a word has roots in 'self – doing' and 'accomplishment' which I argue can be seen as related to praxis (theory informed action).

[12] John Shotter [32] has similar concerns when he poses the question (p. 19): 'why do we feel that our language works primarily by us using it accurately to represent and refer to things and states of affairs in the circumstances surrounding us, rather than by using it to influence each other's and our own behaviour?'

Table 2.1 Explanations of some generalised systems concepts likely to be experienced when encountering a system practitioner (Adapted from [5, 25, 39])

Concept	Explanation
Boundary	The borders of the system, determined by the observer(s), which define where control action can be taken: a particular area of responsibility to achieve system purposes
Communication	1. Communication is understood by some as a simple feedback process (as in a heater with thermostat) involving information but this should not be confused with human communication, which has a biological basis;
	2. From a theory of cognition which encompasses language, emotion, perception and behaviour communication amongst human beings gives rise to new properties in the communicating partners who each have different experiential histories
Connectivity	The relationships between components or elements (including sub-systems) within a system based on factors such as influence and logical dependence
Difficulty	A situation considered as a bounded and well defined problem where it is assumed that it is clear who is involved and what would constitute a solution within a given time frame
Emergent properties	Properties which arise or come into being at a particular level of organisation and which are not possessed by constituent sub-systems. These properties emerge from the operational or relational dynamics between the elements or subsystems that comprise a system
Environment	That which is outside the system boundary and which is coupled with, or affects and is affected by the behaviour of the system; alternatively the 'context' for a system of interest
Feedback	A form of interconnection, present in a wide range of systems. Circularity is inherent where the result of a process is taken as an input to the process so that the process is modified. Feedback may be negative (compensatory or balancing) or positive (exaggerating or reinforcing)
Hierarchy	Layered structure; the location, or embedding, of a particular system within a continuum of levels of organisation. This means that any system is at the same time a sub-system of some wider system and is itself a wider system to its sub-systems
Measure of performance	The criteria against which a system of interest formulated by an observer is judged to have achieved its purpose. Data collected according to measures of performance are used to modify the interactions within the system
Mess	A mess is a set of conditions that produces dissatisfaction. It can be conceptualised as a system of apparently conflicting or contradictory problems or opportunities; a problem or an opportunity is an ultimate element abstracted from a mess
Monitoring and control	Monitoring consists of observations related to a system's performance in the form of prescribed measures or data. When these observations are outside a specified range, action is taken through some avenue of management to remedy or "control" the situation
Networks	An elaboration of the concept of hierarchy which avoids the human projection of 'above' and 'below' and recognises an assemblage of entities in relationship, e.g. organisms in an ecosystem. Networked entities may be totally parallel, embedded, or partially embedded (structurally intersected)

(continued)

Table 2.1 (continued)

Concept	Explanation
Perspective	A way of experiencing which is shaped by our current state and circumstances as these are influenced by our unique personal and social histories, where experiencing is a cognitive act
Purpose	What the system does or exists for from the perspective of someone; the raison d'être of a system of interest formulated by someone and achieved through the particular transformation that has been ascribed
Resources	Elements (e.g. matter, energy or information) which are available either within the system boundary or present outside the system in a manner that the system can access and which enable a desired transformation to occur
System	An integrated whole distinguished by an observer whose essential properties arise from the relationships between its parts; from the Greek 'synhistanai', meaning 'to place together'
System of interest	The product of distinguishing a system in a situation, in relation to an articulated purpose, in which an individual or a group has an interest (a stake); a constructed or formulated system of interest to one or more people, used in a process of inquiry; a term suggested to avoid confusion with the everyday use of the word 'system'
Systemic thinking	The type of thinking that arises from the evolutionary trajectory of cognition. In humans this form of thinking takes place through the systemic action of our own cognitive system in a manner that is not limited to language and logic (background systemic thinking). Within language (i.e. in the foreground) it refers to the understanding of a phenomenon within the context of a larger whole. To understand things systemically literally means to put them into a context, to establish the nature of their relationships
Systematic thinking	Methodical, regular and orderly thinking about the relationships between the parts of a whole or the stages of a process. Systematic thinking usually takes place in a linear, step-by-step manner
Tradition	A network of pre-understandings or prejudices from which individuals, culturally embedded, think and act; how we make sense of our world
Transformation	Changes, modelled as an interconnected set of activities or processes which convert an input to an output which may leave the system (a 'product') or become an input to another transformation. Transformations are sometimes referred to as "processes"
Trap	A term derived through analogy with a lobster pot by Geoffrey Vickers; a way of thinking and acting which is difficult to escape from, and no longer relevant to the changed circumstances
Worldview	That conception or understanding of the world which enables each observer to attribute meaning to what is observed (sometimes the German word Weltanschauung, which refers to both attitude as well as concept, is used synonymously)

can't read signs of failure' by Simon Caulkin [8]. The author is a well known journalist formerly with a major UK newspaper who has an espoused commitment to the use of systems thinking.

Reading 1

Dyslexic Management Can't Read Signs of Failure

Simon Caulkin

The Observer, **Sunday November 25 2007**

The real British disease is the unerring talent for putting together entities that are less than the sum of their parts. The comical inability to think in systems terms – call it management dyslexia – was on dazzling display last week, all over the front and back pages.

First up, the England football team. Management is supposed to amplify effort by providing a creative framework for individual expression that benefits the team. But defeat against Croatia was the reverse, the culmination of unmanagement that over several matches has diminished team effort and turned good players into turnips.

It was the opposite of management that left players individually and collectively bereft. At least the England rugby players, in the World Cup, took the initiative to create their own playing system that, although limited, suited the available talent and took them against the odds to the very brink of triumph.

England's Premiership is the wealthiest football league in the world. Its consistent failure to generate a satisfactory national team is deeply rooted and reflected in other systemic shortcomings. Only one of the top teams, Manchester United, has a British manager; the starting line-up of the Premiership leader, Arsenal, contains just one, sometimes no, English player. Oh, and the new £800m Wembley stadium can't even produce a decent surface to play on. From grass upwards, English football is a system for growing anti-synergies.

Second up, a performance by HM Revenue & Customs that makes it hard to know where to begin – with the IT outsourcing that makes it an expensive extra to separate bank details from other personal data, to senior management's decision to dispense with encryption to Gordon Brown's repeated use of the 'one bad apple' excuse: the leak was the result of one individual's failure to carry out procedures – at the dispatch box.

The spectacle of a general blaming his troops is always distasteful, but in this case is also bankrupt. The HMRC leak is primarily the result not of human error, but poor or non-existent systems design which failed in at least three respects: not segregating sensitive from insensitive information, allowing the two to be sent out together, and omitting to encrypt it. If any of those steps had been followed, the further error, of leaving a junior to decide to put it in the post, would have been harmless. This is called fail-safeing – part of any good systems design.

(continued)

Reading 1 (continued)

And by the way, don't bother with a witch-hunt or a full-scale investigation to find out what went wrong: with the help of readers and the junior HMRC official, this column offers to find the root cause in a day, using a basic problem-solving technique called the 'five whys' (asking 'why' five times over) – and apply the answers to prevent the problem happening again. The five whys are at the heart of continuous improvement which, in turn, is the motor of systemic performance enhancement.

Last week's third outbreak of British anti-synergy syndrome centred on Norfolk and Norwich Hospital. On Wednesday, the hospital went into 'major incident' alert because it was chocker. At one stage, 10 ambulances (nearly half Norfolk's total) were immobilised waiting to unload their patients.

So the hospital's too small, right? Well, hang on a minute. Why was the hospital full? Because of high demand, coupled with high bed-occupancy rates. Why are bed rates so high? Partly because the hospital is 'efficient', operating at occupancy rates of more than 90 per cent. But also because 60 beds are occupied by patients who have finished treatment but can't be discharged. Why can't they be discharged? Because, for financial reasons, the Norfolk Primary Care Trust is busy closing down the community hospitals that would traditionally have taken recovering patients, and social care, as almost everywhere in the country, is utterly inadequate to cope.

And why is demand so high? An epidemic or major accident? Nope. The extra demand comes from within. It is largely generated by NHS Direct which, terrified of making mistakes, routinely directs callers to A&E or their GP – but since GPs are no longer available out of hours, as a result of the government-imposed contracts, that means A&E.

In other words, the Norfolk NHS crisis, like that of HMRC and team England, was self-generated, the result of complete and continuing system-blindness. 'Problems in organisations,' points out Russell Ackoff, one of the first and best systems thinkers, 'are almost always the product of interactions of parts, never the action of a single part.' Treating a single part destabilises the whole and demands more fruitless management intervention; management becomes a consumer of energy, rather than a creator.

Unfortunately, that's the hallmark of 21st century UK management. As last week demonstrated, it still shows no sign of recognising it.

Source: Caulkin, S., 'Dyslexic management can't read signs of failure', The Observer, Sunday November 25th 2007. Copyright Guardian News & Media Ltd 2007.

As you worked through this reading you may have recognised that Simon satisfies my criterion of what it is to be a systems thinker, i.e., he connects, through a social dynamic, with the history of systems thinking and practice, including the main concepts that have been developed and continue to be used (Table 2.1). The social dynamic is played out in the relationship between a journalist and his offering of a systemic explanation about a set of complex situations. In doing this he connects with the history of systems scholarship, via a named scholar (Ackoff) and his use of systems concepts such as emergence, connectedness, purpose (Table 2.1).[13]

Let me say a little more about how systems practice arises in social dynamics through the analogy of family history research. Family history research (FHR), like systems thinking and practice, is a practice open to anyone. FHR could be seen as an inquiry process into who we are. Understanding who we are through constructing narrative explanations about our past creates a new present – perhaps a new sense of identity – and thus, as we accept new explanations, can create different futures. This overall process is something many people are interested in as evidenced by the television programmes that have now developed based on family history research. The main products and processes of FHR involve the construction of lineages. Some stop with the lineage, the family tree, but often the most fascinating part is the stories about people, places, historical events and awareness about the contingency or luck of our own existence. As a form of practice family history research generally starts from the present and works backwards before coming back to the present. For those who engage in it, it invariably changes who they are and thus their future manner of living.

In a manner not dissimilar to family history research, Simon Caulkin referred to Russ Ackoff, an influential scholar in one of the many systems thinking lineages (see below). More importantly though he used stories with concepts based on Ackoff's work to make sense of the situation he was writing about. He thus drew on a particular lineage of systems thinking and applied it to contemporary circumstances.[14] Of course lineages are not static, they evolve and change producing innovative new insights and sometimes conserving unhelpful ideas or ideas no longer relevant to current circumstances.

You may have already realised that my analogy only partly works – in systems thinking and practice there are no genes and thus 'blood ties'. This does not particularly matter – in my experience family history research often reveals how weak

---

[13] When Simon speaks of 'entities that are less than the sum of their parts' he is referring to the systems concept of emergence; Simon's 'why' questions are associated with purpose and the systems notion of layered structure i.e., system, supra or sub-system (see Table 2.1).

[14] It is perhaps fair to say that the most obvious aspect of Caulkin's article is him complaining about a lack, something missing, rather than offering an alternative. This raises the question: "what good did it do?" My own response is to argue the following: (i) can a journalist do more than raise awareness? (ii) might it not be systemically undesirable to offer 'alternatives' developed out of context (i.e., without stakeholders etc.)? and (iii) is not the alternative he offers, perhaps implicitly, the development of systems thinking and practice skills for more effective managing in similar situations? I will return to this issue in Part IV.

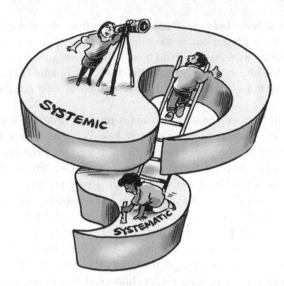

Fig. 2.1 My understanding of the relationship between systemic and systematic, the two adjectives arriving from the word 'system' – the systematic is nested within the systemic or, in other words the systematic is a special case of the systemic; together systemic and systematic form a whole, a unity, known as a duality

'blood ties' actually are[15] – but it does highlight how important social conventions or institutions (including those in politics, hospitals and academia e.g., through curricula, reading lists etc.) are in constituting different intellectual lineages. More importantly in our living we each develop our own intellectual lineage which I shall call our 'tradition of understanding.'[16] I will say more about this shortly.

In my experience the word 'system' is difficult to come to terms with for many people. I know this from first hand experience when, for example, at a dinner party there is often a stunned silence or an awkward initial conversation when, in answer to the question: What do you do? I reply 'Well I am a Professor of Systems'! My explanation in these situations usually varies with context, but behind all my responses there are a few simple distinctions which I depict in Figs. 2.1 and 2.2.

The other distinction that informs my responses is how I understand the relationship between thinking and practice (Fig. 2.2). The terms systems thinking and

---

[15] Using DNA technology it is now easy to determine what geneticists call a 'non-paternity event.' It is reported that in any project involving more than 20 or 30 people there is likely to be 'an oops in it' [23, p. 9].

[16] Following Russell and Ison [29] I use tradition-of-understanding to refer to what arises in our living – our thinking and acting in the moment – based on our individual development and the history of our evolving understanding (ontogeny of understanding) situated in, or coupled to, a cultural context. From the perspective of an observer a culture can also be said to be evolving.

Fig. 2.2 An image of the dynamic relationship between systems thinking and systems practice

systems practice are different ways of being in the same situation. This can be understood as a recursive dynamic much like the relationship between the chicken and egg – they are linked recursively and bring each other forth – speaking metaphorically they can be seen as mirror images of each other. Understood as a recursive dynamic systems thinking and practice can also be described as systems praxis – theory informed practical action.

On the basis of the distinctions depicted in Figs. 2.1 and 2.2 it might be reasonable to understand systems thinking and practice as arising from the dynamics of systemic and systematic thinking and practice operating within a particular set of social dynamics as I have tried to describe. For me these sets of distinctions help to make sense of the word 'systems' when it is used in a wide range of contexts – such as 'systems approaches' or 'systems disciplines' or 'systems lineages.'[17]

2.4 Different Systems Lineages

This book is not about different systems thinking and practice lineages. On the other hand to be able to make the most of the book, you will need some awareness of the territory, a rough road map of the 'systems field'. The material in this section provides a short overview – if you are already familiar with this territory then you

---

[17]Another term is that of 'systemics' which can be understood as an intellectual field – 'an open set of concepts and practical tools useful for gaining a better understandings of and eventual management of complex situations' [15, p. 354].

can pass it over. If you are not familiar with it there are also many books and websites available to explore this territory. When you do so I urge you to try to appreciate where different systems thinkers and practitioners are 'coming from' in what it is they do when they do what they do.

The word system comes from the Greek verb *'synhistanai'*, meaning 'to stand together' (the word 'epistemology' has the same root). A system is a perceived whole whose elements are "interconnected" (Table 2.1). Someone who pays particular attention to interconnections is said to be systemic (e.g. a systemic family therapist is someone who considers the interconnections amongst the whole family; the emerging discipline of Earth Systems Science is concerned with the interconnections between the geological and biological features of the Earth). On the other hand, to follow a recipe in a step-by step manner is being systematic. Medical students in courses on anatomy often take a systematic approach to their study of the human body – the hand, leg, internal organs etc. – but at the end of their study they may have very little understanding of the body as a whole because the whole is different to the sum of the parts, i.e. the whole has emergent properties such as 'life' (Table 2.1). Effective systems practice to change or improve situations of complexity and uncertainty means being both systemic and systematic when appropriate (Fig. 2.1).

Many, but not all, people have some form of systemic awareness, even though they may be unaware of the intellectual history of systems thinking and practice as a field of practical and academic concern (Fig. 2.3).[18] Systemic awareness comes from understanding:

Illustration 2.1

---

[18] There is an argument that all people have some form of systemic awareness, that it is inherent in our nervous system and is just not always recognised as such. People will refer to it as "hunch" or "gut feeling" or "insight" … or just act without noticing how they chose to do what they do [4]. This raises an interesting point about my meaning. I mean awareness of one's systemic thinking when I say "systemic awareness" – my concern is how we become better at, or use more of, systems thinking and practice in our climate-changing world. We cannot do this unless we can cultivate our abilities, however developed and to do so means bringing what systems thinking and practice is into awareness.

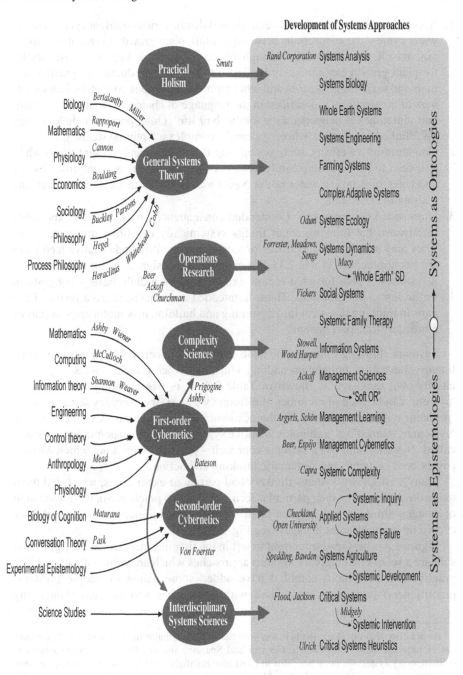

Fig. 2.3 A model of some of the different influences that have shaped contemporary systems approaches and the lineages from which they have emerged (Adapted from [18])

1. 'Cycles', such as the cycle between life and death, various nutrient cycles and the water cycle – the connections between rainfall, plant growth, evaporation, flooding, run-off, percolation etc. Through this sort of systemic logic water availability for plant growth can ultimately be linked to the milk production of grazing animals and such things as profit and other human motivations. Sometimes an awareness of connectivity is described in the language of chains, as in 'the food chain' and sometimes as networks, as in the 'web of life'. Other phrases include 'joined up', 'linked', 'holistic', 'whole systems', 'complex adaptive systems' etc.
2. Counterintuitive effects, such as realising that floods can represent times when you need to be even more careful about conserving water, as exemplified by the shortages of drinking water in the New Orleans floods that followed hurricane Katrina in 2005, and
3. Unintended consequences. Unintended consequences are not always knowable in advance but thinking about things systemically can often minimise them. They may arise because feedback processes (i.e. positive and negative feedback) are not appreciated (Table 2.1). For example the designers of England's motorways did not plan for what is now experienced on a daily basis – congestion, traffic jams, emissions etc. These unintended consequences are a result of the gaps in thinking that went into designing and building new motorways as part of a broader 'transport system'

Systems thinking embraces a wide range of concepts which most systems lineages have as a common grounding. Thus, like other academic areas, 'Systems' has its own language as shown in Table 2.1. It is worth noting that the word 'system' can be used in a number of different ways: (i) the everyday sense as in the 'problem with the system!'; (ii) the academic use of the term as in the phrase a 'system of interest' or 'a complex adaptive system'; (iii) the academic area of study called 'Systems' and (iv) a systems approach – practice or thinking which encompasses both systemic and systematic thinking and action.

Many well-known systems thinkers had particular experiences, which led them to devote their lives to their particular forms of systems practice. So, within the field of systems thinking and practice there are different traditions, which develop or evolve through different lineages (see Fig. 2.3).[19]

Figure 2.3 is best read from right to left in the first instance. Down the right-hand side are a set of contemporary systems approaches which are written about, put into practice and sometimes taught. I have added some names of people (systems practitioners) particularly associated with approaches and lineages though my

---

[19] I do not clam that this depiction is in any way definitive – a major limitation of it is that it does not include the many valid French, German and Spanish, and possibly other, contributions to contemporary systems approaches. This in itself also highlights how the different language communities give rise to intellectual silos. Many people have contacted me after seeing an earlier version of this figure to tell me that something, or someone, is missing. If this is the case then I am apologetic, to a certain extent; but in a way it also helps me to make the point that like all disciplinary fields Systems is not a homogeneous field – how it understands itself is contested. So, please feel free to take this figure and adapt it as you see fit.

choices are far from comprehensive. The approaches are also organised from top to bottom in terms of what I perceive to be common commitments, or tendencies, of a majority of practitioners within the given approaches to seeing systems as entities (ontologies) or heuristic devices (epistemologies).[20]

On the left I identify seven formative clusters that have given rise to these contemporary systems approaches. By following the arrows backwards one can get a sense of some of the different lineages, though rarely are they as simple as depicted here. This figure has many limitations and it is not possible to describe all these influences nor approaches in detail but it does capture a way of understanding the 'Systems field'.

There is a close affinity between systems thinkers and process thinkers. Some historical accounts of systems lineages start with the concerns of organismic biologists who felt that the reductionist thinking and practice of other biologists was losing sight of phenomena associated with whole organisms [3]. Organismic or systemic biologists were amongst those who contributed to the interdisciplinary project described as 'general systems theory' or GST [4]. Interestingly 'systemic biology' is currently enjoying a resurgence [24]. Other historical accounts start earlier – with Smuts' [33] notion of practical holism or even earlier with process thinkers such as Heraclitus who is reputed to have said: "You cannot step into the same river twice, for fresh waters are ever flowing in upon you." Other historical accounts can be found in Checkland [9]; Francois [15]; Flood [12,13]; Jackson [19]; Midgley [22]; Ramage and Shipp [27] or on Principia Cybernetica [26].

Some of the motivation for the 'GST project' in interdisciplinary synthesis can be explained by the realisation in many disciplines that they were grappling with similar phenomena. This project had its apotheosis in the interdisciplinary Macy conferences in the 1940s and 1950s which did much to trigger new insights of a systems and cybernetic nature and subsequently a wide range of theoretical and practical developments [16]. So, although many now argue that GST, as an intellectual project, has not been sustained it has none-the-less left a rich legacy [5].

A good example of how the lineages have operated is that of the relationship between Kurt Lewin, the cybernetic concept of 'feedback' and the everyday concept of 'feedback' [19]. Checkland [9] established a connection with Kurt Lewin's view of 'the limitations of studying complex real social events in a laboratory, the artificiality of splitting out single behavioural elements from an integrated system' [14]. Checkland goes on to say: 'this outlook obviously denotes a systems thinker, though Lewin did not overtly identify himself as such...' (p. 152). A central idea in Lewin's milieu was that psychological phenomena should be regarded as existing

---

[20] I expand on this issue in Chapter 3; some may have chosen to describe this in terms of positivist or constructivist commitments, respectively. My intention in this figure is not to label or classify any approach or practitioner – merely to organise, and reflect on, my experience. How others might order the various systems practitioners/approaches on these dimensions could produce an illuminating conversation. I also want to make it clear that placement along this spectrum is not an attempt to rank on my part.

in a 'field': 'as part of a system of coexisting and mutually interdependent factors having certain properties as a system that are deducible from knowledge of isolated elements of the system' ([10], quoted in Sofer [34]). Whilst Lewin may not have overtly described himself as a systems thinker, he was none-the-less a member of the Macy conferences 'core group'. He attended the first two conferences but died in 1947, shortly before the third conference, and his influence was lost to the group (especially his knowledge of Gestalt psychology). His work was taken up however and has informed the study and description of group dynamics and he is also seen as the founder of 'action research'.

Below GST in Fig. 2.3 the next two clusters are associated with cybernetics, from the Greek meaning 'helmsman' or 'steersman'. The term was coined to deal with concerns about feedback as exemplified by the person at the helm responding to wind and currents so as to stay on course. A key image of first order cybernetics is that of the thermostat-controlled radiator – when temperatures deviate from the optimum, feedback processes adjust the heat to maintain the desired temperature. Major concerns of cyberneticians were that of communication and control (Table 2.1). As outlined by Fell and Russell [11] the first-order cybernetic "idea of communication as the transmission of unambiguous signals which are codes for information has been found wanting in many respects. Heinz von Foerster, reflecting on the reports he edited for the Macy Conferences that were so influential in developing communication theory in the 1950s, said it was an unfortunate linguistic error to use the word 'information' instead of 'signal' because the misleading idea of 'information transfer' has held up progress in this field [5]. In the latest theories the biological basis of the language we use has become a central theme" (see first and second-order communication in Table 2.1).

Fell and Russell [11] go on to describe the emergence of second-order cybernetics in the following terms: "second-order cybernetics is a theory of the observer rather than what is being observed. Heinz von Foerster's phrase, 'the cybernetics of cybernetics' was apparently first used by him in the early 1960s as the title of Margaret Mead's opening speech at the first meeting of the American Cybernetics Society when she had not provided written notes for the Proceedings" [36].

The move from first to second-order cybernetics is a substantial philosophical and epistemological jump as it returns to the core cybernetic concepts of circularity and recursion. These scholars applied the core concept of circularity to itself by recognising that there is a circularity between the observer and their world. An action on the world changes perception of the world which in turn changes the action, again. Action and perception develop as a circularity. This leads to the understanding that observers bring forth their worlds [21,37]. von Foerster [38], following Wittgenstein, put the differences in the following terms: "Am I apart from the universe? That is, whenever I look am I looking through a peephole upon an unfolding universe [the first-order tradition]. Or: Am I part of the universe? That is, whenever I act, I am changing myself and the universe as well [the second-order tradition]." The implications of these two questions are addressed in Chapter 5. It is worth making the point that understandings from second-order cybernetics have been influential in fields as diverse as family therapy and environmental management. Some authors equate a second order cybernetic tradition with radical constructivism although not all agree.

Operations research (OR) is another source of influence on contemporary systems thinking and practice. OR flourished after the Second World War based on the success of practitioners in studying and managing complex logistic problems. As a disciplinary field it has continued to evolve in ways that are mirrored in the systems community.

A set of influences, recently popularised again, have come from the so-called complexity sciences (Fig. 2.3) which is a lively arena of competing and contested discourses. As has occurred between the different systems lineages, there are competing claims within the complexity field for institutional capital (e.g. many different academic societies have been formed with little relationship to each other), contested explanations and extensive epistemological confusion [30]. However, some are drawing on both traditions to forge exciting new forms of praxis (e.g. Mackenzie [20]).

Other recent developments draw on interdisciplinary movements in the sciences, especially in science studies. These include the rise of discourses and understandings about the 'risk' and 'networked' society' [2,7], and associated globalisation which have raised awareness of situations characterised by connectedness, complexity, uncertainty, conflict, multiple perspectives and multiple stakeholdings [17]. It can be argued that this is the reformulation and transformation of an earlier discourse about the nature of situations that Ackoff [1] described as 'messes' rather than 'difficulties' (Table 2.1), Schön [31] as the 'real-life swamp' rather than the 'high-ground of technical rationality' and Rittel and Webber [28], as 'wicked' and 'tame' problems. Schön, Ackoff and Rittel all had professional backgrounds in planning so it is not surprising that they encountered the same phenomena even if they chose to describe them differently.

Unfortunately the systems thinkers and practitioners responsible for the different lineages depicted in Fig. 2.3 are more often than not remembered for a particular method, methodology or technique. These are not insubstantial achievements but an unintended consequence has been to divert attention away from the dynamics of systems thinking and practice as depicted in Fig. 2.2 – or in other words away from systems praxis. Later I draw upon a particular lineage of systems ideas which is concerned with the relationship between a 'framework of ideas', a method, a practitioner and a context or situation. I will argue that this relationship is key to effective systems practice and also the essence of methodology. The point being that the ways in which all combinations of these factors are combined are unique to time and place – much like an actor's performance.

I do not know to what extent Simon Caulkin has immersed himself in the different systems lineages discussed above; in one sense it does not matter. What is clear is that he has a sufficiently profound understanding of some of these lineages to be able to write an insightful article about an important issue. And by his use of the concepts he shows, to me at least, that he understands what he is talking about.

In Reading 1 Caulkin also referred to a form of systems practice which he called 'system design'. I say more about this form of practice in Part III. A test of Simon's effectiveness as a systemic journalist is whether, as a result of reading the article, you now have: (i) a new (or more) systemic understanding that did not exist before of some complex situations, or (ii) an experience in which your own systemic understanding has been affirmed. A key element of this social dynamic is whether you found Simon's explanations satisfying (in the sense of accepting them) or not![21]

2.5   System or Situation?

Simon Caulkin's article (Reading 1) can also be explained in terms of the dynamic depicted in Fig. 1.3 which related changes in understandings to changes in practices for transforming situations. In his article Simon describes situations in which something is at issue and, although not strictly in these terms, he argues that the only way these situations can be transformed for the better is through changes in understandings and practices that are more systemic. But what is it about situations that make them more or less amenable to systemic description and improvement? Is this even a sensible question? In the next chapter (Chapter 3) I want to answer these questions.

References

1. Ackoff, R. L. (1974) Redesigning the future. Wiley: New York.
2. Beck, U. (1992) Risk Society: Towards a New Modernity. Sage: London.
3. Bertalanffy, L. von. (1968 [1940]) The organism considered as a physical system. In L. von Bertalanffy (Ed.), General System Theory (pp. 120–138). Braziller: New York.
4. Bunnell, P. (2009) Personal communication. Systems Ecologist, President of Lifeworks, Vancouver, British Columbia, June 2009.
5. Capra, F. (1996) The Web of Life. HarperCollins: London.
6. Carroll, J.B. (Ed.) (1959) Language thought and reality. Selected writings of Benjamin Lee Whorf. MIT Press: New York.
7. Castells, M. (2004) Informationalism, networks, and the network society: a theoretical blueprint. In M. Castells (Ed.), The Network Society: a Cross Cultural Perspective (pp. 3–48). Edward Elgar: Northampton, MA.
8. Caulkin, S. (2007) Dyslexic management can't read signs of failure. The Observer, Sunday November 25th 2007.
9. Checkland, P.B. (1981) Systems Thinking, Systems Practice. Wiley: Chichester.
10. Deutsch, M. and Krauss, R.M. (1965) Theories in Social Psychology. Basic Books: New York.
11. Fell, L. and Russell, D.B. (2000) The human quest for understanding and agreement. In R.L. Ison and D.B. Russell (Eds.), Agricultural Extension and Rural Development: Breaking out of traditions. Cambridge University Press: Cambridge.
12. Flood, R.L. (1999) Rethinking 'The Fifth Discipline': Learning within the Unknowable. Routledge: London.

---

[21] I will say more about how explanations arise in a social dynamic in Chapter 3.

13. Flood, R.L. (2001) The relationship of systems thinking to action research. In H. Bradbury and P. Reason (Eds.), Handbook of Action Research: Participative Inquiry and Practice (pp. 133–144). Sage Publications: London.
14. Foster, M. (1972) An introduction to the theory and practice of action research in work organizations, Human Relations 25(6), 529–556.
15. Francois, C. (ed) (1997) International Encyclopaedia of Systems and Cybernetics. K. Sauer: Munchen.
16. Heims, S. (1991) Constructing a Social Science for Postwar America: the Cybernetics Group 1946–1953. MIT Press: Cambridge, MA.
17. Ison, R.L., Röling, N. and Watson, D. (2007) Challenges to science and society in the sustainable management and use of water: investigating the role of social learning. Environmental Science & Policy 10(6), 499–511.
18. Ison, R.L. (2008) Systems thinking and practice for action research. In P. Reason and H. Bradbury (Eds.), The Sage Handbook of Action Research Participative Inquiry & Practice (2nd edn) (pp. 139–158). Sage Publications: London.
19. Jackson, M. (2000) Systems Approaches to Management. Kluwer: New York.
20. McKenzie, B. (2006) http://www.systemics.com.au/. Accessed 31 July 2009.
21. Maturana, H. and Poerkson, B. (2004) From Being to Doing: the Origins of the Biology of Cognition. Carl-Auer: Heidelberg.
22. Midgley, G. (2000) Systemic Intervention: Philosophy, Methodology and Practice. Kluwer/Plenum: New York.
23. Olson, S. (2007) Who's your Daddy? Australian Financial Review. Friday 31st August, 9.
24. O'Malley, M.A. and Dupré, J. (2005) Fundamental issues in systems biology. BioEssays 29, 1270–1276.
25. Pearson, C.J. and Ison, R.L. (1997) Agronomy of Grassland Systems, 2nd Edition. Cambridge University Press: Cambridge.
26. Principia Cybernetica (2009) http://pespmc1.vub.ac.be/DEFAULT.html. Accessed 30th July 2009.
27. Ramage, M. and Shipp, K. (2009) Systems Thinkers. Springer: London.
28. Rittel, H.W.J. and Webber, M.M. (1973) Dilemmas in a general theory of planning. Policy Science 4, 155–69.
29. Russell, D.B. and Ison, R.L. (2007) The research-development relationship in rural communities: an opportunity for contextual science. In Ison, R.L. and Russell, D.B. (Eds.), Agricultural Extension and Rural Development: Breaking out of Knowledge Transfer Traditions (pp. 10–31). Cambridge University Press: Cambridge, UK.
30. Schlindwein, S.L. and Ison, R.L. (2004) Human knowing and perceived complexity: implications for systems practice. Emergence: Complexity & Organization 6(3),19–24.
31. Schön, D.A. (1995) The new scholarship requires a new epistemology. Change November/December, 27–34.
32. Shotter, J. (1993) Conversational Realities: Constructing Life Through Language. Sage: London.
33. Smuts, J.C. (1926) Holism and Evolution. Macmillan: London.
34. Sofer, C. (1972) Organizations in Theory and Practice. Heinemann: London.
35. Steirlin, H. (2004) The Freedom to venture into the unknown. In Poerkson, B. (Ed.) The Certainty of Uncertainty. Dialogues introducing constructivism (pp. 153–172). Imprint Academic: Exeter.
36. van der Vijver, G. (1997) Who is galloping at a narrow path: conversation with Heinz von Foerster. Cybernetics & Human Knowing 4(3), 3–15.
37. von Foerster, H. and Poerkson, B. (2004) Understanding Systems. Conversations on Epistemology and Ethics (IFSR International Series on Systems Science & Engineering, 17). Kluwer: New York/Carl-Auu: Heidelberg.
38. von Foerster, H. (1992) Ethics and second-order cybernetics. Cybernetics & Human Knowing 1, 9–19.
39. Wilson, B. (1984) Systems: Concepts, Methodologies and Applications. Wiley: Chichester.
40. Wittgenstein L. (1953) Philosophical Investigations. Basil Blackwell: Oxford.

Chapter 3
Making Choices About Situations and Systems

3.1 Choices that Can Be Made

I want to start this chapter by pointing out that:

1. We are always in situations,[1] never outside them
2. We have choices that can be made about how we see and relate to situations
3. There are implications which follow from the choices we make

Importantly, one of the choices that can be made is to see a situation as a system, but as I will explain, there are many implications of making this choice that can trap the unwary or uninformed! To explain what I mean I am going to go back on my espoused position about definitions and offer one for 'system' which is contained in Reading 2.

This reading [16] comes from a former Open University colleague, Dick Morris, and is concerned with how Systems, in the academic, as well as practical sense, might contribute towards achieving sustainable lifestyles – a situation of considerable complexity! The approach taken is also typical of much Open University Systems teaching of the last 30 years; thus, as a practitioner, Morris exemplifies many aspects of the lineage of systems practice developed at the Open University. As you read try to become aware of the 'elements' of his practice.

---

[1] The word situation has roots in the Latin, *situare*, to place or locate. From *situs*, place, position. It can also be understood as 'to set down' or 'to leave off' [1]. From this comes the 'act of setting or positioning' and the 'extended sense of a state or condition' – in conceptual terms the act of distinguishing a situation could be seen as part of the dynamics of 'bringing forth'.

R. Ison, *Systems Practice: How to Act in a Climate-Change World*,
DOI 10.1007/978-1-84996-125-7_3, © The Open University 2010.
Published in Association with Springer-Verlag London Limited

Thinking About Systems for Sustainable Lifestyles

R.M. (Dick) Morris

The Anglia Schumacher conference is concerned with promoting humane and sustainable lifestyles in the Region.[2] To achieve such lifestyles, we all need to make decisions about a whole complex of interacting requirements, for food, housing, livelihood, health, transport etc., and decisions about one aspect can have unexpected, and perhaps undesired, effects on others. Choosing to work from home can save transport fuel, but could involve even more fuel for home heating! To be effective, we need to consider our whole lifestyle system, not just separate activities. Schumacher's aphorism that "small is beautiful" can be interpreted as referring to small, often local, systems of living.

This word system has become so much a part of our 21st century vocabulary, as in "the Transport system" (when it breaks down), "the Social Security System" (ditto), a "stereo system" etc., that we probably take its use for granted, and do not consider some of the implications of using it. Really to think in terms of systems is not necessarily so easy, but is an essential part of our outlook if we are to develop our world in a sustainable manner. When thinking in terms of systems, we have at least partially to move away from our usual manner of thinking, which has been heavily influenced by the generally science-based model that has characterised European thought, particularly during the last century. Such thinking in science and its partner, technology, has produced enormous strides in our material well-being, but we also recognise that it has also brought some problems. A key feature of classical science has been to work under carefully controlled experimental conditions, looking in detail at one factor at a time. The success of this approach has unintentionally encouraged a widespread popular belief that we can isolate a single cause for any observed event. How often do we see headlines suggesting that childrens' behavioural problems arise from food additives, street crime is the result of shortage of police on the beat, traffic accidents or congestion are the result of inadequate expenditure on the roads etc.? As we read these, we may mentally note reservations about the over-simplification, but all too often, political or societal responses to these concerns are based on such monocausal explanations. It's much easier for a politician or a manager to demonstrate that the supposed single cause is being tackled than to ask the much harder question as to whether it will really produce the desired result.

---

[2] The East Anglia region in the UK which includes the city of Cambridge.

(continued)

Reading 2 (continued)

A classic example arose from the series of rail crashes in England in the first years of this century. Tragically, several people were killed, and the obvious "cause" was problems with the rails. To avoid further loss of life, draconian speed limits were imposed on the trains and repairs to the tracks instigated. This no doubt reduced the chances of further rail accidents, but in the process, persuaded many people to abandon rail travel in favour of their cars. Given that the probability of an accident per kilometre travelled is a couple of orders of magnitude larger for car travel than rail travel, the decisions taken about the railways may actually have increased the number of travel-related deaths and injuries, rather than reduced them. A decision taken about the safety of the railway system may well have had completely the opposite affect to that intended when considered in relation to the wider transport system.

Similar examples could be drawn from any number of situations, highlighting the need to think beyond single cause-effect relations. One of the responses to this has been the movement, particularly in some aspects of medicine, towards so-called holistic methods, which indeed look beyond one-to-one links, to consider the whole range of factors affecting human health such as diet, income, social relations, posture etc. and the complex interactions between these. This approach undoubtedly has its strengths, but there is always a danger that it is impossibly time-consuming and may even conceal or confuse simple solutions. Somewhere between the delightful simplicity of reductionist explanation and the possibly unreal requirements of unrestricted holism, there should be a pragmatic level of discrimination that is both effective and efficient.

This is where the ideas of systems and of systems thinking are valuable. When we start to think about sustainability, it is essential that we ask questions at a range of levels from the local to the global. Questions arise about what aspects of our existence do we want to sustain, how much are we prepared to compromise with others' needs and what unexpected results of our actions might occur. That is, we need to start asking questions about the systems involved in sustainable living. To do this, we need some agreed definitions, and some techniques for thinking about systems. One possible definition of a system, based on one used in the Open University's Technology Faculty, is:

- A collection of entities
- That are seen by someone
- As interacting together
- To do something

The implications of the various elements of this definition are that a system is not a single, indivisible entity, but has component parts (that may themselves be regarded as systems and termed sub-systems) and that the components

(continued)

Reading 2 (continued)

interact with one another to cause change. So, the land, animals, machinery, and organisations involved in supplying our food can (and should) be regarded as a complex, interacting system, rather than just examined in isolation as crops, retail outlets or consumers. Perhaps the most difficult aspect of the definition is the subjective one – the collection of entities is chosen by someone as a system. Different individuals may see different systems in a particular situation. For example, a farm can be seen as a system to produce food, to produce a profit, to maintain a particular landscape, etc. Equally, a supermarket to a consumer is a source of food, whereas to its operator and shareholders it is primarily a source of profit. Both farms and supermarkets can also be seen as part of a wider food supply system, as for example in Fig. 3.1.

Negotiating and choosing an appropriate system for debate and decision-making can be crucial, since we cannot solve all the problems of the world in one go! It is essential to put some boundary round the system we are debating, and different conceptions of a system of interest can also carry with them different criteria for the success or otherwise of that system. Choosing an inappropriate boundary, and with it, inappropriate criteria, can be misleading.

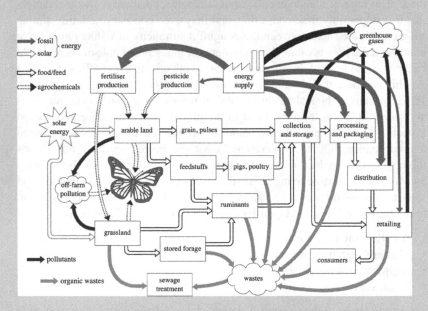

Fig. 3.1 A possible representation of some of the energy and material flows through "a food system" (the butterfly represents biodiversity and landscape!)

(continued)

Reading 2 (continued)

For example, choosing to put a boundary around a system of using animals for food production can suggest that this is grossly inefficient, since it takes about 10 kg of feed to produce one kg of meat. However, if the system is redefined to include the land, then a more interesting measure may be the amount of food produced from the total area of land available. In this situation, some ruminant animals are probably essential to obtain useful human food from those areas that can only grow grass or other plant materials that cannot be used directly by humans. Changing the boundary and the criteria can produce very different conclusions.

So what are the possible systems associated with sustainable living? We are all concerned with this, but with different emphases, timescales, and skills. We are all stakeholders in some sustainable, human-oriented system, but we are unlikely all to have the same vision of what that involves, or what affects what. In order to share our visions, and to debate futures, we need to have some way of explaining what we regard as the system of interest and its key features. We need to have some model of the system which is necessarily simpler than the whole, complex situation itself, but shows what we think are the important aspects. It might be possible to do this in words, but often it is much quicker and more powerful to use some sort of diagram. Words have to flow in a sequential manner to make sense, and one of the features of most systems is that the interactions between entities are often recursive, that is they form loops, where A may affect B, which in turn affects C, but C can also affect A.[3] In such a situation, a diagram can literally be "worth a thousand words"! In the same way that a map highlights a selection of important features of the landscape, an appropriate diagram can make clear the key features of our interpretation of a system. Diagrams can provide the means for sharing different understandings of the world around us and of the potential outcomes of our actions within the multiple, complex systems of which we are a part.

---

[3] The concepts of recursion and circularity are not always well understood – see http://en.wikipedia.org/wiki/Recursion#Recursion_in_plain_English.

(continued)

In the next part of this Reading Morris introduces systems diagramming (e.g. Systems maps and Multiple-cause – alternatively, causal loop – diagrams) as forms of conceptual modelling. Because diagrams can provide a powerful means for communicating different understandings of the world around us and of the potential outcomes of our actions they have become an important ingredient of how systems practice has been taught at the Open

University (UK). Diagramming is an important means for individuals and groups to 'reveal' how they understand a situation, particularly the different elements, patterns of causality and the nature of influences, and eventually what 'systems' might usefully be seen in some situation as part of an inquiry process i.e. the diagram 'captures' someone's thinking and 'mediates' communication about this thinking with others. Morris continues:

Reading 2 (continued)

Two diagrammatic forms that can be useful here are Systems maps and Multiple-cause (alternatively, causal loop) diagrams. An example of each is given below (Fig. 3.2). A systems map uses closed shapes (usually circles or clouds) to show the components that the person drawing the map regards as important in the system that they see in some situation. The spatial relationship between the shapes can be used to highlight some of the structural links between these aspects. So, for example, farms, food processors, the food distribution network and the supermarkets in Fig. 3.2a are all components of a food production sub-system, and might be grouped together on a map of a larger economic, or nation state, system. More dynamic relationships can be represented on a multiple cause diagram, where arrows are used to show where one factor causes another to change, or causes some event to occur (Fig. 3.2b). Such diagrams can be developed into more formal, even computable models of systemic behaviour. However, for many purposes, a diagram alone is more than adequate.[4]

Source: Morris, R.M. (2009) 'Thinking about systems for sustainable lifestyles', *Environmental Scientist*, 18(1), pp. 15–18.

---

[4]This material is based on a talk given as the Schumacher Lecture to GreenChoices, Cambridge and published as Morris [16].

Using systems diagrams for creating and communicating understandings, in order to work towards innovations (i.e., changes) in situations of local human existence, was one objective of the event where Morris presented this paper. Delegates worked in small groups to produce diagrams of some 'relevant systems' involved in supplying human needs. As you can see Morris addressed conceptual and praxis issues in his paper; this is a hallmark of the approach taken at the Open University [8]. I will say more about systems diagramming in Chapter 6.

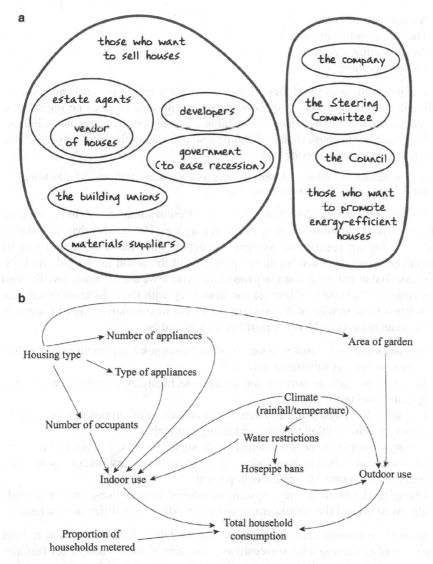

Fig. 3.2 Examples of (**a**) systems map of a system to promote energy efficient houses and (**b**) multiple cause diagram of the factors affecting domestic water use

3.1.1 OU Systems Course Definition of Systems

You will have noted in the reading that Morris [16] argues the need for 'some agreed definitions, and some techniques for thinking about systems'. The definition of a *system* he proposes comes from systems courses developed at the Open University (UK):

- A collection of entities
- That are seen by someone
- As interacting together
- To do something

The implications of this definition are, he suggests, that 'a system is not a single, indivisible entity, but has component parts (that may themselves be regarded as systems and termed sub-systems – see Table 2.1) and that the components interact with one another to cause change'.[5] But he also talks about the 'subjective' element and makes an important point that:

> [T]he collection of entities is chosen by someone as a system. Different individuals may see different systems in a particular situation.

Morris goes on to explore how aspects of different situations could be 'seen' as different types of systems pursuing different purposes. He suggests that 'negotiating and choosing an appropriate system for debate and decision-making can be crucial, since we cannot solve all the problems of the world in one go!' He says: 'It is essential to put some boundary round the system we are debating, and different conceptions of a system of interest can also carry with them different criteria for the success or otherwise of that system.' If I were to summarise what I discern to be the main aspects of Morris's position they would be:

- In many situations it makes sense to avoid monocausal explanations and reductionism as well as unfettered holism
- In situations such as concern for sustainable lifestyles it makes sense to use systems thinking
- It is sensible to agree that a system is: 'A collection of entities that are seen by someone as interacting together and doing something'
- Amongst people in the same situation 'systems' will be perceived differently and thus have different boundaries – i.e. making boundary judgments and thus specifying a system of interest is important
- Changing the boundary to a system of interest is something we can actively choose to do and the criteria employed can produce very different conclusions

In order to communicate our visions, and to debate futures, we need to have some way of explaining what we regard as the system of interest and its key features. He proposes an approach that involves developing a 'model of the system which is necessarily simpler than the whole, complex situation itself, but shows what we think are the important aspects'.

---

[5]Some will argue against the notion that components interacting 'cause' change, but few would dispute that through their interaction change happens. Of contention here is the nature of causality. Likewise it will be of concern to some that all systems are conceptualised "to do something". They may suggest that whilst this is appropriate for designed systems, it is not adequate, for example, for an ecosystem. This is a significant conceptual point with practical implications which I will address later in the book. For the moment the difficulty can be overcome by saying "doing something" rather than "to do something".

3.2 Systems Practice as Process

As I intimated earlier many people either implicitly or explicitly refer to things that are interconnected (exhibit connectivity – Table 2.1) when they use the word 'system'. A common example is the use of 'transport system' or 'computer system' in everyday speech as outlined by Morris in Reading 2. As well as a set of interconnected 'things' (elements) a 'system' can also be seen as a way of thinking about the connections (relationships) between things – hence a process. A constraint to thinking about 'system' as an entity and a process is caused by the word 'system' being a noun – a noun implies something you can see, touch or discover, but in contemporary systems practice more attention can be paid to the process of 'formulating' a 'system' as part of an inquiry process in particular situations. The key elements of this practice are depicted in Fig. 3.3.

Reading 2 makes many important points that will be central to this book, but one aspect that is not addressed explicitly is the potential for confusion between situation and system and between system as 'process' and system as 'thing'. Why is this discussion relevant at this stage you might well ask? Well because in what I call aware systems practice it is important not to fall into a historical trap that has plagued systems scholarship, and held back practice for a long time. The main elements of this trap are depicted in Fig. 3.4. The practitioner at the top sees the world as made up of systems. For example, this position applies to many ecologists who classify, describe and research systems called 'ecosystems'. An unintended consequence of adopting this position is that too often it is assumed that there is agreement about the nature of 'the system' i.e. what it is, what its elements are, and most importantly where the boundaries lie and thus what a change for the

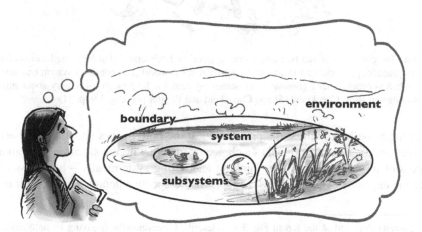

Fig. 3.3 Key elements of systems practice as a process – a system of interest comprising a system (with sub systems), boundary and environment is 'brought forth' or distinguished, by someone as they engage with a particular situation. In this example the woman could be thinking about fish for dinner, and hence her conception of what the system does is to produce fish

Fig. 3.4 The choices that can be made about 'system' and 'situation' that have implications for systems practice: practitioner 1 (*top*) situates systems in the world (i.e., conflates system and situation) whereas practitioner 2 (*bottom*) understands 'systems' to be a means of inquiry about situations (Adapted from Checkland [5] and Checkland and Poulter [6], Fig. 1.9, p. 21)

better might be![6] In contrast the practitioner at the bottom sees situations that are complex and confusing and makes a choice to engage with the situation through a process of inquiry that involves thinking and acting systemically – a process of systemic inquiry. An ecologist who adopts the perspective depicted at the bottom,

---

[6]The position depicted at the top in Fig. 3.4 is described theoretically as giving an ontological status to systems – i.e. proponents of this position, implicitly or explicitly 'see' systems as existing in the world. They may thus claim that the world is systemic! The position depicted at the bottom is described as recognising 'a system' as an epistemological device – i.e. a way or means of engaging with a situation so as to better know or inquire systemically.

for example, would see the concept 'ecosystem' as a particular way of thinking about a complex set of interactions in particular situations or as a device to describe and classify so as to enhance communication and understanding.[7]

My argument is that the effectiveness of systems practice is constrained when a practitioner confuses or uses interchangeably situation and system. With these distinctions in mind I make the following claims about my own understanding of systems practice:

- The place to start is with the situation and not the system
- We have choices we can make about how to characterise and, thus to understand situations and Systems scholars have coined a number of neologisms (such as 'ecosystem', 'mess' or 'difficulty', 'complex adaptive system', 'system of interest', 'learning system' etc.)[8] that can help in making these choices
- Be aware that 'distinguishing, or bringing forth' a system in a situation is a particular way of knowing the situation
- Having made a choice about a way of knowing a situation do not fall into the trap of reifying the choice – be aware of the implications of the choice that is made, and be prepared to make a different choice if that later appears more useful

I will add further claims to this list as the book progresses but now I want to present a model, which if properly understood, challenges the mainstream way in which we have come to understanding situations, such as climate change, which cause us concern. This heuristic model links how we, as human beings, connect or engage with situations, including other people. The model is an attempt to move away from the traditional dualism of subjective (our inner dynamics) and objective (the outer world, or the 'reality' out there) to a dynamic relation between ourselves and the world we assume to exist outside ourselves.[9]

3.3 Practitioner, Framework, Method, Situation

To understand systems practice it is first necessary to understand practice as a particular dynamic. Some of the implications of confusing system with situation can be understood by developing a simple model of practice as a relational dynamic between certain

---

[7]My own interpretation of the conceptual issues at stake in Fig. 3.4 differ from what I understand Checkland's position to be, though perhaps not greatly. It is likely that Checkland would see Fig. 3.3 as depicting a practitioner who sees systems in the world; in contrast I claim it depends on the awareness of the practitioner. If the practitioner acts with awareness that the act of distinguishing a system of interest brings forth an epistemological device then I would claim the end result is similar to that which Checkland claims for the enactment of Soft Systems Methodology as a learning system. Further it makes sense not to assume another's epistemological commitments – but to explore them, as through conversation.

[8]I will expand on this point below – some of these 'neologisms' are described in Table 2.1.

[9]John Shotter [17, p. 6] writes of this dynamic relation as the "formative uses to which 'words in their speaking' are put and upon the nature of relational situations."

Fig. 3.5 A conceptual model which can be applied to many forms of practice (e.g. researching, policy making, leading etc.) comprising a person thinking about a 'real world situation' in which a person or practitioner (P) (who may be the same as the person who is thinking) engaging with a situation (S) with a framework of ideas (F) and a method, M (Adapted from Blackmore et al. [2] and Checkland [4])

elements (Fig. 3.5). Figure 3.5 is what is known as a heuristic device, something that is designed to explore a situation – many of the figures in this book have a heuristic intent i.e., they do not set out to describe or claim 'this is how it is' but are designed for you to use as a way of challenging and developing your own understandings.

In its simplest form practice within the logic of Fig. 3.5 involves a practitioner (P) with a framework of ideas (F), a method or methodology (M) and a situation (S).[10] What more can be said about this conceptualisation? Well I could posit that all practitioners come to situations with existing theoretical frameworks. This idea is captured in a remark attributed to Maynard Keynes: that in his experience 'all men [sic] who claimed to be practical were the victim of some theory 30 years out of date.'[11]

In social research, medical, nursing and policy circles the idea that a theoretical framework can be explicitly chosen is more obvious – for example 'actor network

---

[10] Some refer to this as a 'real world' situation to distinguish it from a conceptual or abstract situation but this is an artificial separation though one that is often useful to help make sense of what is happening. The inverted commas around 'real world' thus denote Checkland's [3] original distinctions between situation and the 'conceptual world' of the researcher/practitioner – this is a distinction to aid praxis, not a commitment, on my part, to a 'reality' independent of an observer. If one is to use the phrase 'real world situation' at all, then my preference is to understand it as all that exists within the large thought bubble, i.e. the practitioner, situation dynamic.

[11] I owe this anecdote to Peter Checkland, though I cannot verify its source. On the other hand I can point to Keynes' claim that 'Practical men, who believe themselves to be quite exempt from any intellectual influence, are usually the slaves of some defunct economist' [13].

theory'[12] or 'transition theory' [7] might be chosen as the theoretical framework from which to answer questions about a particular social situation, or cardiologists may interpret treatment regimes based on either 'metaphors of love' [14] or, more likely, on the latest theories about statins, a particular class of drugs which reduce cholesterol levels.[13]

Method means a way of teaching or proceeding, derived from the Greek 'méthodos', meaning 'pursuit', or to 'follow after'; thus today it commonly means any special procedure or way of doing things [1]. From this the adjectives methodic, or methodical, arise meaning something done according to a method. In research and practice fields there is often confusion between method and methodology – in literal terms the latter means the logos, or logic of method. The place of methodology in every day practice is not as clear as in, say, research practice, but even then confusion often exists. On the other hand whenever we act we usually employ some tool, technique, or method and often sequence these in particular ways. Think about using a street directory to find out where a friend lives: this involves knowing how a directory is organised, knowing how to use an index, reading map coordinates, and then taking the right directions and turns. So when you successfully arrive at your friend's house you could claim that you have mastered a particular method – street directory using.[14,15]

For the moment let's accept that a generic description of practice comprises a practitioner (P) with a history, a tradition of understanding, possibly a chosen framework of ideas (F), a chosen method (M) and a situation (S) in which they practise. Let's further assume that practice is concerned with understanding, discovering, describing or changing some aspect of a situation. Then if Fig. 3.5 is

---

[12] See http://en.wikipedia.org/wiki/Actor-Network_Theory (Accessed 18th January 2009).

[13] On the other hand some researchers, engineers, doctors etc. who are epistemologically naive empiricists may argue, or imagine, that they come to situations as if they were theory free. Equally, social researchers who are theoretically adept may forget that the purposeful choice of any particular theory does not negate the understanding that as human beings with a history they too have traditions of understanding which they bring forth in the moment, and that these, as embodied understandings, may be different to the theories they espouse in moments of rational reflection [11].

[14] In some areas of practice the idea of 'methodology' is associated with rational choice and with this choice a range of methods and techniques become deployed. A choice to use a street directory could also be seen as example of rational choice of a method and might contrast with some who were native to a locale who had an innate sense of direction or ability to read a landscape. I will explain why this line of argument is important in Chapter 7.

[15] When engaging with systems approaches it is easy to focus on method rather than methodology. I argue that methodology rather than being simply the logos, or logic of method, is something that has to be experienced where the key experience is that of the degree of coherence, or congruence between espoused theory and practice (see [9]); in my example of a Street Directory it might be used methodically (i.e. as method) or methodologically (i.e. as methodology). An example of the latter would be if in response to experiencing the Directory as poorly designed a more effective one was developed based on a redeployment of the underlying concepts or the invention of new ones.

considered systemically, as a whole, potential emergent properties of practice can be seen; these include the possibility of:

- Learning about each or all of F, M or S
- Considering the conduct of the practice – the act of connecting F, M and S as a form of performance – e.g. how effective was the practice (first-order effectiveness)?
- Taking a meta or second order perspective on the practice system-environment relationship (as depicted in Fig. 3.5 by the person operating at two levels)[16]

Exploring this heuristic and moving between first and second–order perspectives exemplifies the practice of moving between levels of abstraction that is so important in systems practice. When this skill is mastered it should be apparent that Fig. 3.5 first 'abstracts' both the practitioner and the methodology choice out of the 'situation' and then moves to yet another level of abstraction that enables us to think about the systemic relationships between four factors, P, F, M and S as a means of better understanding what it is that we do when we do what we do! But these are abstractions as we are always in the situation, although at times we may pretend we are not! In Fig. 3.6 I present another way of interpreting these dynamics.

Figure 3.5 can be used to explore other aspects of practice – by introducing more and different actors, e.g. colleagues, other stakeholders, co-researchers etc.; by reflecting on the implications of epistemological awareness, but perhaps most importantly, for becoming aware of the nature of situations in which practice is being conducted. Figure 3.5 used heuristically enables an exploration of what happens when individuals or groups engage with situations and the choices that are possible when this is done knowingly. Evidence suggests that practitioners, including researchers/scientists or policy makers lack a reflexive understanding of their own practice and the rationalities (or epistemologies) out of which they think and act [10–12].

With awareness of the different understandings within the first and second-order cybernetic lineages depicted in Fig. 2.3 four possible forms of practice can be identified (Fig. 3.6). Within the first-order tradition the relationship between F, M and S can be seen as linear or causal or alternatively circular i.e. with feedback processes operating. In both of these depictions the observer/practitioner is not in evidence. Within the second-order tradition the observer/practitioner is always part of the situation and aware that they are as depicted within participatory praxis. But within this tradition there may be different

---

[16] In this book I will use the terms 'meta' and second-order. Sometimes as in this case I will use them interchangeably because whilst similar they convey slightly different meanings and have different histories as terms. 'Meta' means 'beyond, or transcending' i.e., at a higher conceptual level which theoretically relates to the systemic notion of levels, i.e. a system is meta to a sub-system. 'Second order' has a history of use in mathematics and logic (see http://en.wikipedia.org/wiki/Second-order_logic Accessed 5th April 2009) but my usage can be traced to second-order cybernetics or, the cybernetics of cybernetics. Within this conception it has a different operational dynamic – something applied to itself which attains a higher or different conception. Within this framing second and first order phenomena are understood as different but operating as part of a whole, a duality, or totality.

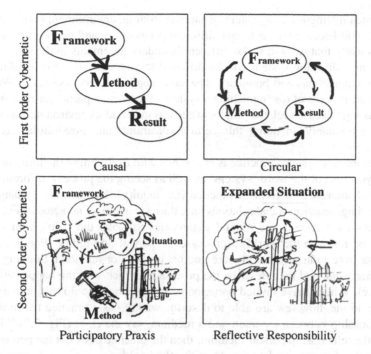

Fig. 3.6 Different practice possibilities within the first and second-order cybernetic traditions of systems practice

levels of awareness: in reflective responsibility the practitioner is aware of, and thus takes responsibility for, their role in creating an F, M, S dynamic or 'performance'.

Drawing on initial work by Peter Checkland [4] and refined through a decade of research training workshops with Ph.D. students at the Open University (UK), I have found the heuristic depicted in Fig. 3.5 useful in opening up a conversational and reflexive space in which practitioners gain insights into the nature of their own practice. I usually start by asking participants to develop a conceptual model of their own practice – they generally find this very hard to do because, I would contend, of the prevalence of the 'mainstream understanding' of situations existing independent of themselves. It is also because most people are very poor at understanding and describing phenomena in dynamic and relational ways; instead the tendency is to rely on linear or monocausal explanations (as noted in Reading 2).

3.4 Bringing Forth Systems of Interest

When teaching systems at the Open University we have found it useful to talk about formulating 'systems of interest' as a means of engaging with complex situations in particular. In the process of distinguishing a system a boundary judgment is made which distinguishes a system of interest from an environment (Fig. 3.3).[17,18] It follows that because we each have different perspectives and interests (histories) then it is likely that we will make different boundary judgments in the same situation, i.e. my education system will be different to yours because we see different elements, connections and boundary. The same applies to 'ecosystems'.[19] What is more, in the process of distinguishing a system of interest, a particular relationship is also distinguished – a relationship between a system and its environment as mediated by a boundary. In doing this certain constraints and possibilities are also created.

Contemporary systems practice is concerned with overcoming the limitations of the everyday use of the word 'system' as well as seeing the process of formulating systems of interest as a form of practice (i.e. inquiry) that facilitates changes in understanding, practice, social relations and, thus, situations as depicted in Fig. 1.3.

Sometimes it takes some mental gymnastics to make the 'flip' from seeing systems as things in the world to processes as part of an inquiry (Fig. 3.4). You might like to test out whether you are good at process, or relational thinking. If so take a short walk and then describe the process by which walking happens (i.e. as a practice). When I have invited participants in workshops to do this activity it is amazing, to me, how few are able to describe walking as a practice that arises in the relationship between a person and a medium, say the floor (Fig. 3.7).[20] If you destroy the relationship person-medium, then the practice does not happen as with the example of the person hanging from the sky hook!

Whilst walking may seem a trivial example it is, I suggest, a profound and not well appreciated insight that has wide applicability. The key point is that walking, like all practices, arises from a relational dynamic which we have come to take for granted. I rarely encounter workshop participants who 'see' this dynamic until it is revealed to them. Instead they tend to see walking as something determined in early childhood or by internal motivational or physiological factors! When participants

---

[17] I use the verb distinguish but need to make clear that the 'system of interest' does not pre-exist the act of making a distinction – in a sense it is a process of bringing forth, like getting a bright idea. Other verbs are possible – formulate, create, invent, generate.

[18] I will say more about the act of making distinctions in later parts of the text – for now I mean the action of distinguishing or discriminating, the noting or making of a difference, the result of which is a difference made or appreciated (following The New Shorter Oxford English Dictionary [19]).

[19] Tansley [18] who did most to establish the concept ecosystem said:

"... the whole method of science... is to isolate systems mentally for the purposes of study, so that the series of isolates we make become the actual objects of our study, whether the isolate be a solar system, a planet, a climatic region, a plant or animal community, an individual organism, an organic molecule or an atom. Actually, the systems we isolate mentally are not only included as parts of larger ones, but they also overlap, interlock and interact with one another. The isolation is partly artificial, but is the only possible way in which we can proceed" (ibid., p. 299–300).

[20] Most people give answers in terms of development of the young human, or in terms of motivation. Common to most answers is that walking arises because of some internal human cause, rather than as a relational dynamic.

Fig. 3.7 How walking happens, or not, as a relational phenomenon between a person and a medium, in this case, a pathway

give these answers they are engaging in deterministic and causal thinking not relational thinking. Others repeat a story they were told – "you learned to walk when you were two" – or they focus on the acquisition of a skill; we rarely consider the relational dynamics involved in any of our skills, we just do them!

The relational thinking I have just described is important to interpreting the general model of practice as described in Fig. 3.5 and the process of distinguishing a system of interest as part of an inquiry (Fig. 3.4). You might like to re-examine these figures in the light of the example of walking as a relational dynamic. This type of relational thinking is central to becoming systemic in what you do.

3.5 Systems Practice – an 'Ideal Type'

In some circles there is a perception that being a systems practitioner is a specialist role mainly carried out by paid consultants but this does not have to be the case. Recent research shows that many people who claim to use systems thinking as part of their practice, whether consultants or managers, do not make this public. This is in addition to those who do it but don't know that that is what they do. Much of systems practice, it would seem, is a 'silent practice'. This creates a conundrum:

- On what basis could we judge whether someone was engaged in systems practice?
- How would we know if they were doing it well?

- How can demand for more and better systems practice be created if it is a silent practice?
- How good is systems practice in improving situations for the better?[21]

At the moment these questions are difficult to answer. In what follows (Chapters 4–8) I build on my general model of practicing (Fig. 3.5), to develop and present a model of a systems practitioner as an 'ideal type'.[22] My aim is to make transparent what occurs in effective systems practice. In creating an ideal type I am not attempting to prescribe or define, but to create, through the use of a particular metaphor, a device to enable reflection and public discussion. Following Krippendorff [15] who argued the need for designers to build their practice through strengthening design discourse, I want to contribute to a more widespread and rigorous discourse about systems practice. This can enable those who claim they are systems practitioners to be more publicly accountable as well as allowing those who would like to develop these skills to know what to do. It may also allow those who are already systems thinkers/practitioners, but are unaware of it, to recognise themselves in a different light.[23]

The 'ideal type' model of a systems practitioner is in part based on my personal experiences of systems practice for managing in situations of complexity. In first proposing this 'ideal type' I was motivated by the question: 'what is it that we do when we do what we do?' In particular, and as depicted in Fig. 3.5, I am concerned with understanding the systemic dynamics of practice in general and more specifically as applied to systems practice for everyday managing – the everyday struggle to keep our heads above water when we try to understand and manage the complexity we experience.[24] This "ideal type", which I introduce in the next chapter, has also been particularly useful as part of our practise at the Open University (UK) as Systems educators and consultants concerned with creating meaningful learning experiences for mature-age students.[25]

---

[21] I raise this point again because I do not want to lose sight of it – but I need several chapters to present what systems practice might be, before addressing this question.

[22] 'Ideal type, also known as pure type, or *Idealtyp* in the original German, is a term most closely associated with sociologist Max Weber (1864–1920). An ideal type is formed from characteristics and elements of the given phenomena, but it is not meant to correspond to all of the characteristics of any one particular case. It is not meant to refer to perfect things, moral ideals nor to statistical averages but rather to stress certain elements common to most cases of the given phenomena'. (See http://en.wikipedia.org/wiki/Ideal_type accessed 19th January 2009)

[23] I would like to think that the trap which currently seems to exist, of no, or little, institutionalised 'demand pull' for systems thinking and practice skills, might be soon overcome.

[24] In this book the idea of practice, or practising, is a general one in that it is something everyone does. The dictionary definition of practice is 'to carry out or perform habitually or constantly... to carry out an action' [19]. Almost everyone has some role in which they practise. Most people occupy a number of roles, in their work or in their community. In these roles it is usual to encounter a number of issues that need dealing with, improving, resolving, or obviating. For example I am a practising father as well as a practising academic.

[25] The median age of students at the Open University UK is 32.

References

1. Barnhart, R. (2001) Chambers Dictionary of Etymology. Chambers: USA.
2. Blackmore, C.P., Ison, R.L. and Jiggins, J. (2007) Social learning: an alternative policy instrument for managing in the context of Europe's water. Environmental Science & Policy 10, 493–498.
3. Checkland, P.B. (1981) Systems Thinking, Systems Practice. Wiley: Chichester.
4. Checkland, P.B. (1985) From optimizing to learning: a development of systems thinking for the 1990s. Journal of the Operational Research Society 36, 757–767.
5. Checkland, P.B. (1999) Soft Systems Methodology: A 30 Year Retrospective. Wiley: Chichester.
6. Checkland, P.B. Poulter, J. (2006) Learning for Action. A short definitive account of soft systems methodology and its use for practitioners, teachers and students. Wiley: Chichester.
7. Frantzeskakia, N and de Haanb, H (2009) Transitions: Two steps from theory to policy. Futures 41(9), 593–606.
8. Ison, R.L. (2001) Systems practice at the United Kingdom's Open University. In J. Wilby and G. Ragsdell (Eds.), Understanding Complexity (pp. 45–54). Kluwer/Plenum: New York.
9. Ison, R.L. and Russell, D.B. (2000) Exploring some distinctions for the design of learning systems. Cybernetics & Human Knowing 7(4), 43–56.
10. Ison, R.L. (2002) Some reflections on a knowledge transfer strategy: a systemic inquiry. In Farming and Rural Systems Research and Extension, Proceedings Fifth IFSA European Symposium, Florence (April).
11. Ison, R.L. (2008a) Methodological challenges of trans-disciplinary research: some systemic reflections. Natures Sciences Sociétés 16, 241–251.
12. Ison, R.L. (2008b) Reprising "wicked problems": social learning, climate change adaptation and the sustainable management of water. Proc. ANZSYS (Australia NZ Systems Society) Conference, Perth, 1–2 December.
13. Keynes, M. The General Theory of Employment, Interest and Money, Book VI, pp. 383–384. http://en.wikipedia.org/wiki/The_General_Theory_of_Employment,_Interest_and_Money Accessed 16th July 2009.
14. Kövecses, Z. (2000) Metaphor and Emotion. Language, culture and body in human feeling. Cambridge University Press: Cambridge.
15. Krippendorff, K. (1995) Redesigning design: An invitation to a responsible future. In P. Tahkokallio and S. Vihma (Eds.), Design – Pleasure or Responsibility? (pp.138–162). University of Art and Design: Helsinki.
16. Morris, R.M. (2005) Thinking about systems for sustainable lifestyles. Open University Systems Society (OUSys) Newsletter No 39 (Autumn), 15–19.
17. Shotter, J. (1993) Conversational Realities: Constructing Life Through Language. Sage: London.
18. Tansley, A.G. (1935) The use and abuse of vegetational concepts and terms. Ecology 16, 284–307.
19. The New Shorter Oxford English Dictionary (1993) Oxford University Press.

Chapter 4
The Juggler: A Way to Understand Systems Practice

4.1 Introduction of the Juggler

It follows from the dictionary definition that a practitioner is anyone involved in practise – in carrying out an action. But as outlined in the previous chapter practice can be understood as a series of elements that are combined to produce a type of performance. If I reflect on my own practice, I am aware that I have a myriad of factors to consider in any given day. What I do is not a simple interaction between practitioner and situation. I experience myself as something of a juggler trying to keep a number of balls in the air as I practise so I am going to employ the metaphor of the systems practitioner as juggler and will focus on four particular balls I think need to be kept in the air for any form of effective systems practice (Fig. 4.1).

Based on my experience, I claim that effective practice involves being aware that all of these four balls need to be juggled – it takes active attention, and some skill, to keep them all in the air. Things start to go wrong if we let any one of them slip. To be an effective practitioner, I find it necessary to continuously think about, and act to maintain, four elements: the processes of being a practitioner, the means we engage with a situation, putting the approach taken into context and managing my own involvement in the situation (Fig. 4.1). The four verbs, the activities, I am drawing your attention to are *being*, *engaging*, *contextualising* and *managing*.

Practice, which is a systemic dynamic, can only be realised through actions, hence my focus on verbs, rather than say nouns. For example, I could have chosen, but rejected, descriptors such as (i) the process, (ii) the approach, (iii) the context and (iv) the manager.

The metaphor of a juggler keeping the four balls in the air is a way to think about what I do when I try to be effective in my own practice. It matches with my experience: it takes concentration and skills to do it well. But metaphors conceal or obscure some features of experience, while calling other features to attention. The juggler metaphor obscures that the four elements of effective practice are related. I cannot juggle them as if they were independent of each other. I can imagine them interacting

R. Ison, *Systems Practice: How to Act in a Climate-Change World*,
DOI 10.1007/978-1-84996-125-7_4, © The Open University 2010.
Published in Association with Springer-Verlag London Limited

Fig. 4.1 For effective practice, what I have distinguished as four balls are juggled. The B-ball symbolises the attributes of Being a practitioner with a particular tradition of understanding. The E-ball symbolises the characteristics ascribed to the 'real-world 'situation that the juggler is Engaging with. The C-ball symbolises the act of Contextualising a particular approach to a new situation. The M-ball is about how the practitioner is Managing their involvement with the situation

with each other through gravitational attraction even when they are up in the air. Further, the juggler can juggle them differently, for example tossing the E ball with the left hand and the B ball with the right hand. These visualisations allow me to say that, in effective practice, the movements of the balls are not only interdependent but also dependent on my actions. Also, when juggling you really only touch one ball at a time, give it a suitable trajectory so that you will be able to return to it while you touch another ball. So it's the way attention has to go among the various domains, a responsible moment of involvement that creates the conditions for continuance of practice.

I'll describe each ball briefly here, and then in the next four chapters expand on the praxis that particularly relates to each. The first ball, Being, is concerned with our own awareness and our ethics of action, thus the responsibility we take as citizens. Though it may manifest differently in various situations, Being is primarily a consequence of the background, experiences and prejudices, or pre-understandings, of being the practitioner. So, to consider the B-ball it is necessary to focus on some of the attributes of the practitioner. One of these attributes is awareness, awareness of self in relation to the balls being juggled and the context for this juggling. The nature of this awareness will be explored.

The second ball is the E-ball – engaging with a 'real world' situation. How a practitioner engages with a situation is not just a property of the situation. The practitioner can choose how to orient and look, and has choices in how to engage. Thus the 'real world' could be experienced as simple or complicated, as a situation or as a system. I will argue that the failure to be aware of the choices we have in juggling the E-ball has given rise, all too often, to policy failure [1] or some other unintended consequence.

The third ball, the C-ball, is concerned with how a systems practitioner puts particular systems approaches into Context for taking action in 'real world' situations. One of the main skills of a systems practitioner is to learn, through experience, to manage the relationship between a particular systems approach and the 'real-world' situation she or he is using it in. Adopting an approach is more than just choosing one of the methods that already exists. This is why I use the phrase 'putting into context', to indicate a process of Contextualisation involved in the choice of approach in relation to situation. Courses that teach about systems approaches are often designed to focus primarily on the C-ball; however, they usually teach how to match an existing systems approach with an area of application.[1]

The final ball the effective practitioner juggles is that of Managing (M). The M-ball is concerned with juggling as an overall performance; managing both the juggling and the desired change in the world. Another way to describe this is as co-managing self and situation. As the term managing is often used to describe the process by which a practitioner engages with a 'real-world' situation it can be considered as a special form of engagement, so later I will explore some of the features associated with what I include in the notion of managing. For example, managing also introduces the idea of change over time, in the situation, the approach and the practitioner – of adapting oneself and one's performance.

I invite you to interpret the juggler metaphor in terms of:

- Your relationship with yourself as a practitioner or your sensibilities of being a practitioner
- The choices you are envisaging about a situation as you and other stakeholders perceive it (i.e. your mode of engaging with a situation of interest)
- Your manner of adapting your practice to the circumstances (contextualising)
- How you plan to perform your practice through the act of managing the overall activity

---

[1] For too long in my view the Systems field has been plagued by method and methodology wars – incessant arguing about the virtues, or otherwise, of particular methods and methodologies, often in the form of a product offered by a consulting group. Unfortunately, this has constrained both the institutionalisation of Systems within academic life as well as drawing attention away from the praxis of systems as described here. As I do not want to perpetuate this unhelpful situation I want to make it clear that in juggling the C-ball it is not just about choice of a method or application of a method but, rather, the question of how a method or methodology can mediate the emergence of situation-improving action. 'Putting into context' could also be understood as a form of bringing forth (as per Maturana – see Proulx [7]) or as a form of context sensitive design [4].

Having introduced all of the key elements of the juggler metaphor I will now provide some general background as to why I consider these four balls to be important. I will also introduce a Reading to exemplify how a particular systems practitioner engages in their 'juggling' and draw out some practical implications. In Parts II and III of the book, I will give examples of my own systems practice, that is my own juggling.

Juggling, as practice, results from a set of relationships. A juggler is a person or living system in a particular context, with body positioned so as to give support from the floor, and in my use of the metaphor, four different balls. If any of these things is taken away, the juggler, the connection to the floor or all the balls then juggling will not arise as a practice (as with my example of walking). In some situations an audience might also be important, especially if juggling for money or another form of performance. If I chose to see this situation as if it were a 'system for making money through a juggling performance' then taking away the audience would destroy the 'system of interest', the interconnected set of relationships that was envisioned. But there's more to this set of relationships than meets the eye. Take the juggler for example, s/he's both a unique person and also part of a lineage of organisms or 'living systems'. All 'living systems' have an evolutionary past, which means biologically we are essentially the same, but an individual developmental past that is unique to each individual. For humans this means we each have a unique set of experiences so that one person's world is always different from another person's world. We humans never truly 'share' common experiences because this is biologically impossible. We can however communicate with each other about our experiences.

Before I introduce Reading 3, which exemplifies a form of systems practice, and which can be used to tease out aspects of 'juggling' I want to take a short conceptual detour. When I introduced the idea of thinking of systems practice in terms of juggling I referred to it as a metaphor. In the past I have been happy enough with this description but I am no longer convinced it is totally appropriate. The essence of how metaphor works is to express one thing in terms of something else, such as in the phrase 'the office is a warzone'. This metaphor invites us to think of the office **as if** it were a warzone; in operational terms it provides a sort of gestalt which is often the basis of innovation, i.e. it takes one to a new place. At no stage do we consider the office to be an actual warzone. Following this logic of how metaphors work my use of 'systems practice as juggler' if it were to be regarded as a metaphor would take you from 'systems practice' as one thing to juggling as something else. But this dynamic is an abstraction divorced from our doing, our actions – thinking, feeling, experiencing – which has led Humberto Maturana with Kathleen Forsythe to coin the term isophor to explain the dynamic of experiencing the same thing through another means – in this case experiencing systems practice by the doing of juggling. To put it rather simplistically by doing some juggling you begin to feel what it would be like to be doing systems practice.[2] On the other hand the balls themselves could still be regarded as metaphors.[3]

---

[2] It was Forsythe's paper on Cathedrals of the Mind [3, p. 175] that led Maturana to invent the term isophor [2]. However, I am not sure the metaphor/isophor distinction is an either/or choice – we can choose to see it as a metaphor – which reveals some things – or to see it as an isophor which reveals different things.

[3] Humberto Maturana (personal communication, 19th August 2009) explains what he meant and what he means with the word "isophor" in the following way: 'The notion of metaphor invites to

4.2 An Example of Systems Practice as Juggling

Having introduced the 'juggler' as a means to appreciate systems practice as well as the idea that there are choices to make about situations as part of a relational dynamic (as discussed in Chapter 3 and depicted in Fig. 3.5) I would now like to introduce a reading (Reading 3) which exemplifies some further aspects of systems practice in comparison to earlier readings. You will gain much from this reading in its own right but as I am using readings for a particular purpose I would like to invite you to do a little extra work as you read. To reiterate, my purpose is to create the circumstances where you can better appreciate what systems practitioners do when they do what they do. Of course I also want you to make connections with your own life and to begin the process of adding variety to your own systems practice.

My invitation is that as you engage with this reading [5], please take a particular approach by attending to the following questions:

1. What is the situation and its nature?
2. Who are the main 'actors' in the situation?
3. What is at issue?
4. What different ways of understanding and/or engaging with the situation are described by the author?
5. What does the article reveal about the author?
6. Given your current understanding of the idea of juggling in relation to systems practice what can you say about Donella Meadows' juggling?
7. Faced with the same or a similar situation would you think about it similarly or differently?
8. What does this article reveal about your own ways of thinking and acting?

There is much to be gained from this reading other than answers to my questions. But through your attempt to answer them you will gain the experience of an inquiry process based on the metaphor of the juggler. This inquiry process could also, of course, be applied to the earlier readings as well as to readings in later chapters.

---

understand something by proposing an evocative image of a different process in a different domain. With the metaphor you liberate the imagination of the listener by inviting him or her to go to a different domain and follow his or her emotioning. When I proposed the notion of isophor to Kathleen [Forsythe] I wanted it to refer to a proposition that takes you to another case of the same kind in another domain. So, with an isophor you would not liberate the imagination of the listener but you would focus his or her attention on the configuration of processes or relations that you want to grasp. In these circumstances, the fact that a juggler puts his or her attention on the locality of the movement of one ball as he or she plays with them, knowing how to move at every instant in relation to all the other balls, shows that the whole matrix of relations and movements of the constellation of balls is accessible to him or her all the time. So, juggling is an isophor of the vision that one must have of the operational-relational matrix in which something occurs to be able to honestly claim that one understands it. That is, juggling is an isophor of the vision that one wants to have to claim that one understands, for example, a biological or a cultural happening'.

Reading 3

Places to Intervene in a System

Donella H. Meadows

Folks who do systems analysis have a great belief in "leverage points." These are places within a complex system (a corporation, an economy, a living body, a city, an ecosystem) where a small shift in one thing can produce big changes in everything. The systems community has a lot of lore about leverage points. Those of us who were trained by the great Jay Forrester at MIT have absorbed one of his favorite stories. "People know intuitively where leverage points are. Time after time I've done an analysis of a company and I've figured out a leverage point. Then I've gone to the company and discovered that everyone is pushing it in the wrong direction!"

The classic example of that backward intuition was Forrester's first world model.

Asked by the Club of Rome to show how major global problems poverty and hunger, environmental destruction, resource depletion, urban deterioration, unemployment, are related and how they might be solved, Forrester came out with a clear leverage point:

Growth. Both population and economic growth. Growth has costs among which are poverty and hunger, environmental destruction the whole list of problems we are trying to solve with growth!

The world's leaders are correctly fixated on economic growth as the answer to virtually all problems, but they're pushing with all their might in the wrong direction.

Counterintuitive. That's Forrester's word to describe complex systems. The systems analysts I know have come up with no quick or easy formulas for finding leverage points. Our counter intuitions aren't that well developed. Give us a few months or years and we'll model the system and figure it out. We know from bitter experience that when we do discover the system's leverage points, hardly anybody will believe us.

Very frustrating. So one day I was sitting in a meeting about the new global trade regime, NAFTA and GATT and the World Trade Organization. The more I listened, the more I began to simmer inside. "This is a HUGE NEW SYSTEM people are inventing!" I said to myself. "They haven't the slightest idea how it will behave," myself said back to me. "It's cranking the system in the wrong direction: growth, growth at any price!! And the control measures these nice folks are talking about, small parameter adjustments, negative feedback loops, are PUNY!"

(continued)

Reading 3 (continued)

Suddenly, without quite knowing what was happening, I got up, marched to the flip chart, tossed over a clean page, and wrote: "Places to Intervene in a System," followed by nine items:

9 Numbers (subsidies, taxes, standards)
8 Material stocks and flows
7 Regulating negative feedback loops
6 Driving positive feedback loops
5 Information flows
4 The rules of the system (incentives, punishment, constraints)
3 The power of self-organization
2 The goals of the system
1 The mindset or paradigm out of which the goals, rules, feedback structure arise

Everyone in the meeting blinked in surprise, including me. "That's brilliant!" someone breathed. "Huh?" said someone else. I realized that I had a lot of explaining to do.

In a minute I'll go through the list, translate the jargon, give examples and exceptions. First I want to place the list in a context of humility. What bubbled up in me that day was distilled from decades of rigorous analysis of many different kinds of systems done by many smart people. But complex systems are, well, complex. It's dangerous to generalize about them. What you are about to read is not a recipe for finding leverage points. Rather it's an invitation to think more broadly about system change. That's why leverage points are not intuitive.

9. Numbers

Numbers ("parameters" in systems jargon) determine how much of a discrepancy turns which faucet how fast. Maybe the faucet turns hard, so it takes a while to get the water flowing. Maybe the drain is blocked and can allow only a small flow, no matter how open it is. Maybe the faucet can deliver with the force of a fire hose. These considerations are a matter of numbers, some of which are physically locked in, but most of which are popular intervention points.

Consider the national debt. It's a negative bathtub, a money hole. The rate at which it sinks is the annual deficit. Tax income makes it rise, government expenditures make it fall. Congress and the president argue endlessly about the many parameters that open and close tax faucets and spending drains. Since those faucets and drains are connected to the voters, these are politically charged parameters. But, despite all the fireworks, and no matter which party is in charge, the money hole goes on sinking, just at different rates.

(continued)

Reading 3 (continued)

The amount of land we set aside for conservation. The minimum wage. How much we spend on AIDS research or Stealth bombers. The service charge the bank extracts from your account. All these are numbers, adjustments to faucets. So, by the way, is firing people and getting new ones. Putting different hands on the faucets may change the rate at which they turn, but if they're the same old faucets, plumbed into the same system, turned according to the same information and rules and goals, the system isn't going to change much. Bill Clinton is different from George Bush, but not all that different.

Numbers are last on my list of leverage points. Diddling with details, arranging the deck chairs on the Titanic. Probably ninety-five percent of our attention goes to numbers, but there's not a lot of power in them. Not that parameters aren't important, they can be, especially in the short term and to the individual who's standing directly in the flow. But they RARELY CHANGE BEHAVIOR. If the system is chronically stagnant, parameter changes rarely kick-start it. If it's wildly variable, they don't usually stabilize it. If it's growing out of control, they don't brake it.

Whatever cap we put on campaign contributions, it doesn't clean up politics. The Feds fiddling with the interest rate haven't made business cycles go away. (We always forget that during upturns, and are shocked, shocked by the downturns.) Spending more on police doesn't make crime go away.

However, there are critical exceptions. Numbers become leverage points when they go into ranges that kick off one of the items higher on this list. Interest rates or birth rates control the gains around positive feedback loops. System goals are parameters that can make big differences. Sometimes a system gets onto a chaotic edge, where the tiniest change in a number can drive it from order to what appears to be wild disorder.

Probably the most common kind of critical number is the length of delay in a feedback loop. Remember that bathtub on the fourth floor I mentioned, with the water heater in the basement? I actually experienced one of those once, in an old hotel in London. It wasn't even a bathtub with buffering capacity; it was a shower. The water temperature took at least a minute to respond to my faucet twists. Guess what my shower was like. Right, oscillations from hot to cold and back to hot, punctuated with expletives. Delays in negative feedback loops cause oscillations. If you're trying to adjust a system state to your goal, but you only receive delayed information about what the system state is, you will overshoot and undershoot.

Same if your information is timely, but your response isn't. For example, it takes several years to build an electric power plant, and then that plant lasts, say, thirty years. Those delays make it impossible to build exactly the right number of plants to supply a rapidly changing demand. Even with immense effort at forecasting, almost every electricity industry in the world experi-

(continued)

Reading 3 (continued)

ences long oscillations between overcapacity and undercapacity. A system just can't respond to short-term changes when it has long-term delays. That's why a massive central-planning system, such as the Soviet Union or General Motors, necessarily functions poorly.

A delay in a feedback process is critical RELATIVE TO RATES OF CHANGE (growth, fluctuation, decay) IN THE SYSTEM STATE THAT THE FEEDBACK LOOP IS TRYING TO CONTROL. Delays that are too short cause overreaction, oscillations amplified by the jumpiness of the response. Delays that are too long cause damped, sustained, or exploding oscillations, depending on how much too long. At the extreme they cause chaos. Delays in a system with a threshold, a danger point, and a range past which irreversible damage can occur, cause overshoot and collapse.

Delay length would be a high leverage point, except for the fact that delays are not often easily changeable. Things take as long as they take. You can't do a lot about the construction time of a major piece of capital, or the maturation time of a child, or the growth rate of a forest. It's usually easier to slow down the change rate (positive feedback loops, higher on this list), so feedback delays won't cause so much trouble. Critical numbers are not nearly as common as people seem to think they are. Most systems have evolved or are designed to stay out of sensitive parameter ranges. Mostly, the numbers are not worth the sweat put into them.

8. Material stocks and flows

The plumbing structure, the stocks and flows and their physical arrangement, can have an enormous effect on how a system operates. When the Hungarian road system was laid out so all traffic from one side of the nation to the other had to pass through central Budapest, that determined a lot about air pollution and commuting delays that are not easily fixed by pollution control devices, traffic lights, or speed limits. The only way to fix a system that is laid out wrong is to rebuild it, if you can. Often you can't, because physical building is a slow and expensive kind of change. Some stock-and-flow structures are just plain unchangeable.

The baby-boom swell in the US population first caused pressure on the elementary school system, then high schools and colleges, then jobs and housing, and now we're looking forward to supporting its retirement. Not much to do about it, because five-year-olds become six-year-olds, and sixty-four-year-olds become sixty-five-year-olds predictably and unstoppably. The same can be said for the lifetime of destructive CFC molecules in the ozone layer, for the rate at which contaminants get washed out of aquifers, for the fact that an inefficient car fleet takes ten to twenty years to turn over.

The possible exceptional leverage point here is in the size of stocks, or buffers. Consider a huge bathtub with slow in and outflows. Now think about a small one with fast flows. That's the difference between a lake and a river.

(continued)

Reading 3 (continued)

You hear about catastrophic river floods much more often than catastrophic lake floods, because stocks that are big, relative to their flows, are more stable than small ones. A big, stabilizing stock is a buffer.

The stabilizing power of buffers is why you keep money in the bank rather than living from the flow of change through your pocket. It's why stores hold inventory instead of calling for new stock just as customers carry the old stock out the door. It's why we need to maintain more than the minimum breeding population of an endangered species. Soils in the eastern US are more sensitive to acid rain than soils in the west, because they haven't got big buffers of calcium to neutralize acid. You can often stabilize a system by increasing the capacity of a buffer. But if a buffer is too big, the system gets inflexible. It reacts too slowly. Businesses invented just-in-time inventories, because occasional vulnerability to fluctuations or screw-ups is cheaper than certain, constant inventory costs, and because small-to-vanishing inventories allow more flexible response to shifting demand.

There's leverage, sometimes magical, in changing the size of buffers. But buffers are usually physical entities, not easy to change. The acid absorption capacity of eastern soils is not a leverage point for alleviating acid rain damage. The storage capacity of a dam is literally cast in concrete. Physical structure is crucial in a system, but the leverage point is in proper design in the first place. After the structure is built, the leverage is in understanding its limitations and bottlenecks and refraining from fluctuations or expansions that strain its capacity.

7. Regulating negative feedback loops

Now we're beginning to move from the physical part of the system to the information and control parts, where more leverage can be found. Nature evolves negative feedback loops and humans invent them to keep system states within safe bounds. A thermostat loop is the classic example. Its purpose is to keep the system state called "room temperature" fairly constant at a desired level. Any negative feedback loop needs a goal (the thermostat setting), a monitoring and signaling device to detect excursions from the goal (the thermostat), and a response mechanism (the furnace and/or air conditioner, fans, heat pipes, fuel, etc.).[4]

---

[4] This claim may warrant critical scrutiny – a common fallacy here is the idea that a goal is needed, i.e. the signalling device that responds to above or below is all that matters. However, we interpret the "goal" based on the result. We create a goal with our thermostat setting, but the 'thermostat system' itself has no "goal". In other words, the goal is only a heuristic invention, not part of the system (unless the observer/designer with a goal in mind is included), but about our relation to the mechanism, it represents our value.

(continued)

A complex system usually has numerous negative feedback loops it can bring into play, so it can self-correct under different conditions and impacts. Some of those loops may be inactive much of the time, like the emergency cooling system in a nuclear power plant, or your ability to sweat or shiver to maintain your body temperature. One of the big mistakes we make is to strip away these emergency response mechanisms because they aren't often used and they appear to be costly. In the short term we see no effect from doing this. In the long term, we narrow the range of conditions over which the system can survive.

One of the most heartbreaking ways we do this is in encroaching on the habitats of endangered species. Another is in encroaching on our own time for rest, recreation, socialization and meditation.

The "strength" of a negative loop, its ability to keep its appointed stock at or near its goal, depends on the combination of all its parameters and links, the accuracy a rapidity of monitoring, the quickness and power of response, the directness and size of corrective flows.

There can be leverage points here. Take markets, for example, the negative feedback systems that are all but worshipped by economists, and they can indeed be marvels of self-correction, as prices vary to keep supply and demand in balance. The more the price, the central signal to both producers and consumers, is kept clear, unambiguous, timely, and truthful, the more smoothly markets will operate. Prices that reflect full costs will tell consumers how much they can actually afford and will reward efficient producers. Companies and governments are fatally attracted to the price leverage point, of course, all of them pushing in the wrong direction with subsidies, fixes, externalities, taxes, and other forms of confusion. The REAL leverage here is to keep them from doing it. Hence anti-trust laws, truth-in-advertising laws, attempts to internalize costs (such as pollution taxes), the removal of perverse subsidies, and other ways of leveling market playing fields.

The strength of a negative feedback loop is important RELATIVE TO THE IMPACT IT IS DESIGNED TO CORRECT. If the impact increases in strength, the feedbacks have to be strengthened too.

A thermostat system may work fine on a cold winter day, but open all the windows and its corrective power will fail. Democracy worked better before the advent of the brainwashing power of centralized mass communications. Traditional controls on fishing were sufficient until radar spotting and drift nets and other technologies made it possible for a few actors to wipe out the fish. The power of big industry calls for the power of big government to hold it in check; a global economy makes necessary a global government.

(continued)

Reading 3 (continued)

Here are some other examples of strengthening negative feedback controls to improve a system's self-correcting abilities: preventive medicine, exercise, and good nutrition to bolster the body's ability to fight disease, integrated pest management to encourage natural predators of crop pests, the Freedom of Information Act to reduce government secrecy, protection for whistle blowers, impact fees, pollution taxes and performance bonds to recapture the externalized public costs of private benefits.

6. Driving positive feedback loops

A positive feedback loop is self-reinforcing. The more it works, the more it has power to work some more.

The more people catch the flu, the more they infect other people. The more babies are born, the more people grow up to have babies. The more money you have in the bank, the more interest you earn, the more money you have in the bank. The more the soil erodes, the less vegetation it can support, the fewer roots and leaves to soften rain and runoff, the more soil erodes. The more high-energy neutrons in the critical mass, the more they knock into nuclei and generate more.

Positive feedback loops drive growth, explosion, erosion, and collapse in systems. A system with an unchecked positive loop ultimately will destroy itself. That's why there are so few of them.

Usually a negative loop kicks in sooner or later. The epidemic runs out of infectable people, or people take increasingly strong steps to avoid being infected. The death rate rises to equal the birth rate, or people see the consequences of unchecked population growth and have fewer babies. The soil erodes away to bedrock, and after a million years the bedrock crumbles into new soil, or people put up check dams and plant trees.

In those examples, the first outcome is what happens if the positive loop runs its course, the second is what happens if there's an intervention to reduce its power.

Reducing the gain around a positive loop, slowing the growth, is usually a more powerful leverage point in systems than strengthening negative loops, and much preferable to letting the positive loop run.

Population and economic growth rates in the world model are leverage points, because slowing them gives the many negative loops, through technology and markets and other forms of adaptation, time to function. It's the same as slowing the car when you're driving too fast, rather than calling for more responsive brakes or technical advances in steering.

The most interesting behavior that rapidly turning positive loops can trigger is chaos. This wild, unpredictable, unreplicable, and yet bounded behavior

(continued)

Reading 3 (continued)

happens when a system starts changing much, much faster than its negative loops can react to it.

For example, if you keep raising the capital growth rate in the world model, eventually you get to a point where one tiny increase more will shift the economy from exponential growth to oscillation. Another nudge upward gives the oscillation a double beat. And just the tiniest further nudge sends it into chaos.

I don't expect the world economy to turn chaotic any time soon (not for that reason, anyway). That behavior occurs only in unrealistic parameter ranges, equivalent to doubling the size of the economy within a year. Realworld systems do turn chaotic, however, if something in them can grow or decline very fast. Fast-replicating bacteria or insect populations, very infectious epidemics, wild speculative bubbles in money systems, neutron fluxes in the guts of nuclear power plants. These systems are hard to control, and control must involve slowing down the positive feedbacks.

In more ordinary systems, look for leverage points around birth rates, interest rates, erosion rates, "success to the successful" loops, any place where the more you have of something, the more you have the possibility of having more.

5. Information flows

There was this subdivision of identical houses, the story goes, except that the electric meter in some of the houses was installed in the basement and in others it was installed in the front hall, where the residents could see it constantly, going round faster or slower as they used more or less electricity.

Electricity consumption was 30 percent lower in the houses where the meter was in the front hall.

Systems-heads love that story because it's an example of a high leverage point in the information structure of the system. It's not a parameter adjustment, not a strengthening or weakening of an existing loop. It's a NEW LOOP, delivering feedback to a place where it wasn't going before.

In 1986 the US government required that every factory releasing hazardous air pollutants report those emissions publicly. Suddenly everyone could find out precisely what was coming out of the smokestacks in town. There was no law against those emissions, no fines, no determination of "safe" levels, just information. But by 1990 emissions dropped 40 percent. One chemical company that found itself on the Top Ten Polluters list reduced its emissions by 90 percent, just to "get off that list."

(continued)

Reading 3 (continued)

Missing feedback is a common cause of system malfunction. Adding or rerouting information can be a powerful intervention, usually easier and cheaper than rebuilding physical structure.

The tragedy of the commons that is exhausting the world's commercial fisheries occurs because there is no feedback from the state of the fish population to the decision to invest in fishing vessels. (Contrary to economic opinion, the price of fish doesn't provide that feedback. As the fish get more scarce and hence more expensive, it becomes all the more profitable to go out and catch them. That's a perverse feedback, a positive loop that leads to collapse.)

It's important that the missing feedback be restored to the right place and in compelling form. It's not enough to inform all the users of an aquifer that the groundwater level is dropping. That could trigger a race to the bottom. It would be more effective to set a water price that rises steeply as the pumping rate exceeds the recharge rate.

Suppose taxpayers got to specify on their return forms what government services their tax payments must be spent on. (Radical democracy!) Suppose any town or company that puts a water intake pipe in a river had to put it immediately DOWNSTREAM from its own outflow pipe. Suppose any public or private official who made the decision to invest in a nuclear power plant got the waste from that plant stored on his/her lawn.

There is a systematic tendency on the part of human beings to avoid accountability for their own decisions. That's why there are so many missing feedback loops, and why this kind of leverage point is so often popular with the masses, unpopular with the powers that be, and effective, if you can get the powers that be to permit it to happen or go around them and make it happen anyway.

4. The rules of the system (incentives, punishments, constraints)

The rules of the system define its scope, boundaries, degrees of freedom. Thou shalt not kill. Everyone has the right of free speech. Contracts are to be honored. The president serves four-year terms and cannot serve more than two of them. Nine people on a team, you have to touch every base, three strikes and you're out. If you get caught robbing a bank, you go to jail.

Mikhail Gorbachev came to power in the USSR and opened information flows (glasnost) and changed the economic rules (perestroika), and look what happened.

Constitutions are strong social rules. Physical laws such as the second law of thermodynamics are absolute rules, if we understand them correctly. Laws, punishments, incentives, and informal social agreements are progressively weaker rules.

(continued)

Reading 3 (continued)

To demonstrate the power of rules, I ask my students to imagine different ones for a college. Suppose the students graded the teachers. Suppose you come to college when you want to learn something, and you leave when you've learned it. Suppose professors were hired according to their ability to solve real-world problems, rather than to publish academic papers. Suppose a class got graded as a group, instead of as individuals.

Rules change behavior. Power over rules is real power.

That's why lobbyists congregate when Congress writes laws, and why the Supreme Court, which interprets and delineates the Constitution, the rules for writing the rules, has even more power than Congress.

If you want to understand the deepest malfunctions of systems, pay attention to the rules, and to who has power over them.

That's why my systems intuition was sending off alarm bells as the new world trade system was explained to me. It is a system with rules designed by corporations, run by corporations, for the benefit of corporations. Its rules exclude almost any feedback from other sectors of society. Most of its meetings are closed to the press (no information, no feedback). It forces nations into positive loops, competing with each other to weaken environmental and social safeguards in order to attract corporate investment. It's a recipe for unleashing "success to the successful" loops.

3. The power of self-organization

The most stunning thing living systems can do is to change themselves utterly by creating whole new structures and behaviors. In biological systems that power is called evolution. In human economies it's called technical advance or social revolution. In systems lingo it's called self-organization.

Self-organization means changing any aspect of a system lower on this list, adding or deleting new physical structure, adding or deleting negative or positive loops or information flows or rules. The ability to self-organize is the strongest form of system resilience, the ability to survive change by changing.

The human immune system can develop responses to (some kinds of) insults it has never before encountered. The human brain can take in new information and pop out completely new thoughts.

Self-organization seems so wondrous that we tend to regard it as mysterious, miraculous. Economists often model technology as literal manna from heaven, coming from nowhere, costing nothing, increasing the productivity of an economy by some steady percent each year. For centuries people have regarded the spectacular variety of nature with the same awe. Only a divine creator could bring forth such a creation.

In fact the divine creator does not have to produce miracles. He, she, or it just has to write clever RULES FOR SELF-ORGANIZATION. These rules

(continued)

govern how, where, and what the system can add onto or subtract from itself under what conditions.

Self-organizing computer models demonstrate that delightful, mindboggling patterns can evolve from simple evolutionary algorithms. (That need not mean that real-world algorithms are simple, only that they can be.) The genetic code that is the basis of all biological evolution contains just four letters, combined into words of three letters each. That code, and the rules for replicating and rearranging it, has spewed out an unimaginable variety of creatures.

Self-organization is basically a matter of evolutionary raw material, a stock of information from which to select possible patterns, and a means for testing them. For biological evolution the raw material is DNA, one source of variety is spontaneous mutation, and the testing mechanism is something like punctuated Darwinian selection. For technology the raw material is the body of understanding science has accumulated. The source of variety is human creativity (whatever THAT is) and the selection mechanism is whatever the market will reward or whatever governments and foundations will fund or whatever tickles the fancy of crazy inventors.

When you understand the power of self-organization, you begin to understand why biologists worship biodiversity even more than economists worship technology. The wildly varied stock of DNA, evolved and accumulated over billions of years, is the source of evolutionary potential, just as science libraries and labs and scientists are the source of technological potential. Allowing species to go extinct is a systems crime, just as randomly eliminating all copies of particular science journals, or particular kinds of scientists, would be.

The same could be said of human cultures, which are the store of behavioral repertoires accumulated over not billions, but hundreds of thousands of years. They are a stock out of which social evolution can arise. Unfortunately, people appreciate the evolutionary potential of cultures even less than they understand the potential of every genetic variation in ground squirrels. I guess that's because one aspect of almost every culture is a belief in the utter superiority of that culture.

Any system, biological, economic, or social, that scorns experimentation and wipes out the raw material of innovation is doomed over the long term on this highly variable planet.

The intervention point here is obvious but unpopular. Encouraging diversity means losing control. Let a thousand flowers bloom and ANYTHING could happen!

Who wants that?

(continued)

Reading 3 (continued)

2. The goals of the system

Right there, the push for control is an example of why the goal of a system is even more of a leverage point than the self-organizing ability of a system.

If the goal is to bring more and more of the world under the control of one central planning system (the empire of Genghis Khan, the world of Islam, the People's Republic of China, WalMart, Disney), then everything further down the list, even self-organizing behavior, will be pressured or weakened to conform to that goal.[5]

That's why I can't get into arguments about whether genetic engineering is a good or a bad thing. Like all technologies, it depends upon who is wielding it, with what goal. The only thing one can say is that if corporations wield it for the purpose of generating marketable products, that is a very different goal, a different direction for evolution than anything the planet has seen so far.

There is a hierarchy of goals in systems. Most negative feedback loops have their own goals, to keep the bath water at the right level, to keep the room temperature comfortable, to keep inventories stocked at sufficient levels. They are small leverage points. The big leverage points are the goals of entire systems.

People within systems don't often recognize what whole-system goal they are serving. To make profits, most corporations would say, but that's just a rule, a necessary condition to stay in the game. What is the point of the game? To grow, to increase market share, to bring the world (customers, suppliers, regulators) more under the control of the corporation, so that its operations become ever more shielded from uncertainty. That's the goal of a cancer cell too and of every living population. It's only a bad one when it isn't countered by higher-level negative feedback loops with goals of keeping the system in balance. The goal of keeping the market competitive has to trump the goal of each corporation to eliminate its competitors. The goal of keeping populations in balance and evolving has to trump the goal of each population to commandeer all resources into its own metabolism.

I said a while back that changing the players in a system is a low-level intervention, as long as the players fit into the same old system. The exception to that rule is at the top, if a single player can change the system's goal.

---

[5] From a second-order cybernetic perspective it could be claimed that Donella, like most people, was at this point in her thinking not aware that Control and Goal are both human concepts grounded in our ability to imagine and desire a particular configuration, with the goal being a description of the condition of a system under the configuration of being subject to actions named control. It's our belief that we can choose to do or not do the actions that turns them into a "control".

(continued)

Reading 3 (continued)

I have watched in wonder as, only very occasionally, a new leader in an organization, from Dartmouth College to Nazi Germany, comes in, enunciates a new goal, and single-handedly changes the behavior of hundreds or thousands or millions of perfectly rational people. That's what Ronald Reagan did. Not long before he came to office, a president could say, "Ask not what government can do for you, ask what you can do for the government," and no one even laughed. Reagan said the goal is not to get the people to help the government and not to get government to help the people, but to get the government off our backs. One can argue, and I would, that larger system changes let him get away with that. But the thoroughness with which behavior in the US and even the world has been changed since Reagan is testimony to the high leverage of articulating, repeating, standing for, insisting upon new system goals.

1. The mindset or paradigm out of which the system arises

Another of Jay Forrester's systems sayings goes: It doesn't matter how the tax law of a country is written. There is a shared idea in the minds of the society about what a "fair" distribution of the tax load is. Whatever the rules say, by fair means or foul, by complications, cheating, exemptions or deductions, by constant sniping at the rules, the actual distribution of taxes will push right up against the accepted idea of "fairness."

The shared idea in the minds of society, the great unstated assumptions, unstated because unnecessary to state; everyone knows them, constitute that society's deepest set of beliefs about how the world works. There is a difference between nouns and verbs. People who are paid less are worth less. Growth is good. Nature is a stock of resources to be converted to human purposes. Evolution stopped with the emergence of *Homo sapiens*. One can "own" land. Those are just a few of the paradigmatic assumptions of our culture, all of which utterly dumbfound people of other cultures.

Paradigms are the sources of systems. From them come goals, information flows, feedbacks, stocks, flows.

The ancient Egyptians built pyramids because they believed in an afterlife. We build skyscrapers, because we believe that space in downtown cities is enormously valuable. (Except for blighted spaces, often near the skyscrapers, which we believe are worthless.) Whether it was Copernicus and Kepler showing that the earth is not the center of the universe, or Einstein hypothesizing that matter and energy are interchangeable, or Adam Smith postulating that the selfish actions of individual players in markets wonderfully accumulate to the common good.

People who manage to intervene in systems at the level of paradigm hit a leverage point that totally transforms systems.

(continued)

Reading 3 (continued)

You could say paradigms are harder to change than anything else about a system, and therefore this item should be lowest on the list, not the highest. But there's nothing physical or expensive or even slow about paradigm change. In a single individual it can happen in a millisecond. All it takes is a click in the mind, a new way of seeing. Of course individuals and societies do resist challenges to their paradigm harder than they resist any other kind of change.

So how do you change paradigms? Thomas Kuhn, who wrote the seminal book about the great paradigm shifts of science, has a lot to say about that. In a nutshell, you keep pointing at the anomalies and failures in the old paradigm, you come yourself, loudly, with assurance, from the new one, you insert people with the new paradigm in places of public visibility and power. You don't waste time with reactionaries; rather you work with active change agents and with the vast middle ground of people who are open-minded.

Systems folks would say one way to change a paradigm is to model a system, which takes you outside the system and forces you to see it whole. We say that because our own paradigms have been changed that way.

0. The power to transcend paradigms

Sorry, but to be truthful and complete, I have to add this kicker. The highest leverage of all is to keep oneself unattached in the arena of paradigms, to realize that NO paradigm is "true," that even the one that sweetly shapes one's comfortable worldview is a tremendously limited understanding of an immense and amazing universe.

It is to "get" at a gut level the paradigm that there are paradigms, and to see that that itself is a paradigm, and to regard that whole realization as devastatingly funny. It is to let go into Not Knowing.

Illustration 4.1

(continued)

Reading 3 (continued)

People who cling to paradigms (just about all of us) take one look at the spacious possibility that everything we think is guaranteed to be nonsense and pedal rapidly in the opposite direction. Surely there is no power, no control, not even a reason for being, much less acting, in the experience that there is no certainty in any worldview. But everyone who has managed to entertain that idea, for a moment or for a lifetime, has found it a basis for radical empowerment. If no paradigm is right, you can choose one that will help achieve your purpose. If you have no idea where to get a purpose, you can listen to the universe (or put in the name of your favorite deity here) and do his, her, its will, which is a lot better informed than your will.

It is in the space of mastery over paradigms that people throw off addictions, live in constant joy, bring down empires, get locked up or burned at the stake or crucified or shot, and have impacts that last for millennia.

Back from the sublime to the ridiculous, from enlightenment to caveats. There is so much that has to be said to qualify this list. It is tentative and its order is slithery. There are exceptions to every item on it. Having the list percolating in my subconscious for years has not transformed me into a Superwoman. I seem to spend my time running up and down the list, trying out leverage points wherever I can find them. The higher the leverage point, the more the system resists changing it – that's why societies rub out truly enlightened beings.

I don't think there are cheap tickets to system change. You have to work at it, whether that means rigorously analyzing a system or rigorously casting off paradigms. In the end, it seems that leverage has less to do with pushing levers than it does with disciplined thinking combined with strategically, profoundly, madly letting go.

Source: Meadows, D.H. (1997) 'Places to Intervene in a System', *Whole Earth*, Winter.

I like this Reading a lot. Why? Because it reveals the passion, enthusiasm and conviction, as well as the analytical and conceptual rigour, of the author. It also, as with previous readings, provides concepts and examples that I can build into my own systems practice. Importantly though, the paper gives many rich insights into Donella Meadows' form of systems practice. Let me expand on this point by providing responses (R), from my perspective, to the questions (Q) I posed earlier (of course I do not expect you to have responded in the same way. Nor are there right or wrong answers):

Q. What is the situation and its nature?

R. In a first-order sense the situation is the (then) new global trade regime, NAFTA and GATT and the World Trade Organisation – some would claim this was 'the problem' but it is clear from Donella's own behaviour that the nature of the situation was (and still is) contested. "This is a HUGE NEW SYSTEM people are inventing!" she said to herself. Possibly she assumed the situation to be a complex system much as she does for 'a corporation, an economy, a living body, a city, an ecosystem.' More specifically however the situation was Donella's participation in a meeting where this important, but complex matter was being discussed and where the conceptualisations and explanations were at odds with her own: "I said to myself. 'They haven't the slightest idea how it will behave,' myself said back to me. 'It's cranking the system in the wrong direction: growth, growth at any price!! And the control measures these nice folks are talking about, small parameter adjustments, negative feedback loops, are PUNY!'" It was her experiences in this meeting that triggered the actions and reflections which she writes about in the paper.

Q. Who are the main 'actors' in the situation?

R. These are not made clear; we do not know how Donella came to be in the meeting or who else was present. She does suggest a main set of actors when she says: 'It [NAFTA, GATT] is a system with rules designed by corporations, run by corporations, for the benefit of corporations. Its rules exclude almost any feedback from other sectors of society.' Clearly governments, policy makers, etc. are involved. These are the actors that come to attention if I focus on the situation as something independent of Donella. If I take on board the dynamic depicted in Fig. 3.5 other actors become apparent such as (i) Jay Forrester, of MIT, the 'founding father' of 'systems dynamics'[8], (ii) the 'systems community' and 'systems folks' who do (iii) systems analysis – thus systems analysts.

Illustration 4.2

Q. What is at issue?

R. A simplistic response would be to say the design of free trade agreements and other institutional arrangements to foster globalisation and
international trade. This is certainly at issue but it is more than this as
when Donella says: "my systems intuition was sending off alarm bells
as the new world trade system was explained to me. It is a system with
rules designed by corporations, run by corporations, for the benefit of
corporations. Its rules exclude almost any feedback from other sectors of
society. Most of its meetings are closed to the press (no information, no
feedback). It forces nations into positive loops, competing with each
other to weaken environmental and social safeguards in order to attract
corporate investment. It's a recipe for unleashing 'success to the successful' loops." To me this account amounts to a common, but often
unacknowledged, phenomenon – a contestation over what constitutes a
valid explanation and/or design for something. It seems to me highly
possible that for Donella the underlying assumptions, concepts and
explanations that were dominating the discussion in the room were, in
her experience, totally inadequate.

Q. What different ways of understanding and/or engaging with the situation
are described by the author?

R. In many ways this question gets to the crux of the article. The emotion
needed to jump up and explicate her intervention strategies reveal that
she was emotionally connected to, or engaged with, the situation
(which, as I will explain later, I view as inescapable but desirable when
acknowledged and reflected upon, though in academic circles emotion
is often sublimated or censored). As I outlined above she chooses the
concept of 'system' or 'complex system' as a means to engage with the
situation intellectually, hence the title of her article 'places to intervene
in a system'; I counted 96 uses of the word 'system', one use of 'systematic', none of 'systemic'. Occasionally the 'problem' metaphor is
apparent (four mentions). If I use the ten strategies to intervene, that she
outlines, to analyse her own intervention in the meeting (understood by
me as a complex, possibly conflictual, situation), then my sense is that
she was intervening at levels 0 to 2 and possibly 4 (rules). What success, if any, her intervention had in this meeting is unstated but we
know pretty much that the thinking behind international trade and globalisation politics and policy has remained the mainstream view and
still persists in early 2009 despite the global financial collapse and, for
me, the validity of her arguments. Yet her example in the publication
has altered the views of many readers.

In this article Donella uses and exemplifies many key systems concepts
some of which are also described in Table 2.1. She thus connects with the
theoretical and historical background of Systems in doing what she does

and makes explicit linkages to the systems dynamics lineage depicted in Fig. 2.3 through the work of Jay Forrester.

Q. What does the article reveal about the author?

R. The article has been written around a key experience of the author's: jumping up in a room full of people and using a flip chart to explicate how her understanding of the situation differs from the 'mainstream conversation'. This particular action is an excellent example of juggling the B-Ball – the author's own being (I will say more about this when I discuss the B-Ball in the next chapter). Importantly for me, she acknowledges her experience within the article – it could have been written as a rather dry and theoretical paper which was based on '10 ways to intervene in a system' but it wasn't. This article as well as the action of outlining the intervention points in front of what might have been a hostile audience is testimony to how well the author knew her material – how capable and immersed she was within the systems practice lineage of Systems Dynamics. Her examples also demonstrate how her systems dynamics training and understandings enable her to see or recognise issues, concerns, opportunities of a systemic nature in many situations.

Q. Given my current understanding of the idea of juggling in relation to systems practice what can I say about Donella Meadow's juggling?

R. Well it sounds like it was quite a performance! As I outlined above there is good evidence of her juggling the B-ball – her own being. Whether she chose the best way to Engage with the situation is unclear – but it does seem that her audience were prepared to listen and make sense of what she said. So she was certainly using the E-ball, not least through the ten different ways of engaging with a situation (a phrasing I prefer to 'intervening in a system'). As is common in the systems dynamics lineage she chose to see the situation as 'a system' (i.e. as depicted in Fig. 3.4). This is a choice that we can make but as I have outlined it has implications. Her 'performance' as much as we can tell seems to have been appropriately contextualised (the C-Ball) – her systems understandings were powerfully brought into play to illuminate a complex situation. No doubt it was dramatic – but whether effective in the longer term is an open question which probably has little to do with the adequacy of the ideas or explanations for the situation. It is also a matter of contextualising where to look. There is perhaps least evidence of how Donella juggled the M-ball in the initial situation – but her paper is testimony to effective 'managing' of the process of creating a readable and engaging narrative.

At this stage the juggler is my metaphor, not yours, and you may experience my answer as a 'forced-fit'; in some ways it is. As I write I imagine a conversation with Donella in which she might argue that her ten interventions are the balls that she juggles in her practice. Alternatively she may have rejected the metaphor as inadequate. In asking you to engage with these questions and to use the metaphor as an inquiry device my main aim is to help you stand back from the detail of this reading (usually 'content' or 'results' are the main focus) and appreciate it as a practice dynamic, a form of performance.

Q. Faced with the same or a similar situation would I think about it similarly or differently?

R. I think the end result of my own thinking about the situation (i.e. globalisation) would be very similar to Donella's. I would use or draw upon many of the same systems concepts (e.g. self-organisation) but perhaps deploy them differently as well as using other systems concepts. My own experience of the systems dynamics lineage is not nearly as strong as Donella's and so I would not be able to use the thinking in the way that she did in the meeting – though in similar circumstances I could draw on my own understandings to make similar points. I will certainly add these ten points or variations of them to my own systems practice repertoire. As I outline in my response to the next question I would choose to use some of the concepts and language differently for both practical and theoretical reasons.

Q. What does this article reveal about my own ways of thinking and acting?

R. In my response I am going to focus on the insights I have gained from this article about my own systems thinking and explore some of the differences and similarities with the author. In doing so I want to make it clear that I am not criticising or being critical of this article although what I write could be seen as a critique. My differences in thinking relate to:

- The way in which 'system' is conflated with 'situation' (what I mean can be understood by looking at Fig. 3.4) and, as a consequence
- The absence of attention to who brings forth a system by what means – in systems theoretical terms this is the issue of who participates in making boundary judgements about what is in or outside a system of interest
- The focus on goals rather than desired outcomes or questions of purpose
- Some of the implications of certain metaphors e.g. "leverage points", "systems analysis", "analysing a system", "intervening in a system"
- Anxieties as to how misunderstandings could arise about the nature of 'information' and her use of, and role for, 'rules'

This is not such a long list so we have much in common. If I had to highlight some of the things I particularly like in this paper at the top of my list would be:

- The need for awareness that transcends paradigms – "It is to 'get' at a gut level the paradigm that there are paradigms, and to see that that itself is a paradigm, and to regard that whole realisation as devastatingly funny. It is to let go into Not Knowing" or the recognition that 'there is no certainty in any worldview' and that systems practice is concerned with 'disciplined thinking combined with strategically, profoundly, madly letting go'

- Excellent accounts of the nature and importance of positive and negative feed-back processes (see also Table 2.1)
- Sound arguments for paying much more attention to creating the circumstances for self-organisation

The ten intervention points are also a useful device to think about where my own focus has been in developing and using my systems practice. My assessment is that it has been at levels 0 to 5, though not exclusively — something you will no doubt find reflected in this book.

In relation to my own ways of acting I find I relate readily to the sense of frustration that led Donella to jump up in the meeting, triggered I suggest by her experience of the failure by others to engage systemically with complex situations. My experience is that it is necessary to act in this way from time to time to maintain one's equilibrium but that in the main it is not the most useful form of practice. Instead, I am now guided by the following question: How is it that I could create the circumstances whereby others could engage with this situation systemically? In Part III, I will discuss this under the rubric of the 'design of learning systems'.

4.3 Reflecting on Reflections

Effective systems practice incorporates reflective practice, including an awareness of how questions are framed and answered. With this in mind I would like to draw your attention to some aspects of the questions I posed in section 4.2. Considered in isolation some could be seen to be about the situation Meadows found herself in (1, 2, 3). The next set of questions (4, 5) were more to do with Donella and her mode of engagement with the situation. So questions 1 – 5 illuminate the dynamic depicted in Fig. 3.5 but they do so through the question-responder's history, or traditions of understanding and this is different to yours and Donella's (to which we have no direct access). Finally questions 6 – 8 were more focused on you and your understandings, from a position outside, or meta as depicted in Fig. 3.5. Your answers to questions 1 – 5 rely heavily on your interpretive skills and thus, on the tradition of understanding from which you interpret. Historically and as part of daily life there is a tendency to focus only on questions and concerns associated with 'the situation', i.e. questions of the 1 – 3 type. Important as these are they are only part of the systemic dynamic that underlies systems practice that is captured by the juggler metaphor. The practitioner with a chosen framework of ideas (F in Fig. 3.5) is well illustrated in the Reading by the commitment of the author to the systems dynamics lineage of seeing systems in the world (in the sense shown in Fig. 3.4) – so this is part of her engaging with the situation [6]. Becoming aware of these distinctions and levels is part of improving your own system practice.

Having provided a set of answers to the questions I posed for Reading 3, I want to draw attention back to my primary purpose in offering readings. That is to provide you with vicarious experiences that give insights into what is entailed in doing systems practice – your own systems practice – through the lens of the question: what is it that we do when we do what we do? To do this I want to briefly return to discussing systems practice as an 'ideal type', as introduced in Chapter 3. In the next four chapters (5–8), I will further unpack the 'juggler metaphor'.

So what constitutes an Ideal Systems Practitioner? First, this persons is able to draw on their experience of different systems traditions to enact a systems approach, or approaches, in managing 'real world' situations. Understandably I will not be overly concerned here with approaches to practise other than systems approaches, and will not be making any extravagant claims that any given systems approach is better than any other forms of practice (but I will return to the issue of effectiveness in Part IV). I will, however, develop arguments that support four claims. These are:

- Systems practice has particular characteristics that make it qualitatively different to other forms of practice
- An effective and aware systems practitioner can call on a greater variety of options for doing something about complex 'real-world' situations than other practitioners do
- Being able to deploy more choices when acting so as to enhance systemically desirable and culturally feasible change has important ethical dimensions
- Our individual and collective capabilities to think and act systemically are under-developed and this situation is a strategic vulnerability for us, as a species, at a time when concerns are growing for our continued existence in a co-evolutionary, climate change world

These are important claims. They will structure most of the argument made in the rest of Part II through the vehicle of the juggler isophor.

References

1. APSC (Australian Public Service Commission) (2007) Tackling Wicked Problems. A Public Policy Perspective. Australian Government/Australian Public Service Commission: Canberra.
2. Bunnell, P. (2009) Personal communication. Systems Ecologist, President of Lifeworks, Vancouver, British Columbia, May 2009.
3. Forsythe, K. (1986) Cathedrals in the mind: the architecture of metaphor in understanding learning. In Trappl, R. (Ed.), Cybernetics and Systems '86: Proceedings of the Eighth European Meeting on Cybernetics and Systems Research, organized by the Austrian Society for Cybernetic Studies, held at the University of Vienna, Austria, 1–4 April 1986. D. Reidel: Dordrecht 285–292.
4. Ison, R.L., Blackmore, C.P., Collins, K.B. & Furniss, P. (2007) Systemic environmental decision making: designing learning systems. Kybernetes 36 (9/10), 1340–1361.

5. Meadows, D. H. (1997) Places to Intervene in a System. Whole Earth, Winter.
6. Meadows, D. H. (2008) Thinking in Systems. A primer. Earthscan: London.
7. Proulx, J. (2008) Some differences between Maturana and Varela's theory of cognition and constructivism. Complicity: An International Journal of Complexity & Education 5(1), 11–26 – www.complexityandeducation.ca
8. Ramage, M. and Shipp, K. (2009) Systems Thinkers. Springer: London.

Chapter 5
Juggling the B-Ball: Being a Systems Practitioner

5.1 Accepting Different Explanations

As I write, I imagine this ball is shiny and thus acts as a mirror reflecting an image of the juggler. The properties of the juggler as systems practitioner come under the spotlight in this section. In choosing the word 'being' I am deliberately playing, metaphorically, with different meanings of being – one of which is, of course, 'human being'. Some of the features of being human include self-consciousness, language, emotions and the capacity to reason, rationalise and reflect. Human beings also live with a desire for explanations they find satisfying. You may have had the experience of a child repeatedly asking why? how? and then stopping after you have given a particular answer. The child finally finds your explanation satisfying – it makes sense within the child's world – and the child no longer needs to ask (Fig. 1.2).[1]

Perhaps you have experienced explanations that did not satisfy at all. If you are aware of this occurring do you remember what it felt like? By this I mean, were you in touch with your emotions when you became aware that a particular explanation was satisfying or dissatisfying? By asking this question, I am saying it is legitimate to acknowledge your emotions – they are part of living and need not be ignored. What is often not apparent to us is that "calm" is an emotion or that "insistence" is an emotion, including the "insistence on rationality"! We are

---

[1] Following Maturana [27] I understand the social dynamics of explanations to be a key aspect of being human and which begins to be learnt in early childhood. Maturana says (ibid p. 148) that an 'explanation is an answer to a question about the origin of some particular experience of the observer that is asked in such a way that it explicitly or implicitly demands an answer that satisfies the following two conditions: (i) the answer must consist in the proposition of a mechanism or process that, if it were allowed to operate, would give rise in the observer as a result of its operation the experience that she or he wanted to explain; (ii) the generative mechanism or process proposed as an answer must be accepted as doing what it claims to do by an observer, who could be the same person that proposes it, because it satisfies some other condition that he or she puts in his or her listening.'

R. Ison, *Systems Practice: How to Act in a Climate-Change World*,
DOI 10.1007/978-1-84996-125-7_5, © The Open University 2010.
Published in Association with Springer-Verlag London Limited

ALWAYS in one emotion or another! We only note the odd/extreme ones, such as fear, anger or joy as worth comment. The fact that I raise this at all reflects deeply held convictions and commitments in particular professions and academic traditions as characterised in Table 5.1. Take for example Sir Geoffrey Vickers who became a noted systems thinker when he chose in his retirement to try to make sense of his professional life [43–45]. He coined the term 'appreciative system' to describe the cycles of decision-making about fact and value that were for him key to relationship maintaining, or breaking, in the daily flux of managing [7, 46]. It is clear from his personal correspondence that Vickers found it very difficult to let go of the rationality of his culture and time, born as he was into late Victorian society. For him and many of his generation poetry proved to be the one culturally

Table 5.1 Some contrasting features between the traditional western conception of the disembodied person with that of an embodied person [21, pp. 552–557]

Traditional Western conception of the disembodied person	The conception of an embodied person
The world has a unique category structure independent of the minds, bodies or brains of human beings (i.e. an objective world)	Our conceptual system is grounded in, neutrally makes use of, and is crucially shaped by our perceptual and motor systems
There is a universal reason that characterises the rational structure of the world. Both concepts and reason are independent of the minds, bodies and brains of human beings	We can only form concepts through the body. Therefore every understanding that we can have of the world, ourselves and others can only be framed in terms of concepts shaped by our bodies
Reasoning may be performed by the human brain but its structure is defined by universal reason, independent of human bodies or brains. Human reason is therefore disembodied reason	Because our ideas are framed in terms of our unconscious embodied conceptual systems, truth and knowledge depend on embodied understanding
We can have objective knowledge of the world via the use of universal reason and universal concepts	Unconscious, basic-level concepts (e.g. primary metaphors) use our perceptual imaging and motor systems to characterise our optimal functioning in everyday life – it is at this level at which we are in touch with our environments
The essence of human beings, that which separates us from the animals, is the ability to use universal reason	We have a conceptual system that is linked to our evolutionary past (as a species). Conceptual metaphors structure abstract concepts in multiple ways, understanding is pluralistic, with a great many mutually inconsistent structurings of abstract concepts
Since human reason is disembodied, it is separate from and independent of all bodily capacities: perception, bodily movements, feelings emotions and so on	Because concepts and reason both derive from, and make use of, the sensorimotor system, the mind is not separate from or independent of the body (and thus classical faculty psychology is incorrect)

acceptable outlet for non-rational and even systemic thinking. So, I argue that an ideal systems practitioner is able to include an awareness of their emotions as well as their rational ideas – in fact I would say that these are inextricably connected.

Juggling is a particularly apt isophor (see Chapter 4) in regard to one's being because good practice results from centering your body and connecting to the floor. So juggling arises from a particular embodiment. Effective juggling is thus an embodied way of knowing associated with our lived history as human beings.[2] Lakoff and Johnson [21] argue that in the Western world, the most common sense of what a person is arises from a false philosophical view, that of disembodied reason, that has influenced almost all of the professions [21]. They contrast this with what they term an "embodied person" (Table 5.1). For example, in medicine until quite recently the brain was seen as quite distinct from the body – the mind-body dualism – whereas the brain is only one part of a much larger cognitive system, a network of molecular relations that we have variously isolated as if they were independent, for example as the nervous, endocrine and immune systems [31].

I have started this chapter by offering some explanations that are for many outside of the mainstream. My invitation is to consider seriously the explanations I have offered about (i) the nature of explanations, (ii) the role and place of emotions, (iii) embodied knowing.

There is a rigorous and increasing evidence base for these explanations some of which I will draw upon in the following sections. In the sections that follow I address four aspects of being a systems practitioner. The first is being aware of the constraints and possibilities of the observer. The second is understanding understanding and knowing knowing. The third is learning, and the fourth is being ethical.

5.2 Being Aware of the Constraints and Possibilities of the Observer

The essence of a systems approach is that of seeing the world in a special way – and by 'seeing' I mean more than just vision. This immediately prompts the question of what is meant by the phrase 'seeing the world'. Because we live so intimately with the world of objects and people and phenomena, we tend to think our own way of seeing the world is the only way, or even of thinking, 'Well that is my view because the world is like that'. Actually, your perception and cognition is special in several separate ways:[3]

---

[2] I will say more about embodiment later in this chapter – at this stage it is worth noting that embodiment is a term I take from theoretical and practical concerns in a number of disciplines about embodied or embedded cognition, a position in cognitive science, for example, stating that intelligent behaviour emerges out of the interplay between brain, body and world (See http://en.wikipedia.org/wiki/Embodiment. Accessed 16th July 2009).

[3] I am grateful to Peter Roberts for his original work on which this is based.

1. If your vision is not impaired, you see your surroundings using only light of wavelengths between 380 and 780 nm. Bees, for example, see flowers using wavelengths much less than 380 nm. You have quite a small visual window on the world

2. With normal hearing you hear frequencies of sound between 20 and 20,000 Hz (Hertz). Bats use sound waves of higher frequency than 20 kHz, which we cannot hear

3. Your ability to detect odours is vastly inferior to a dog's. A dog's 'smell world' is vastly richer than its visual world

4. Research on colour perception in the 1960s showed that colour was not something that is fixed in the world, but is a property of our own unique biology and histories. This led one of the researchers involved to change the question he was concerned with from 'how do I see colour?' to 'what happens in me when I say that I see such a colour?'[4]

5. The language you have learned steers you into categorising your world in ways you are largely unaware of, just as a fish is unaware of the water it is immersed in throughout its life. Sometimes it is possible to become aware of this when speaking another language – when immersed in the other language the experience is sometimes like being a different person[5]

6. Your physiological, hormonal and emotional dynamics are interrelated and affect how you experience the world. This includes neuropeptide dynamics and aspects of the functioning of your nervous system as these play a role in cognition; hormonal events such as menstruation and the release of natural endorphins during exercise are also involved

From birth, unless circumstances are extraordinary, people live in language and a particular culture or cultures. An individual's social or cultural history creates constraints and possibilities to their ways of 'seeing' the world. For example, the role the culture of the society in which you have developed determines what you see as well as how you can respond in any flow of relationships. Your culture also determines what you implicitly accept as your perceptions and emotions. So the ways you learn to see manners, relationships and behaviours is dependent on how people around you see and act. A consequence is that we get caught in a trajectory and tend to articulate it ever more finely, rather than change it.

'Institutional arrangements' are a particular aspect of living in a culture that is worthy of highlight because the influences are so pervasive. These are the norms and rules invented by people which we are born into and which are sometimes modified and often added to. They operate in the family (e.g. dinner time), community (e.g. town council) and our governance (e.g. road rules, parliaments etc.).

---

[4]My reference to history makes it sound like colour perception is learned – and partly it is. We learn the names of nuances of wavelengths that are relevant in our culture. But there is also "relativistic colour coding" that means we see colour by ratios of stimulation of cones, not by wavelength of light. We don't work the same way as the instruments we invent to record colour.

[5]This may often involve developing and using different muscles, i.e. differences in our body.

Fig. 5.1 A metaphorical account of the way theories (planet on telescope) determine what we see in the world. The mischief makers represent what happens implicitly with any theory [30]

What is particularly important is that 'institutions' are a backdrop to all that we do and often we are not aware of them, particularly the forms of understanding that have been built into them at different historical moments.

Another special subset of culture is the particular explanations we accept for things we experience. The 'theoretical windows' through which we interpret and act are always with us regardless of whether we are aware of them or not. Figure 5.1 provides a metaphorical account of this phenomenon. The theory or explanation you accept will determine what you see and thus the meaning you will give to an experience. Think here of the fundamentally different cosmology, the set of explanations for the origin and evolution of the universe, developed by the Mayan civilisation in South America that was entirely coherent but so different to our own Western cosmology.[6] This phenomenon is sometimes described as the theory dependency of facts.[7]

The ways in which perception and cognition work raise important questions that are relevant to practice. Take for example the image depicted in Fig. 5.2. When you

---

[6] We usually vilify other cosmologies as "myths" because we know better now.

[7] See http://en.wikipedia.org/wiki/Philosophy_of_science: 'most observations are theory-laden – that is, they depend in part on an underlying theory that is used to frame the observations. Observation involves perception as well as a cognitive process. That is, one does not make an observation passively, but is actively involved in distinguishing the thing being observed from surrounding sensory data. Therefore, observations depend on some underlying understanding of the way in which the world functions and that understanding may influence what is perceived, noticed or deemed worthy of consideration.' (Accessed 25th January 2009).

Fig. 5.2 Individuals are likely to perceive different things in this image, depending in part on their prior experience [1]

look at this Figure you may see a young woman wearing a necklace looking away, or you may see an old woman with a big nose, looking down. Some of you will see both, one after the other. What you perceive in relation to this image raises two important questions:

1. What is experience? In this example some people experienced a young woman whilst others experienced an old woman yet both looked at the same image. It is thus possible to claim that experience arises by making a distinction – if you are unable to distinguish a young woman then you have no experience of one!
2. Is it possible to decide on which interpretation, the young woman, the old woman or merely the ink on the paper, is correct? In other words do we reject those people who see only an old woman as being 'wrong'?

The following story also illuminates what I mean by experience. The story relates to an incident towards the end of a flight I was making from Johannesburg to East London in the new province of the Eastern Cape, South Africa. I was doing a consultancy just after the first multiracial elections, which was a time of good will and enthusiasm and general optimism. As the plane taxied up the tarmac towards the terminal, I experienced my South African colleague, in the seat next to me, as becoming agitated and tense. I looked out the window and could not distinguish anything that might have been the cause of his distress. When I enquired, my colleague pointed to some seemingly innocuous cement pillars, which he explained were the remains of gun emplacements left over from the state of emergency in the apartheid era. Until he pointed them out I had not seem them. Because my colleague's history was different to mine he had seen what I could not see, that is, his observation consisted of distinctions that I had not made. Furthermore, my colleague's distinctions altered his own mental, emotional and physiological state – they altered his being.

My colleague made distinctions that I was unable to make and thus he experienced something I did not. The act of making a distinction is quite basic to what it is to be human. When we make a distinction we split the world into two parts: this and that.[8] We separate the thing distinguished from its background. We do that when we distinguish a system from its environment as depicted in Fig. 3.3.

It is worth remembering that using the word system is actually a shorthand for specifying a system in relation to an environment.[9] In process terms, this is the same as drawing a circle on a sheet of paper. When the circle is closed, three different elements are brought forth at the same time: an inside, an outside and a border (in systems terminology, a boundary).[10] In daily life we have developed all sorts of perceptual shortcuts that cause us to forget this is what we do – we live, most of the time, with our focus on one of these three elements: the inside, the outside, or the border. Heinz von Foerster [47] observed that biologically we cannot focus on both sides of a distinction at the same time. He then proposed that the descriptions we make say more about ourselves than about the world (situations) we are describing.

While the old woman – young woman example is now well known, the implications that flow from it are not. This simple example demonstrates that in the moment of experience we cannot distinguish between perception and illusion and that we do not see that we do not see [24]. Intentional illusions are a mocking, jesting play with our sense of perception, a role that Fig. 5.2 and other variations of it, play well. But illusions also operate in daily life; we become aware of this phenomenon when we think we see a friend, a lover, a rival and experience a sudden change in our emotions at the prospect of meeting them, only to find some moments later that we were mistaken because the person we glimpsed turns out to be someone else. It is ironic that we pay money to go and see illusionists, and marvel at their artistry,

---

[8]And at the same time we have specified a domain (a framing) in which that particular split makes sense. This is probably much more significant than the particular split because people will discuss the boundary, but never realise they have already accepted the domain, or they will think they speak of the same split, but it won't be the same if they have done it in different domains [3].

[9]In systems theoretical terms the word 'environment', as "that which surrounds", is the correct term but it has no relation to the so-called physical environment – it is an abstract or conceptual notion that only arises whenever a system of interest is distinguished by someone.

[10]The concept circle is also brought forth in this action – but in every day life most of us have from an early age, as part of our tradition of understanding, the concept circle and thus when we distinguish circles in everyday life the boundary usually disappears from view. Mainly we only become aware of most borders when something is ambiguous.

yet remain unaware that illusion is also part of daily life. Perhaps we do this because acknowledging the interplay between perception and illusion is too challenging. For systems practice the existence of this phenomenon is challenging in a number of ways:

1. It draws attention to what is involved in the process of modelling of which diagramming is a subset (as discussed in Reading 2). It raises the question of whether we model some part of the world or model our mental models, that is, our set of accepted distinctions, of some part of the world
2. It challenges the certainty of some practitioners who claim they are objective or they are right in a way that justifies the way they practise
3. It reminds us that our perspective is always partial and a product of our cognitive history. Thus, when forming a system of interest, the question of 'perspective, whose perspective?' is crucial[11]
4. It invites awareness of the constraints and possibilities of the observer as the B-ball is juggled in practise

The properties and role of the observer have been largely ignored in science and everyday culture despite Werner Heisenberg's finding in 1927 that the act of observing a phenomenon is an intervention that alters the phenomenon in ways that cannot be inferred from the results of the observation. This is the essence of Heisenberg's uncertainty principle, which limits the determinability of elementary events.[12] The story of how the observer came into focus is an interesting story in the history of Systems and its associated field of cybernetics as discussed in Chapter 2, Section 2.4 [11–13, 23, 48].

Being aware of the constraints and possibilities of the observer enhances the behavioural repertoire we have at our disposal. Because we are able to communicate with one another, and because we live within cultures, we can take shortcuts: it makes sense most of the time to act as if we are independent of the world around us.[13] Sometimes it also makes sense to act *as if* systems existed in the world and *as if* we could be objective. But remember, the two small words *as* and *if* are important in the context of our behaviour when managing our own systems practice. These two small words always go together as in "we act *as if*". From my perspective it is always a shortcut when they are left out.

---

[11] This is not BAD, it is how we and language are constituted. We manage just fine, unless we fall into the trap of certainty. And knowing this is how we operate opens up the option for new perspectives that better fit our situation.

[12] See von Foerster [49]; people assume Heisenberg's uncertainty principle only pertains to elementary events but a more useful framing might be to say he invented it for that, but it has far wider application!

[13] We would be so hampered in doing our daily tasks if ALL the time we had to think like this! Reflecting alters our path, but then for a while we just want to engage. It can be thought of as an ongoing dynamic between engaging and reflecting.

5.3   Understanding Understanding and Knowing Knowing

In my experience the second-order cybernetic explanation provided for the observer is challenging for many people. When I attend workshops where these ideas are expressed for the first time, some people become angry.[14] It is profoundly disturbing to have the basis for your understanding of the world challenged. It seems important to explore these issues, however, because in my experience, it gives access to new and practical explanations. I have already acknowledged that some explanations may be dissatisfying but, in the end, that is all they are – just explanations. If they are not satisfying they need not be accepted.

Relatively recent findings in cognitive science which are not widely appreciated, challenge some widely held 'common sense' notions. Take information for example. Many people assume that individuals would be better decision makers if they had better information. However, any nervous system, including the human one, is closed in its operation, that is, all it is is a detector of its own changing internal configurations. This means that 'information' arises more from the history of the operations of each person's unique nervous system than from 'messages' from the environment (this relates also to Table 5.1). Unfortunately this does not stop bureaucrats and others wasting a lot of money on 'information campaigns', as in this example:

> Despite a significant evidence base showing that behaviour change can rarely be achieved by trying to drive it with information, many programmes and projects remain essentially information driven, even where the organisations concerned have commissioned analysis which noted that such an approach is unlikely to work. For example the UK Government Energy Savings Trust is running a web based campaign to get individuals to pledge to cut personal carbon by 20%, driven by a list of information on why this is a good idea, and actions you can take. [35]

It is possible to use this example to place the process of cognition in a different light to the dominant paradigm prevailing in psychology, computer science and in the common-sense view (Table 5.1). The prevailing paradigm has been described as the information-processing model of the mind [34]. Since about 1950, when the notion of "information" as a driving force was introduced, the prevailing view in cognitive science has been that the nervous system picks up information from the environment and processes it to provide a representation of the outside world in our brain. But we can now say instead, to paraphrase Francisco Varela [42], that the nervous system is closed, without inputs or outputs, and its cognitive operation reflects only its own organisation. Because of this, we project our constructed information – or our meaning – on to the environment, rather than the other way around. This is much like Fig. 5.1, except this time the image of the planet is contained in

---

[14] This phenomenon is likely to be of interest to systems educators – Pille Bunnell [4] in response to my observation says: "I wonder how much that is the approach to the explanation per se. Though I admit it takes me about 25 h of class time to build up to a point where no one becomes 'angry' though some (less than 10% of the class) will not accept this in the long term. Others do take it as a fundamental shift."

our nervous system rather than the lens of the telescope. It implies our interactions with the 'real world', including other people, can never be deterministic, there are no unambiguous external signals.

However, it is clear from lived experience that we do not live an 'anything goes' existence in which we constantly imagine the world into existence – we live always coupled to the materiality and regularities of our world. Our interactions consist of various non-specific triggers, which individuals understand strictly according to their own internal structural dynamics [11]. This has profound implications for how human communication is understood – it is not signal or information transfer but, as I discuss below, something much more fascinating and complex.

5.3.1 Living Within a Network of Conversations

Within the line of reasoning I have been developing it has been argued that:

> we human beings exist, are realised as such, in conversations. It is not that we use conversations, we are a flow of conversations. It is not that language is the home of our being but that the human being is a dynamic manner of being in language, not an entity that has an existence independent of language, and which can then use language as an instrument for communication [26].

This is a particular feature of our evolutionary trajectory along perhaps with a few other species (e.g. whales and dolphins). In the opening to his book 'Conversational Realities', John Shotter provides several quotes which capture the essence of my line of argument [40, p. 1]:

- 'The primary human reality is persons in conversation' [14, p. 58]
- 'Conversation flows on the application and interpretation of words, and only in its course do words have their meaning' [52, p. 135]
- 'Conversation understood widely enough, is the form of human transactions in general' [22, p. 197]
- 'If we see knowing not as having an essence, to be described by scientists or philosophers, but rather as a right, by current standards, to believe, then we are well on the way to seeing conversation as the ultimate context within which knowledge is to be understood' [33, p. 389][15]

These explanations are a long way from the 'mainstream' understanding in our society about the nature of language, and thus human communication and action. But much insightful research has been ignored or become 'unfashionable'. The extensive studies by Benjamin Whorf on these matters in the 1920s and 1930s led him

---

[15]By including this quote I want to make it clear that I am not interpreting it as "my right to believe whatever I like and declare that as knowledge" but because it offers a perspective that differs widely from the mainstream understanding of what most scientists do when they claim scientific knowledge.

to conclude 'that the structure of a human being's language influences the manner in which he [sic] understands reality and behaviour in respect to it' [5, p. 23].

For example, when the word nature is used in modern Western discourse it is often used in such a way that leads us to live as if we human beings are outside nature. The concept 'nature' thus structures who we are and what we do. In some indigenous, non-western languages the term or concept does not exist. John Shotter uses another example, that of the words 'I love you' to explore how language operates [40, p. 3]. He discusses the circumstances that might give rise to someone saying these words and the implications they carry with their saying. Once uttered these words change irrevocably the whole character of the future relational dynamics between two people and in the process they change the being of those in the conversation: 'so the world of those in love is different from those who are not – in other words, they are different in their ways of being'.

I am tempted to conclude that human beings are yet to understand the implications of living in language or, put another way, have failed to see how language can act as a social technology that mediates our understandings and practices and thus our relationship with the biophysical world.[16] Collectively we have failed to realise that we do not use language but that language uses us in ways we are yet to appreciate and master.[17] By "using us" I mean in the sense that one dance partner may "use" another. Obviously, this view has implications for juggling the 'B-ball' and thus for the nature of systems practice.

Appreciating how living in language works, and why diversity and difference in manners of living in language are important, has major implications for how we, as a species, adapt with changing climate. Nick Evans [10], a linguist at the Australian National University who studies Australian Aboriginal languages, makes the point that:

> 'Say a community goes over from speaking a traditional Aboriginal language to speaking Creole.[18] Well let's just use talking about the natural world as an example. You leave behind a language where there's very fine vocabulary for the landscape. Inside the language there's a whole manual for maintaining the integrity of the landscape, for managing it, for using it, for looking for stuff. All that is gone in a creole'. Unfortunately the scale of this issue is poorly appreciated – there are 6-7000 languages that remain in use around the world and at least 40 percent of these will probably by mid-century be replaced by a few meta-languages – English, Spanish, Mandarin [10, 28].

In the sense that I use it conversation can be literally understood, following its Latin roots, as *con versare*, meaning to turn together [37]. Our living in language is both an evolutionary and embodied phenomenon that is intimately connected to the dynamics in our emotions or, as Maturana describes it, our emotioning.

---

[16] I return to the issue of social technologies in Section 5.5.

[17] The evidence for this set of claims can be found in the early work of Benjamin Whorf who found, for example, that 'the strange grammar of Hopi gives rise to different modes of perceiving and conceiving' (see Carroll [5, p. 17]).

[18] Creole in this context means a blend or hybrid of English and the speakers original language.

5.3.2 Thinking and Acting Based on Our Tradition of Understanding

Understanding does not depend on information or, put another way, information does not generate understanding.[19] The notion that we exist in language and co-construct meaning in human communication, much as dancers co-construct the tango or samba on the dance floor, suggests the need to consider on what basis we might accept that understanding has occurred? Asking this question is like opening a Pandora's box. It raises all sorts of questions that we take for granted, like: What is learning? What is understanding? How do we know what we know?

In Fig. 5.3 I depict some of the dynamics that come into play when we do what we do. The figure depicts a person (a living being) over time; as unique human beings we are each part of a lineage and our individual history is a product of both ontogeny, which refers to the biological growth and development of the individual

cultural and
biological
history

tradition
of
understanding

Fig. 5.3 A conceptual model of a practitioner who brings forth their tradition of understanding as they lay down the path of their walking (doing). All humans have a personal history within a wider cultural and biological history. A systems practitioner, for example, will from some moment have a history of thinking, acting or understanding systemically[20]

---

[19] But what is "information"? Is there such a thing at all? I claim it is not a thing, but an arising within – from the Latin roots '*in formare*'; for example, following Bateson, famous for the phrase 'the difference that makes a difference', information could be said to arise when someone distinguishes a difference in two domains. In this context information is an explanatory principle.

[20] Finding an image to depict the concepts that I consider important has been a challenge – this figure owes much to a Michael Leunig cartoon for inspiration.

organism, and our phylogeny, which refers to biological and socio-cultural development, that is our evolution. Thus an individual at any moment is result of both their own life and the history of their culture: living in language, relationships, emotional dynamics etc. Together the ontogeny and phylogeny (that is, both the individual's biological and cultural development) form what I will call a tradition. A tradition is those aspects of our past that create an individual's history, it is the history of an individual's being in the world.

Traditions are important because our models of understanding grow out of traditions; traditions are thus not only ways to see and act in the world but operationally traditions also conceal that aspect of human cultural and biological history that has not been part of our own experience [36, 37].

As I explain further what I mean by a tradition of understanding it might be useful to look at Figs. 5.3, 5.4 and 5.5 together, as a set. What is important to grasp is that when I speak of 'tradition of understanding' I am not using it as a label for something that existed in the past. Instead, as depicted in Fig. 5.3, it arises in our living and becomes the path on which our living, our doing, unfolds. This path never has exactly the same features as we learn, change etc. And whilst I have sited the origin of this path from the head, as I have said it is an embodied phenomenon, not just a product of a disembodied mind!

Figure 5.4 looks at the dynamics I am concerned with from a different perspective. In Fig. 5.4a the practitioner lives with changing understandings of situations over time – as we age and society changes around us. This comes about through accepting and rejecting explanations, being exposed to new experiences etc. We may be more or less aware of how we have engaged with situations in the past – perhaps with more certainty or different convictions than when we were younger (Fig. 5.4b). Through reflection on our past engagement, we might claim that our understandings have changed. This reflection often leads us to realise that even our current understanding may change, that our future self may look back on the present moment and see that which we do now as somewhat limited or immature. Thus we may be willing to accept that no understanding is final, or inherently right, understanding always evolves.

With growing awareness we may also appreciate that when we engage with situations from our current understanding(s) – it need not be just one understanding – we are inherently making connections (engaging) with a 'real-world' situation through the act of making a distinction – selecting a situation from a more or less knowable broader context (Fig. 5.4c).[21] Based on the distinctions made in the situation, the practitioner can probe, or construct, the history of a situation thus becoming aware

---

[21] As I have outlined "making a distinction" is the basis of new experience but over time the act of making a distinction falls into the background as we discern a situation according to an accepted distinction. For example, if we have already distinguished "chair" we can sit on it, pick it up, invite someone else to sit in it… without having to distinguish it as a chair each time! The reason distinctions are so important is that once made they are taken for granted. This makes operational sense but when we are not aware that this is what we do we can end up in traps of our own making.

Fig. 5.4 Changing stages of awareness in practice (a is at top, d at bottom). (**a**) How understandings of situations change over time associated with our personal history. (**b**) A reflexive turn – appreciating how our current understandings differ, yet arise from past understandings of situations. (**c**) Awareness that our focus of attention – the situation is always part of a broader context, which includes (**d**) how understandings that are suitable can be incorporated into a method which over time is reified becoming a social technology or institution in the present which constrains how we think and act because it is no longer suited to the situation

of how social technologies have reified understandings at earlier historical moments (Fig. 5.4d). Thus we have an opportunity to not only reflect on how our tradition of understanding has developed throughout our life, but also how the cultural context that colours or situates our understanding has changed over time.

As I have outlined earlier if those in the social dynamic, including the practitioner themselves, bring forth systems understandings in what they do, then we could claim that this particular practitioner is a systems practitioner.

Figures 5.3 to 5.5 are elaborations of the dynamics at play depicted in earlier Figures, especially Figs. 3.3 and 3.5. The main new elements in the figures are that of history – our own history – and that of the process of constructing histories and thus understandings of situations.[22] Underpinning each figure is an appreciation that any living system only lives in an ever unfolding present as if we existed on an ever rolling wave (Fig. 5.5). From this understanding the "past" and the "future" are merely different ways of living in the present.

Figures 5.3 to 5.5 establish a basis for the processes of being and engaging – two of the balls being juggled. To reiterate, I use the word 'tradition' in a specific way to mean an individual's history of making distinctions as part of their living. It thus follows that a tradition is not 'some thing' but what is enacted or brought forth as we live our traditions of understandings (Fig. 5.3). Another way of describing a tradition is as our experiential history because experiences arise in the act of making a distinction in relation to ourselves. To experience in the way that I mean requires language – if we did not live in language we would simply exist in a continuous present not 'having experiences'. Because of language we are able to reflect on what is happening or in other words we create an object of what is happening and name it 'experience'.

On the wavefront of the present

Fig. 5.5 Our being is like living on an ever rolling wave – we are always in the present

---

[22]As outlined earlier, for simplicity these figures don't show a key human dynamic, namely our conversations with other people, which of course greatly influence how our understanding develops, how we "learn". The 'real world' situation too is an abstraction – we are never apart from or independent of such situations.

Let me try to exemplify what I mean in Figs. 5.3–5.5 by considering the main fictional character Smilla, in Peter Høeg's novel 'Miss Smilla's Feeling for Snow' [16]. Smilla was born and spent her early years in Greenland. Her mother was an Inuit and an expert hunter. Being half Danish, Smilla subsequently pursued a Western education that built on her earlier experiences. She became an expert in the qualities of ice and snow. It was her understanding of the many different qualities of ice and snow that enabled her to solve the murder of an Inuit child around which the story is built. Her understanding also enabled her to survive in the snow and icy water when pursued by the murderers.

As an author, Høeg has grounded the distinctions Smilla is able to make about snow and ice in the history and culture of the Inuit people. Inuit culture is set against the background of continuous snow and ice; survival depends on being able to 'read' the snow and ice in detail. This detail can reveal, for example, how long ago the wolf left its footprints and whether the ice will support the weight of a dog team. The distinctions the Inuit make assume their importance because of the actions they allow. They arise as embodied ways of knowing and acting in which knowledge is not separate from action. The distinctions Smilla, or other Inuit make, are not distinctions I could make except that, having read the book, I could claim to know a little about some of the different categories of ice and snow. But would I, in similar circumstances, be able to escape from the murderers and solve the case based on what I claim to know? The answer is no, because the distinctions Smilla makes are invisible to me in my living. They are not part of my growing up and thus not something I could bring forth, my tradition of understanding, in my living. That would remain the case until such times as the distinctions about snow and ice quality and colour became embodied through my actions: for example running on snow until I became competent to do so without falling or working with snow and thus learning to make the distinctions about what kind of snow can be built into a shelter.

From this example, the only connection I can make with my own history is that of the act of 'making categories'. This is because I have learnt the process of making

Illustration 5.1

Fig. 5.6 A British dinosaur: making a connection with an example of a particular history of symbolic representation (Source: The Open University)

categories and if I came to understand the different types of snow I could probably develop a set of categories. In contrast to Smilla, this is a rather poverty-stricken form of knowing about snow!

A systems practitioner always engages with a 'real world' situation by making distinctions which are grounded in his or her personal history of making distinctions. Based on the distinctions he or she makes, the practitioner can probe the history of the situation, much like an archaeologist, historian or anthropologist, to reveal those dynamics which pertain to the distinctions he or she has made. It is possible to connect with a particular history whenever we make sense of a distinction in relation to its particular historical context. For example, if you look at Fig. 5.6 and you are British then you are likely to have little difficulty making sense of the distinction 'British dinosaur' in relation to a history of symbolising British culture through the image of a bulldog and the union flag. But if you are not British this history is unlikely to be part of your 'being'.[23]

So, after reading Peter Høeg's novel I could claim to know about the different categories of snow by listing or categorising them. Perhaps after visiting Greenland with an Inuit guide I could claim to know the kinds of snow if I could distinguish them in snow. To claim to understand snow would require me to be able to explain how, when and where different kinds are formed or found, and what implications they have for various activities, my own and other animals. In all cases the categories or distinctions would need to be grounded within the historical context, including ways of acting, of the Inuit, much as I have tried to do above. However I would need to do much more to embody these distinctions in my practices in the snow, that is, to know, or practise in, snow (as distinct from knowing only categories). To know snow I would need to be able to claim, through evidence of effective action,

---

[23] To explain a little further for those who are not British – prominent British leaders such as Winston Churchill often had or were photographed with a 'British bulldog' – a breed developed in Britain. This breed came to symbolise a form of British nationalism – if you were British – or a form of oppression if you were subject to British imperial rule. The Union flag itself is a flag generated as part of nation building from the flags of constituent UK nations – England, Scotland, Wales and Northern Ireland.

that I had brought forth my distinctions, i.e. they emerged as part of my own tradition of understandings. A test would be that under similar circumstances to Smilla my behaviour had similar results.

My explanation of knowing (or practising) and understanding in the Smilla example exemplifies how I interpret the two axes depicted in Fig. 1.3. The process of transforming a situation involves both individual and collective shifts in understandings and practices (knowing) which arise, through conversation, in joint action. Wenger [51] refers to this as a shared repertoire which is important when groups need to cooperate. How learning is understood, either individually or socially (as in groups), is key to effective action.

5.3.3   Learning and Effective Action

I claim that learning is concerned with the processes leading to effective action in a particular domain. Learning is distinguished when an individual, or another observer, recognises that they can perform what they were unable to perform before. Following Reyes and Zarama [32], I claim learning is an assessment made by an observer based on a change in an observed capacity for action. From this perspective, learning is not about ideas stored in our mind, but about action, action which is always set in a social dynamic. So what makes an action effective? Reyes and Zarama [32] make the following claims:

> Assessments change through history.... A major blindness we often observe in people is the almost exclusive attention they pay to learning particular skills as a way to become effective and successful in the future. However, they do not pay much attention to the fact that the standards to assess effectiveness in the future may be very different from the ones used today.... Actions by themselves never generate effectiveness. Only actions that comply with existing social standards can produce it.... A good example... is the importance granted today to ecological concerns. Based on historical changes in standards of effectiveness, procedures that were considered extremely effective in the past are now discarded because they do not meet ecological standards.

Within this logic 'effectiveness' is associated with the process of conserving (or enacting) over time viable behavioural repertoires. This historical pattern of changes in what constitutes effectiveness is made in our social communications – it can be referred to as discourses in the social sciences. Making judgements about effectiveness is something we do every day when we say, 'He is a good footballer' or 'She is a good manager'. Implicit in these statements are some measures of performance against which effectiveness was judged. An example might be that of a father whose standards of effectiveness are different to his daughter's when, after listening to a CD, he says, 'She is a good singer' and she goes "What?"!

To be highly competent in practice, any practice, requires that the learning be embodied – incorporated in the body itself (see Table 5.1). This is clear if we watch an Olympic hurdler or any other consummate athlete or performer or a very good teacher articulate a difficult concept or deliver a stimulating lecture. Every learning involves an alteration of the learner's body to perform the newly-learned actions.

This takes time, many, many repetitions of the action each one becoming smoother and smoother as the elements of the action are coordinated as one continuous and elegant action. This of course is what we mean when we say the athlete or performer is "practicing". The same is true for things that we do in other ways. To be effective in performing mental arithmetic, the person has to do that over and over again, that is practise. Thus, to be effective, one must practise whatever the activity, and the practice "embodies" the action.

For example, I have an aspiration to embody a wider repertoire of systems practices and the only way I can do that is engage in them. I think I have a long way to go, but I have also experienced systems practitioners who meet many of the criteria of the 'ideal' model that I am developing. Of course it is not feasible for one person to embody all the myriad different skills that people might embody; each one takes a substantive amount of practice. The same is true for the range of possible forms of systems practice. It is certainly unlikely that one person might embody all the different practices associated with the different systems lineages depicted in Fig. 2.3, even though they may understand the concepts associated with most of them.

I have claimed that central to juggling the B-ball is how we know about our knowing (which is the subject of epistemology) and that, following Maturana [25], all knowing is doing. This suggests a recursive relationship between being and doing just as there is a recursive relationship between theory and practice, though it is better expressed as theorising and practicing. The same applies also for ethics and epistemology. All have implications for the relationship between being and doing, which I will return to in Chapter 13. But now I want to turn to ethics which is the next element of the B-ball that I want to consider.

5.4 Being Ethical

Heinz von Foerster [48], citing philosopher Ludwig Wittgenstein, claimed that 'ethics cannot be articulated'. Further, 'it is clear that ethics has nothing to do with punishment and reward in the usual sense of the terms'. Nevertheless, he contends, there are desirable and undesirable conditions inherent in any action and in the consequences of the action. These conditions, he argues, must reside in the action not in some form or code abstracted from action. As outlined in Chapter 2, von Foerster went on to consider the epistemological choice we all must make in terms of the following questions:

• Am I apart from the universe? Whenever I look, am I looking as through a peep-hole upon an unfolding universe?
• Am I part of the universe? Whenever I act, am I changing myself and the universe as well?

He then went on to say:

Whenever I reflect on these two alternatives, I am surprised again and again by the depth of the abyss that separates the two fundamentally different worlds that can be created by such a choice. Either to see myself as a citizen of an independent universe, whose regularities, rules and customs I may eventually discover, or to see myself as the participant in a conspiracy [in the sense of collective action], whose customs, rules and regulations we are now inventing.

The ethical way forward, von Foerster argues, is to always try to act to increase the number of choices available. By this he seeks in his own practices to act in ways that do not limit the activities of other people: 'Because the more freedom one has, the more choices one has, and the better chance that people will take responsibility for their own actions. Freedom and responsibility go hand in hand.' [50, p. 37] [24]

One of the examples of how von Foerster's concept of choice, and thus ethics, operates, that I find illuminating was related to me by Humberto Maturana [26]. He spoke of how, when his young grandson accidently fell in a swimming pool, he naturally reacted very quickly, but open to the circumstances realised that his grandson was capable of 'saving himself'. Another person less open to the circumstances may have immediately rescued the boy, and thus deprived him of choice and thus responsibility. The choice for the grandson in this context came through an enhanced behavioural repertoire, and thus more embodied options for the future, i.e. enhanced capacity for the recursive operation of responsibility with response-ability. Thus Maturana's ethical action took into account more than just safety and he considered the consequences of his behaviour on the future wellbeing of the child. One of the reasons ethical behaviour cannot be specifically prescribed is that the extent of implications is always dependent on the immediate situation.

This same awareness of a set of wider implications underpins my own experience of re-conceptualising my role as an academic from 'teacher' to 'facilitator of learning'. All too often the former leads to practices which take responsibility away from the learner; in my experience the best learning occurs when a learner is able to take responsibility for their own learning and the facilitator works to create the circumstances in which the learner can be response-able. This applies equally to this book and your own developing systems practice. It is one thing for me to try to create a context, through the auspices of this book, for you to begin to take responsibility for developing your systems practice. But it is another for me to imagine the myriad of contexts within which a reader might be situated. None-the-less the responsibility/response-ability ethic only becomes meaningful if, in understanding systems practice, I conceptualise it always as a systems practitioner in context relationship.

---

[24]By choice von Foerster means enhancing the opportunity to consider the implications of acting one way or another. This does not happen if one is given rules. Rules preclude consideration of alternatives because the action is specified. Treated superficially the concepts espoused by von Foerster can be easily misunderstood, as seems to have been the case with the UK Labour Government who, when Tony Blair was PM, made much of 'increased choice', as in say the operation of the National Health Service. Simplistically they seem to have reduced the concept of choice to the provision of options in contexts that did not matter to patients – the choices that were offered were not choices that mattered to those who used the health service, i.e. the choices were not grounded in the circumstances of patient's own lives. As Seddon observes [38, p. 15] 'to say people have choice when they are in no position to make one is disingenuous.'

For systems practice to flourish contexts in which a practitioner is response-able need to be created. In Part III (Chapter 9) I identify four contextual factors that undermine the emergence of responsible systems practice.

Von Foerster's questions are highly relevant to the ethics of systems practice. Ethics within systemic practice[25] can be seen as operating on multiple levels. Like the systems concept of hierarchy (Table 2.1), what each of us perceives to be good at one level might be bad at another – and we may be unaware that the level we are operating at is different to other stakeholders in the situation (typically confusion over what, why, how). Because an epistemological position (e.g. a choice about how we understand situations) is available to be chosen, rather than taken as a given, the choice involves taking responsibility.[26] The choices made have ethical implications; my aim here is to expand the awareness of choice.

Within traditional, mainstream, practice ethics and values are generally not addressed as a central theme. If there is no awareness, those who follow the mainstream view are not integrated into the change process because the practitioner or researcher takes an objective stance that excludes ethical considerations. But recourse to objectivity can be a means of avoiding responsibility (see also Maturana [25]). Moral codes are also inadequate; they either have implicit in them an objective "right" view of the world or, if first generated by a thoughtful leader as a guideline, they become institutionalised to support a power structure of enforcers or other moralists.

A practical tool for acting ethically is to be aware of the language used in a conversation. For example, by turning away from statements that begin with 'That is the way it is!' A claim such as this, which involves entering a conversation convinced you are right or that your perspective is the only valid one, limits the choices available to those who wish to pursue a conversation. A reframing to 'In my experience I find...' offers more choices and, through the introduction of the word 'I', it involves taking responsibility as long as in the saying it is experienced as authentic by those who listen. Of course this does not mean that in the unfolding conversational dynamic you have to agree with the perspective of others on offer! Nevertheless, progress in human affairs generally involves some form of agreement which I understand as achieving accommodations[27] between those with different perspectives, rather than a consensus (see also Chapter 11).

Another example from practice of attempting to be ethical is shown in Box 5.1. This example comes from work conducted by members of the Open Systems Research Group[28] with the English and Welsh Environment Agency.

---

[25] The distinction between systemic and systematic practice was addressed in Chapter 2; I also return to these distinctions at the end of Part II.

[26] This applies only if we don't remain in the default position of thinking there is no other. Usually people do not choose their position, they don't even know they have one!

[27] The action of adjusting or adapting to enable purposeful action amongst those with different understandings and interests.

[28] Based at The Open University, UK.

Box 5.1   An Example of a Statement of Ethics[29]

Background

This is not a code of ethics. It is a statement of how we intend to work together in undertaking a collaborative research project. It is thus about our ambitions concerning joint practice, our ways of relating and conversing.

There are important distinctions to be made about the connection between research and ethical practice [15]. For example, a distinction can be made between 'fixed rule language games' like the judicial system, and 'emergent rule language games' [53, p. 1107], as in contract law. We consider, for the purpose of this research, ethics in research to be an emergent language game, that is, rather than being captured in pre-specified codes, ethics arise in lived experience. The meaning of what counts as ethical for that conversation is brought forth in the conversation and in the practices that emerge from these encounters (e.g. in writing documents, papers, reporting on what has happened). Thus 'it is working on developing a Code [of Conduct] that foregrounds ethical issues rather than the code itself' [15]. From this perspective our project will not predefine what our Ethics are but will carry this as an ongoing conversation throughout our joint activity. It will not however be implicit – we will always come back to it in our joint engagements.

The mutuality of conversations is also recognised by Clandinin and Connelly [9] who claim that the conversational form in qualitative research is marked by:

- Equality among participants
- Flexibility to allow participants to establish the form and topics important to their inquiry
- Listening
- Probing in a situation of mutual trust, and caring for the experiences described by the other

We agree with this as a starting point in our on-going activities.

An Ethical Agenda

The agenda below is an invitation for us to talk about these matters, not a set of imperatives:

- We will be open with each other about our mutual expectations
- We will commit to reflective practice

---

[29]Agreed between The Open University (Open Systems Research Group – incorporating the SLIM-UK Project), the Environment Agency (EA) funded project 'WFD (Water Framework Directive) and River Basin Planning Project – Social Learning'.

(continued)

- We acknowledge that certain contractual obligations will need to be codified and we will work to ensure that these keep open, as much as possible, the space for our ethical conversation
- We will discuss and agree practices relating to writing-up, reporting and presenting about our joint activities and we will be clear with each other about matters of confidentiality and anonymity
- In our practices we will acknowledge the work of others and invite others with whom we work to accept the same practice

An implication of the 'contract' reported in Box 5.1 is that ethics stands on our emotions as they arise in our living. Ethics do not stand on rationality. An ethical conversation arises in the awareness and care that what is being discussed has consequences for others.

5.5   Constraints and Possibilities Associated with Our 'Being'[30]

5.5.1   Technology as Mediator of Our Being

It seems to me that as human beings no matter how hard we try, we seem always to be caught up in some form of historical 'framing' that presents constraints and possibilities to what we can do. The most basic and pervasive of these is language which, as I indicated earlier, I will call a 'social technology'. Ever since humanoids began to use tools, much as a chimp can use a stick to extract ants from a nest, our being began to be mediated by technologies. Technologies are thus ubiquitous – they provide a pervasive background to our development and on-going daily life. They are not part of us – but neither are they separate from us. Technologies mediate who we are and what we do by shifting the constraints and possibilities for what we can do and since we know according to how we do, they also change our conception of the world.[31]

---

[30]My phrasing here is inspired by the work of Ceruti [6, p. 117] who said the contemporary evolutionary sciences do not just emphasise the notions of possibility and constraint as key notions in the explanation of evolutionary phenomena. They show the existence of a natural history of possibilities and constraints, of a history of reciprocal co-production of possibility and constraints. This 'decisive change consists in placing at the foundation of the evolutionary sciences the notion of constraint and not the notion of cause'.

[31]As in something between, connected but not directly.

Fig. 5.7 A metaphor for the technological framings of our existence from which it is sometimes difficult to escape

Figure 5.7 is a metaphor for what I mean. Technologies are so pervasive and so frequently in the background yet so influential on our lives that it is as if we are living within a large ball. This "ball" enables most of our actions most of the time but at particular times – say if trying to go over rough ground or up or down a hill – can prove particularly difficult, even dangerous (as in significant technological failure or breakdown). But it is only at these moments that we begin to notice that we are in the ball, a sort of technological cocoon, because for the rest of the time we are like goldfish, oblivious to the water in which we exist.

As humans spread across the Earth they took with them fire and other technologies which remain with us in various forms today. Fire created warmth, enabled cooking and could be used to manage the landscape. If one travels it is possible to see many variants of technological lineages such as those for heating or cooling. Thus in many tropical or humid areas (e.g. Florida) humans have expanded their presence through the development of air-conditioning based on relatively cheap fossil fuel energy. This, in the case of Florida, has enabled an expansion of the number of older people who can live in what would have been, without air conditioning, an inhospitable environment. Climate change, through systemic effects associated with factors such as rising energy costs, technology failures associated with increasing temperatures and increased hurricane frequency may challenge the ongoing viability of the forms of human activity that have evolved in Florida. In future people may no longer be able to live as they have due to breakdown of the people-technology relationship.

Air conditioning is likely to be widely understood as a form of technology. However, most people are unlikely to think of 'language' and 'road rules' under the same label. For this reason I use the term social technologies and argue (in the next section) that it makes sense to understand language and other rules, norms etc.

within an intellectual framing informed by the studies of the history and philosophy of technology as well as that of new institutional economics [29, 41]. My purpose is to draw attention to (i) the relational nature of technology; (ii) how understandings become embedded in technologies at particular historical moments and can remain embedded, and thus 'out of sight, out of mind', even though the historical understandings may no longer be accepted or useful; (iii) how our own being is a product of these dynamics. My ambition is to make these dynamics and their implications more transparent as we go about doing what we do!

5.5.2 The Role of Social Technologies

By social technologies I mean technologies that are often invisible because they are embedded in daily practices, including our language and use of numbers. As outlined in Part 1 (Chapter 1) an example based on numbers is the practice of giving scores as exam marks – this was 'invented' in the late eighteenth century but today it seems unthinkable that we would not quantify scores for exams. A good example in my work at The Open University (UK) and in other settings is the use of 'an agenda' and 'minutes' as part of regular meetings. At some historical moment someone invented these practices and they have been conserved as practices over time even though on some occasions our meetings might be more productive and creative if we did not employ an agenda and minutes. Because I am aware of the ways in which an agenda and minute taking can help or hinder a meeting I sometimes use other techniques to structure a meeting e.g. a SWOT (Strengths, Weaknesses, Opportunities, Threats) analysis can be used as an alternative – it structures discussion and leaves behind a written record.

Social technologies are distinct from artefacts such as an air conditioner, hammer or a computer considered in isolation, which is what we usually think about when technology is mentioned. They are characterised by a set of relationships in which the technology plays a mediating role just as the hammer does in Fig. 5.8.

Examples of social technologies include 'cost-benefit analysis', 'white papers', 'public inquiries', 'carbon emission schemes', 'environmental management schemes', 'regulatory impact assessment'... the list is seemingly endless. There are many more examples which are less obvious such as the format of committee meetings, protocols for ministerial decision-making, templates for public consultation, procedures for hiring consultants, etc. Figure 5.9 depicts just some of the social technologies that can be seen to be involved in an 'Israeli water management system'. It also refers to 'desalination plants' and the like, which I will call artifactual or engineered technologies.[32] This systems map was developed as part of a multi-perspective, systemic inquiry into the social and institutional aspects of water managing in Israel.

---

[32] The word 'artifact' is revealing in this context. It can be understood as an 'art-in-fact' and thus raises questions about the nature of social facts – an epistemological divide suggests itself.

Fig. 5.8 When researchers and others talk about the need for new tools (e.g. a hammer) they usually fail to recognise that the situation of concern, and thus what they desire, is a relational dynamic between people (a hammerer), a tool, a practice (hammering) and a situation (frustration with a computer?). The dynamic also produces something we can describe as a result or an effect, i.e. something is hammered!

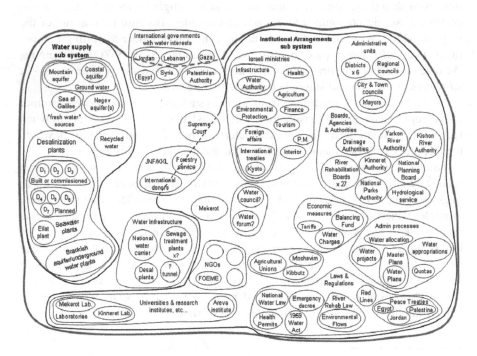

Fig. 5.9 One person's understanding of the Israeli water managing situation depicted using a systems map of the artifactual and social technology elements of a water managing system

In my terms management, or decision making becomes a social technology when it is made up of procedures and rules designed to standardise behaviour, or in other words, practice becomes reduced to sets of techniques used routinely without awareness of the origins of, and implications of the use of, such techniques, the role of the practitioner and the need for contextual understanding about the situation.

My use of the term 'social technologies' is very close to what some economists, particularly institutional economists, refer to as 'institutions'. There are multiple uses and interpretations of the term 'institution' and in English, it is often used interchangeably with 'organisation' as in a 'business institution' or a 'business organization'. Following North [29] I use the term institution more broadly than organisation to describe an 'established law, custom, usage, practice, organisation or other element in the political or social life of a people'.[33] The Oxford English Dictionary defines an institution as 'a regulative principle or convention subservient to the needs of an organised community'. Institutions can be policies and objectives, laws, rules, regulations, organisations, policy mechanisms, norms, traditions, practices and customs [41]. Institutions influence how we think and what we do. In contrast, an organisation can be understood as a hierarchy or network of behaviour and roles to elicit desired individual behaviour and coordinated actions in response to a set of rules and procedures.

In my use of the term 'social technologies' I wish to draw attention to three matters. Firstly how 'institutions' are usually named as 'things' independent of the context of practice in which there is always a relationship between the 'thing' (the hammer in Fig. 5.8), someone (the hammerer) and the context of practice. For example, a lot of research funds go to develop tools for decision making without thinking of how the tool will be put into practise. I would go further and claim that social technologies can orchestrate a performance in the sense that hammering can be considered a performance. Figure 5.8 depicts how technology (in this case an artefact), a person and a situation interact to give rise to a practice – the practice of hammering. The practice arises (is an emergent property) from the interaction of the hammerer, the hammer and the hammered. Hammering as a form of practice would make little sense without each of these elements interacting and aligning with an espoused purpose, e.g. to destroy the computer or to vent one's anger.[34] The same logic applies to 'projects' or 'public inquiries' as to hammering. It should be possible to explore the history of any social technology and to ask if the thinking and circumstances that gave rise to it are still valid today, i.e. does it remain a valid social technology?

---

[33]See http://en.wikipedia.org/wiki/New_institutional_economics for more background (Accessed 24th April 2009).

[34]I am not claiming that all action and thus practice is purposeful – very often the attribution of purpose, as with intention, arises in reflection, not as a precursor to what we do. In other words we do what we do and then seek a way of describing why. But we have also evolved a manner of living or conversing about 'what is to be done' of which rationality is but one facet. This book is based on my experience that enhancing the quality of talking about what might be done benefits the nature and quality of what is done – but I do not make a claim that in the moment we do what we claim we were going to do, or that we have even done what we claimed.

Secondly, whilst I argue that new governance arrangements are needed that are more systemic and adaptive for managing in a climate-change world (Chapter 1), I am also conscious that any such innovation will duplicate the process that gives rise to a social technology. A line of argument follows that could posit that what I am actually talking about is "antisocial technology" in that when technologies and routinised practice become embedded or reified beyond awareness of their history and their role of mediation then they subvert the social nature of being human, i.e. our living in a manner in which the other arises as a legitimate other.

Thirdly I want to claim that there are useful insights to be gained from understanding institutions through the lens of history and through the philosophy of technology studies.[35]

5.6   An Example of Juggling the B-Ball

In my experience many who claim they are systems practitioners on closer inspection actually engage in writing about the value of systems thinking. As Bunnell [2] has observed by 'doing this they report explanations that they themselves find satisfying and would have others enact'. Effective practice in both of these domains (acting and writing about acting) is highly desirable, but there is a need to be clear what form of practice is being enacted in particular contexts. The following anecdote exemplifies one of the main reasons why I think juggling the B-ball is important for systems practice. The story relates to two practitioners who were able to connect with the history of organisational complexity ideas. It describes the process they chose to take in response to a highly specific organisational-development tender document couched in traditional ways [39]:

> Our first decision was to challenge the tender document.... When asked to present our proposals to the tender panel we ignored the presenter/audience structure in which the room had been arranged by drawing chairs up to the table and conversing with the client group. We began a discussion about the way those present were thinking about organisational and cultural change and emphasised the unknowability of the evolution of a complex organization in a complex environment. Instead of offering workshops or programmes we proposed an emergent, one step at a time contract... to discover and create opportunities to work with the live issues and tasks that were exercising people formally and informally in the working environment... we were subsequently told that the panel's decision to appoint us was unanimous.

When reflecting on this experience Patricia Shaw made the following comments:

> We were told by one of the directors, 'Everyone else made a presentation based on knowing what to do. You were the only ones who spoke openly about not knowing while still

---

[35] For example see the body of work by Don Ihde including Technics and Praxis [17]; Experimental Phenomenology: An Introduction [18]; Technology and the Lifeworld: From Garden to Earth [19]; Philosophy of Technology: An Introduction [20].

being convincing. It was quite a relief'.[36] Our success in interesting the client group in working with us seemed to be based on:

1. Making it legitimate in this situation not to be able to specify outcomes and a plan of action in advance, by so doing we made 'not knowing' an intelligent response
2. Pointing out the contradictions between the messy, emergent nature of our experience of organizational life and the dominant paradigm of how organizations change through the implementation of prior intent

This approach helped to contain the anxiety of facing the real uncertainties of such a project together. It was an example of contracting for emergent outcomes.

How do I interpret this story? It shows that how we think about the world: our theories and models that are a result of experience, even if implicit, determine what we do in the world. Our theories predispose us to engage with 'real-world' situations in particular ways. Unlike the other consultants, Patricia Shaw and her colleague did not respond to the tender as if it were a problem for which they had the answer.

This approach is potentially able to encompass all of the complexity in the situation. It is also able to bring forth the multiple perspectives through the engagement of all the actors in the situation.[37] They used conversations, interviews and even drama to achieve this. This allows outcomes to emerge from the process rather than being defined in the form of a plan with outcomes specified in advance. Sometimes highly specific plans that are not renegotiated iteratively as the environment changes are called blueprints and the process called blueprint planning. Shaw and her colleague approached their task as an unfolding process of 'engaging' in which all parties were learning or co-constructing new meanings in the situation [39]. You will, of course, recognise that the behaviour of Shaw and her colleague is not appropriate in all contexts, although I think the approach could be used more. In the case of an engineer responding to some specific request that required precise technical specifications another response may have been appropriate. Being aware, or becoming aware of our being, I argue, increases the repertoire of possible actions available to a systems practitioner.[38]

---

[36] Those experienced in this sort of practice will understand that in similar situations some would prefer to act in the acceptance of a fiction that those presenting their ideas actually know what to do. Such an attitude is fostered in some of the main discourses on 'leadership'.

[37] Sometimes "I don't know" can be an abrogation of responsibility but on other occasions it can entail the acceptance of a deeper responsibility. The consultants in this example presumably demonstrated the latter in their initial interview – and evidently followed through. It is easy to pretend to engage in a collaborative process or to follow a prescription for participation and not really engage deeply.

[38] Patricia Shaw and I at one period discussed extensively our approaches to practice and found that we had quite profoundly different preferences. My own is to do much more planning or designing at an early stage after becoming as aware of the context as is possible. But then to be open to the unfolding circumstances in the practice situation and act in an emotion that is free to change, not one that tries to hold onto the plan or design. My argument for this form of practice is that it enhances one's behavioural repertoire in the moment providing the practitioner with more variety and choices.

Being aware, or not, of the issues I have raised in this chapter creates the initial starting conditions for engaging with situations; I turn to the implications of juggling the E-ball in the next chapter.

Before moving on to the E-ball it is important to acknowledge that Shaw and her colleagues make no claims to being systems practitioners – in fact they have rather publicly rejected systems approaches, though as I have argued elsewhere, and with them personally, what they reject is their own misunderstanding of the diversity and differences within systems thinking and practice (as depicted in Fig. 2.3).[39] In offering this example of systems practice I take responsibility for the claim that I make, based on my own experience of them doing what they do; for me they are acting systemically because they create the circumstances for emergence and they act with (some) awareness of their own being in the situation.

References

1. http://www.braingle.com/brainteasers/teaser.php?id=26745;op=0;comm=1 (Accessed 3rd August 2009).
2. Bunnell, P. (2003) Personal communication. Systems Ecologist, President of Lifeworks, Vancouver, British Columbia, June 2003.
3. Bunnell, P. (2008) Editor's Foreward. The Origin of Humanness in the Biology of Love. In H.R. Maturana and G. Verden-Zoller (eds), The Origin of Humanness in the Biology of Love. Imprint Academic: Exeter.
4. Bunnell, P. (2009) Personal communication. Systems Ecologist, President of Lifeworks, Vancouver, British Columbia, June 2009.
5. Carroll, J.B. (Ed.) (1959) Language thought and reality. Selected writings of Benjamin Lee Whorf. MIT Press: New York.
6. Ceruti, M. (1994) Constraints and Possibilities. The evolution of knowledge and the knowledge of evolution. Gordon and Breach: New York.
7. Checkland, P.B. and Casar, A. (1986) Vickers' concept of an appreciative system: a systemic account. Journal of Applied Systems Analysis 13, 3–17.
8. Checkland, P.B. and Poulter, J. (2006) Learning for Action: a Short Definitive Account of Soft Systems Methodology and its Use for Practitioners, Teachers and Students. Wiley: Chichester.
9. Clandinin, D.J. and Connelly, F.M. (1998) Personal experience methods. In N.K. Denzin and Y.S. Lincoln (Eds.), Collecting and Interpreting Qualitative Materials. Sage: London.
10. Evans, N. (2009) Dying words: Endangered languages and what they have to tell us. Wiley-Blackwell: Maldon, MA.
11. Fell, L. and Russell, D.B. (2000/2007) The human quest for understanding and agreement. In R.L. Ison and D.B. Russell (Eds.), Agricultural Extension and Rural Development: Breaking out of Traditions (pp. 32–51). Cambridge University Press: Cambridge.
12. Glanville, R. (1995a) A (cybernetic) musing: Control 1, Cybernetics & Human Knowing 3(1), 47–50.
13. Glanville, R. (1995b) A (cybernetic) musing: Control 2, Cybernetics & Human Knowing 3(2), 43–46.
14. Harré, R. (1983) Personal Being: A Theory for Individual Psychology. Basil Blackwell: Oxford.

---

[39]For a well-articulated critique see Checkland and Poulter [8, p. 148–153].

15. Helme, M. (2002) Appreciating metaphor for participatory practice: constructivist inquiries in a children and young peoples justice organisation. PhD Thesis, Systems Discipline, The Open University: UK.
16. Høeg, P. (1994) Miss Smilla's Feeling for Snow. Flamingo: London.
17. Ihde, D. (1979) Technics and Praxis. Reidel: Dordrecht.
18. Ihde, D. (1986) Experimental Phenomenology: An Introduction. Putnam: New York.
19. Ihde, D. (1990) Technology and the Lifeworld: From Garden to Earth. Indiana University Press: Bloomington, IN.
20. Ihde, D. (1993) Philosophy of Technology: An Introduction. Paragon: New York.
21. Lakoff, G. and Johnson, M. (1999) Philosophy in the Flesh: The Embodied Mind and Its Challenge to Western Thought. Basic Books: New York.
22. MacIntyre, A. (1981) After Virtue. Duckworth: London.
23. Mahoney, M. (1988) Constructive meta-theory 1: Basic features and historical foundations, International Journal of Personal Construct Psychology 1, 1–35.
24. Maturana, H. and Varela F. (1987) The Tree of Knowledge: The biological roots of human understanding. New Science Library, Shambala Publications: Boston, MA.
25. Maturana, H. (1988) Reality: The search for objectivity or the quest for a compelling argument, Irish Journal of Psychology 9, 25–82.
26. Maturana, H. (1996) Personal Communication. Professor Emeritus, University of Chile, 1996.
27. Maturana, H.R., Verden-Zöller, G., and Bunnell, P. (Ed.). (2008) The Origin of Humanness in the Biology of Love. Imprint Academic: Exeter.
28. Monaghan, P. (2009) Never mind the whales, save the languages. The Australian. Wednesday 24th June, 36.
29. North, D. (1990) Institutions, Institutional Change and Economic Performance. Cambridge University Press: Cambridge.
30. Open University (2000) Managing complexity A systems approach, Course Code T306. The Open University: Milton Keynes.
31. Pert, C.B. (1997) Molecules of Emotion. The science of mind-body medicine. Simon & Schuster: New York.
32. Reyes, A. and Zarama, R. (1998) The process of embodying distinctions: a re-construction of the process of learning, Cybernetics & Human Knowing 5, 19–33.
33. Rorty, R. (1980) Philosophy and the Mirror of Nature. Basil Blackwell: Oxford.
34. Rosch, E. (1992) Cognitive psychology. In J.W. Hayward and F.J. Varela (Eds.), Gentle Bridges: Conversations with the Dalai Lama on the sciences of mind. Shambala Publications: Boston, MA.
35. Rose, C., Dade, P. and Scott, J. (2007) Research into Motivating Prospectors, Settlers and Pioneers to Change Behaviours that Affect Climate Emissions. http://www.campaignstrategy.org/articles/behaviourchange_climate.pdf (accessed 11 August 2009).
36. Russell, D.B. and Ison, R.L. (2000) The research-development relationship in rural communities: an opportunity for contextual science. In R.L. Ison and D.B. Russell (Eds.), Agricultural Extension and Rural Development: Breaking out of Traditions (pp. 10–31). Cambridge University Press: Cambridge.
37. Russell, D.B. and Ison, R.L. (2007) The research-development relationship in rural communities: an opportunity for contextual science. In R.L. Ison and D.B. Russell (Eds.), Agricultural Extension and Rural Development: Breaking out of Knowledge Transfer Traditions (pp. 10–31). Cambridge University Press: Cambridge, UK.
38. Seddon, J. (2008) Systems Thinking in the Public Sector: the failure of the reform regime, and a manifesto for a better way. Triarchy Press: Axminster.
39. Shaw, P. (2002) Changing Conversations in Organizations. A Complexity Approach to Change. Routledge: London.
40. Shotter, J. (1993) Conversational Realities: constructing life through language. Sage: London.
41. SLIM (2004) The role of conducive policies in fostering social learning for integrated management of water. Slim Policy Briefing No. 5. See http://slim.open.ac.uk. Accessed 31 July 2009.

42. Varela, F. (1979) Principles of Biological Autonomy. North Holland: New York.
43. Vickers, G. (1965) The Art of Judgement: a study of policy making. Chapman and Hall: London.
44. Vickers, G. (1970) Freedom in a Rocking Boat: changing values in an unstable society. Penguin: Harmondsworth.
45. Vickers, G. (1983) Human Systems are Different. Harper and Row: London.
46. Vickers, J. (Ed.). (1991) Rethinking the Future: The Correspondence between Geoffrey Vickers and Adolf Lowe. Transaction Publishers: London.
47. von Foerster, H. (1984) Observing Systems. Systems Publications: Salinas, CA.
48. von Foerster, H. (1992) Ethics and second-order cybernetics, Cybernetics & Human Knowing 1, 9–19.
49. von Foerster, H. (1994) Foreword. In Ceruti, M. Constraints and Possibilities: The knowledge of evolution and the evolution of knowledge. Gordon and Breach: Lausanne.
50. von Foerster, H. and Poerkson, B. (2002) Understanding Systems. Conversations on Epistemology and Ethics. IFSR International Series on Systems Science and Engineering, Vol 17. Kluwer/Plenum: New York.
51. Wenger, E. (1998) Communities of Practice: Learning, meaning and identity. Cambridge University Press: Cambridge, UK.
52. Wittgenstein, L. (1981) Zettel. Basil Blackwell: Oxford.
53. Wittgenstein, L. (1999) Philosophical Investigations. Blackwell: Oxford.

Chapter 6
Juggling the E-Ball: Engaging with Situations

6.1 Naming Our Experiences

In this chapter I am concerned with juggling the E-ball, how we choose to engage with situations as part of systems practice. The word 'engage' in its original meaning meant 'to pledge' from which 'the sense of promise to marry', or 'betrothed', originates. This historical meaning connects in a metaphorical way with the sense that I am using 'engaging with situations' – to give what is juggled its full expression. My sense of the mediaeval understanding of 'granting one's pledge' is of an act of distinguishing and committing to someone from the many – to single them out in a way that creates a new form of relationship, a different experience. In the act of distinguishing and committing a form of choice is made against a background of many possible choices. Following this line of argument I could claim that when a young woman, or old woman or both is distinguished in the example I gave in the last chapter (Fig. 5.2) the result is a pledge to one or other descriptions (or explanations) embodied in the process of naming our experience, i.e. old woman, young woman etc. It is this dynamic that I want to unpack and make practical in this chapter.

To illustrate how I interpret the E-ball I will argue that we all have choices about how to engage with situations. Further, that when we experience situations with particular characteristics it makes sense to engage with these in particular ways. My main concern will be with the process of engaging with situations through the practice of naming these situations in particular ways.[1] I will point out the limitations of naming these situations as 'problems' and explore what can be gained by thinking of them as 'complex', or 'contested', or 'wicked' or 'messy'. Thus I am going to elaborate further on the dynamics of practice depicted in Fig. 3.5; of course a primary concern I have is to illustrate how the E-ball might be juggled by a systems practitioner.

The practice of inventing a new name and giving it to a situation in response to an experience, or set of experiences, arises from the act of making a distinction (i.e., naming an experience). In this process new categories can be created as well as

---

[1] In this chapter I do not propose to develop explanations about how changes in our emotional dynamics accompany the acts of distinguishing and naming.

R. Ison, *Systems Practice: How to Act in a Climate-Change World*,
DOI 10.1007/978-1-84996-125-7_6, © The Open University 2010.
Published in Association with Springer-Verlag London Limited

new identities.[2] Much of the time we use names coined by others to name our experiences but in some instances new names are coined; the process of doing this can be called neologising, a practice that gives rise to a new expression as well as a new underlying logic. A recent example of this is the invention of the term 'complex adaptive system' but there are earlier examples which I will explore.

6.1.1 Naming Situations as 'Wicked Problems'

Systems scholars and practitioners, along with others, have commonly engaged in neologising.[3] One that is better known than most is the 'wicked problem'. In 2007, the Australian Public Service Commission (APSC) [3], concerned about the growing number of seemingly intractable policy 'problems' produced a very thoughtful review of 'wicked problems'. In this review they described 'wicked problems' as problems that

> go beyond the capacity of any one organisation to understand and respond to, and [where] there is often disagreement about the causes of the problems and the best way to tackle them…. Usually, part of the solution to wicked problems involves changing the behaviour of groups of citizens or all citizens. Other key ingredients in solving or at least managing complex policy problems include successfully working across both internal and external organisational boundaries and engaging citizens and stakeholders in policy making and implementation.

They go on to say that 'wicked problems require innovative, comprehensive solutions that can be modified in the light of experience and on-the-ground feedback' and that 'wicked problems' 'can pose challenges to traditional approaches to policy making and programme implementation'. In my experience this last claim about policy making and implementation is both valid and profound. In a foreword to the APSC paper, the Australian Public Service Commissioner makes the very powerful point that: 'It is important, as a first step, that wicked problems be recognised as such. Successfully tackling wicked problems requires a broad recognition and understanding, including from governments and Ministers, that there are no quick fixes and simple solutions'. This is a strong statement that does not mince words; it could be seen as a challenge to the way that Australian democracy and associated bureaucracies function and could be also seen as a call to action in the light of climate change and other situations seen as 'wicked problems'.

The authors of the APSC paper identified a number of situations which they chose to describe as 'wicked problems'. These included [3]:

---

[2]Here I include the act of coining new names to describe phenomena as well as that of naming new children – thus I consider a wide range of situations to which the same underlying process applies.

[3]I say this with awareness that what I am doing is what I describe! Neologism – literally new logic or new coinage.

Climate change – because it is a 'pressing and highly complex policy issue involving multiple causal factors and high levels of disagreement about the nature of the problem and the best way to tackle it. The motivation and behaviour of individuals is a key part of the solution as is the involvement of all levels of government and a wide range of non-government organisations (NGOs).'

Obesity – because it is 'a complex and serious health problem with multiple factors contributing to its rapid growth over recent decades. How to successfully address obesity is subject to debate but depends significantly on the motivation and behaviour of individuals and, to a lesser degree, on the quality of secondary health care. Successful interventions will require coordinated efforts at the federal, state and local government levels and the involvement of a range of NGOs.'

Indigenous disadvantage – because it 'is an ongoing, seemingly intractable issue but it is clear that the motivation and behaviour of individuals and communities lies at the heart of successful approaches. The need for coordination and an overarching strategy among the services and programmes supported by the various levels of government and NGOs is also a key ingredient.'

Land degradation – because it 'is a serious national problem. Given that around 60% of Australia's land is managed by private landholders, it is clear that assisting and motivating primary producers to adopt sustainable production systems is central to preventing further degradation, achieving rehabilitation and assisting in sustainable resource use. All levels of government are involved in land use as is a range of NGOs.'

Using the above examples I feel sure you will not find it difficult to name some situations you have encountered as 'wicked problems'. The term 'wicked problems' was coined by Horst Rittel and Melvin Webber in 1969 [25]. At the time Rittel was Professor of the Science of Design and Webber Professor of City Planning at the University of Berkeley in California; these scholars observed that there was a whole realm of 'social planning problems' that could not be treated successfully with traditional linear, analytical approaches. At the time they contrasted 'wicked problems' with 'tame problems' (Box 6.1).

Box 6.1 Some features of wicked and tame problems

Wicked problems

Wicked problems are described as 'ill-defined, ambiguous and associated with strong moral, political and professional issues. Since they are strongly stakeholder dependent, there is often little consensus about what the problem is, let alone how to resolve it. Furthermore, wicked problems won't keep still: they are sets of complex, interacting issues evolving in a dynamic social context. Often, new forms of wicked problems emerge as a result of trying to understand and solve one of them' (see http://www.swemorph.com/wp.html).

(continued)

Box 6.1 (continued)

The following list summarises some of the main features (http://en.wikipedia.
org/wiki/Wicked_problem):

1. There is no definitive formulation of a wicked problem
2. Wicked problems have no stopping rule
3. Solutions to wicked problems are not true-or-false, but better or worse
4. There is no immediate and no ultimate test of a solution to a wicked
   problem
5. Every solution to a wicked problem is a "one-shot operation"; because
   there is no opportunity to learn by trial-and-error, every attempt counts
   significantly
6. Wicked problems do not have an enumerable (or an exhaustively
   describable) set of potential solutions, nor is there a well-described set
   of permissible operations that may be incorporated into the plan
7. Every wicked problem is essentially unique
8. Every wicked problem can be considered to be a symptom of another
   problem
9. The existence of a discrepancy representing a wicked problem can be
   explained in numerous ways. The choice of explanation determines the
   nature of the problem's resolution
10. The planner has no right to be wrong (planners are liable for the conse-
    quences of the actions they generate)

Tame problems

According to Conklin [11, p. 11] these:

1. Have a relatively well-defined and stable problem statement
2. Have a definite stopping point, i.e. we know when the solution or a solution
   is reached
3. Have a solution which can be objectively evaluated as being right or wrong
4. Belong to a class of similar problems which can be solved in a similar
   manner
5. Have solutions which can be tried and abandoned

Sources: http://en.wikipedia.org/wiki/Wicked_problem (Accessed 21 July 2008);
http://www.swemorph.com/wp.html (Accessed 21 July 2008).

I think that the invention of "wicked problems" is a good example of creating
neologisms in response to particular experiences. In trying to answer the question:
What is it that Rittel and Webber did when they did what they did? I find it insightful

to ask 'what experiences did they have that led them to coin these neologisms?' My question can be answered in part by reading the original paper by Rittel and Webber [25]. I will spend some time exploring their work because the 'wicked and tame problems' distinction offers a good case study of how neologisms become conserved (i.e., taken up and used), or not, over time. Through this dynamic some terms seem to take on a life of their own independent of their history, and more importantly, independent of an awareness of the nature of the experiences that were had, when the term was coined.[4] Other terms have little impact – they are not taken up, or only taken up selectively. As I examine Rittel and Webber's paper I will be looking for possible answers to the question: How did they imagine these terms being used in practice? By posing this question I want to draw attention to how a neologism, a new concept, can become employed, or not, as a social technology.

Rittel and Webber's paper, published in the journal Policy Sciences in 1973 was originally presented to the Panel on Policy Sciences of the American Association for the Advancement of Science, Boston, in December 1969. Drawing on their paper it is possible to characterise their concerns; they said:

1. 'There seems to be a growing realization that a weak strut in the professional's support system lies at the juncture where goal-formulation, problem-definition and equity issues meet. Goal-finding (central to planning) is turning out to be an extraordinarily obstinate task'
2. 'We are now sensitised to the waves of repercussions generated by a problem-solving action directed to any one node in the network, and we are no longer surprised to find it inducing problems of greater severity at some other node. And so we have been forced to expand the boundaries of the systems we deal with, trying to internalise those externalities'
3. 'we are calling them "wicked" not because these properties are themselves ethically deplorable. We use the term "wicked" in a meaning akin to that of "malignant" (in contrast to "benign") or "vicious" (like a circle) or "tricky" (like a leprechaun) or "aggressive" (like a lion, in contrast to the docility of a lamb). We do not mean to personify these properties of social systems by implying malicious intent. But then, you may agree that it becomes morally objectionable for the planner to treat a wicked problem as though it were a tame one, or to tame a wicked problem prematurely, or to refuse to recognise the inherent wickedness of social problems'
4. 'The difficulties attached to rationality are tenacious, and we have so far been unable to get untangled from their web. This is partly because the classical paradigm of science and engineering – the paradigm that has underlain modern professionalism – is not applicable to the problems of open societal systems'
5. 'The systems-approach "of the first generation" is inadequate for dealing with wicked-problems. Approaches of the "second generation" should be based on a model of planning as an argumentative process in the course of which an image of the problem and of the solution emerges gradually among the participants, as a product of incessant judgment, subjected to critical argument'

---

[4]This suggests a potential 'reflexive tool' – namely, when considering any term, to think 'what circumstances might have given rise to this term?' [7].

I gain the following insights from my inquiry into Rittel and Webber's paper:

- They were explicitly concerned with the process of problem formulation, particularly with who participates in formulating 'problems' (i.e. equity) and how and by whom goals are articulated[5]
- In what might be regarded as an early appreciation of the nature of networks they recognised that action at one node may induce unintended consequences at another node. They are implicitly referring to positive and negative feedback processes (Table 2.1) and the idea of unintended consequences that arise through connectedness, or its breakdown
- They used the term wicked in a playful way, exploring different metaphors, whilst at the same time recognising the seriousness of such situations
- They raised at least two implications for practice 1. avoiding treating 'wicked problems' as tame 'or to tame a wicked problem prematurely, or to refuse to recognise the inherent wickedness of social problems' and 2. the need to develop a second generation systems approach that operates in language (as an argumentative process) amongst stakeholders to form an image of the problem as 'a product of incessant judgement, subjected to critical argument'
- They recognised a very difficult context for 'the adoption' of their understandings, claiming rational approaches [technical rationality] to be tenacious and unhelpful and supported by 'the classical paradigm of science and engineering – the paradigm that has underlain modern professionalism'…[which] 'is not applicable to the problems of open societal systems'

Like many academic papers Rittel and Webber's is written in the normal academic style and they do not, for example, ground any of their claims in personal experience through one is left in no doubt that they have had these experiences.[6] They also say little about practice that may lie beyond the labelling of situations as wicked or tame. I imagine that for Rittel and Webber having policy makers recognise these 'type' of situations as 'wicked problems' was probably half the battle. Perhaps they felt that the naming, or representation, of these 'type' of situations as 'wicked problems' was their most important task.

But naming a situation as a 'wicked problem' is not the same as acting to improve the situation! Doing something in these situations that most would agree leads to a change for the better is a much greater challenge and in this regard it is tempting to conclude that little has changed since 1969 – the classical paradigm remains pervasive (though human-induced climate change could act as a tipping point) and, as yet, a second generation systems approach has not taken hold in policy and governance circles, i.e., systems explanations, and hence practices are not valued in this context.

---

[5]It is worth noting that they say 'goal finding', not 'goal setting' – this is a subtle but important distinction which has practical implications.

[6]For example there is no use of the first person – the author's experience is abstracted, or written, out of the situation.

Maturana [20] makes the point that when we accept a different explanation our world changes. Unfortunately, as the APSC paper [3] makes clear, the world of policy makers seems not to have changed in response to the naming of 'wicked problems' 40 years ago. It suggests that the explanations about 'wicked problems' are not widely accepted in policy circles. As a consequence it is also unlikely that the APSC authors' recognition that:

> tackling wicked problems also calls for high levels of systems thinking... thinking [that] helps policy makers to make the connections between the multiple causes and interdependencies of wicked problems that are necessary in order to avoid a narrow approach and the artificial taming of wicked problems. Agencies need to look for ways of developing or obtaining this range of skills[7]

is well appreciated or accepted in policy circles. If I stand back from the specifics of Rittel and Webber's paper some interesting questions emerge such as "How, if at all, have the notions of 'wicked and tame problems' entered our understandings and practices?" Unfortunately I have not undertaken or seen a study that sets out to answer these questions in a systematic way. My experience suggests that the 'technical rationality' identified as constraining by Rittel and Webber is still tenacious and that the professionalism of engineers and planners has changed little towards skills for managing 'wicked problems'.[8]

The APSC paper I referred to earlier can be seen as part of a lineage of attempts to introduce understandings about 'wicked' and 'tame' problems' into policy and governance discourse. The practices they suggest are outlined in Box 6.2.

Box 6.2 Practices for tackling wicked problems

Tackling wicked problems is an evolving art but one which seems to at least require:

- holistic, not partial or linear thinking. This is thinking capable of grasping the big picture, including the interrelationships between the full range of causal factors underlying the wicked problem. Traditional linear approaches to policy formulation are an inadequate way to work with wicked policy problems as linear thinking is inadequate in encompassing their complexity, interconnections and uncertainty

(continued)

---

[7] It might be concluded that the issue now is that policy makers have not found a way to develop a new range of skills to match the conception of the problems as wicked? Perhaps the change that has happened in policy makers is an awareness that they don't have the skills, nor do they know how to go about getting them. Perhaps they are stuck?

[8] I have no doubt that in many local settings understandings of wicked and tame problems have been generated but my perception is that they are still not widely appreciated and incorporated into everyday discourse – hence the APSC paper [3].

Box 6.2 (continued)

- innovative and flexible approaches. It has been argued that the public sector needs more systematic approaches to social innovation and needs to become more adaptive and flexible in dealing with wicked problems
- the ability to work across agency boundaries. Wicked problems go beyond the capacity of any one organisation to understand and respond to, and tackling them is one of the key imperatives that makes being successful at working across agency boundaries increasingly important. This includes working in a devolved way with the community and commercial sectors
- increasing understanding and stimulating a debate on the application of the accountability framework. It is important that pre-set notions of the accountability framework do not constrain resolution of wicked problems
- effectively engaging stakeholders and citizens in understanding the problem and in identifying possible solutions
- additional core skills. The need to work across organisational boundaries and engage with stakeholders highlights some of the core skills required by policy and programme managers tackling wicked problems – communication, big picture thinking and influencing skills and the ability to work cooperatively
- a better understanding of behavioural change by policy makers. This needs to be core policy knowledge because behavioural change is at the heart of many wicked problems and influencing human behaviour can be very complex
- a comprehensive focus and/or strategy. Successfully addressing wicked policy problems usually involves a range of co-ordinated and interrelated responses given their multi-causal nature and that they generally require sustained effort and/or resources to make headway
- tolerating uncertainty and accepting the need for a long-term focus

Source: APSC [3].

Unfortunately little is said in the APSC paper about how and where to develop these skills, or capabilities, nor whether current institutional settings are conducive to such practices being enacted.[9]

---

[9]In a presentation to the International Society for Systems Sciences annual conference in Brisbane, Australia in July 2009, entitled "Delivering performance and accountability – intersections with 'wicked problems'", Lynelle Briggs [6], the Australian Public Service Commissioner made the point that removing unnecessary obstacles to innovation, to improve the quality of outcomes in complex and uncertain policy areas and developing more variegated accountability and performance management arrangements, better suited to new modes of policy implementation were needed (see http://www.apsc.gov.au/media/briggs150709.htm Accessed 11 August 2009).

From my own perspective there is also an unfortunate entailment in the term 'wicked problem', namely the concept 'problem'. Later I will expand upon why I have come to see the use of the word 'problem' as problematic – for now let me say that my main concern is about what becomes reified whenever the term 'problem' is used.[10] And my concern is driven by my understanding of the process of reifying problems as well as the question of who participates in naming, or reifying, 'the problem' in a given context.

In this introductory section about juggling the E-ball, I have explored a particular dynamic that arises when we reflect on our experiences and create objects (neologisms, categories etc.) from that experience.[11] This is what Rittel and Webber did when they coined the terms 'wicked' and 'tame problems'. Before I go on to explore some of the potential unintended consequences of this practice, and the implications for juggling the E-ball, I want to ask: 'Did others have similar experiences to Rittel and Webber and, if so, what did they do in response to these experiences?'

6.1.2 Naming Experiences in Similar Ways

At about the time Rittel and Webber were publishing their work, academics in the newly formed Systems Department at The Open University (OU) were developing their first teaching programmes. One of their challenges was to invent how to teach Systems (as there was little experience to go on) and how to do it in a supported open learning mode where students studied at a distance. Naturally they engaged with the published literature and explored the ideas of those who were generally accepted as Systems scholars. In doing this they came across the work of Russ Ackoff and neologisms that he had created, that of *messes* and *difficulties* (see Box 6.3). These distinctions were incorporated into OU Systems courses and have been in almost every Systems course since.[12] With the value of hindsight and after a little investigation it is possible to see that Ackoff, Rittel and Webber, as well as some others – see below – had experiences of a particular type that led them to coin these new terms to describe the situations in which their experiences arose.[13]

---

[10]I sometimes speak of 'wicked situations' as a hybrid term that acknowledges the lineage of ideas coming from the work of Rittel and Webber.

[11]The traps associated with neologising and reifying extend, in my view, to the practices of categorising and typologising as well. The act of categorisation is very common – in research practice the development of typologies is also a frequent form of practice. Typologies and categorisations can themselves become reified; the circulation of the products of reification in academic discourse in particular leads us to lose sight of how these "things" came into existence and, further, the validity or viability in contemporary circumstances, of their on-going use.

[12]The distinction between wicked and tame problems also influenced the teaching of Design at The Open University – see Cross [12, pp. 134–144].

[13]I do not know to what extent Ackoff, Rittel, Webber and later Schön, were influenced by each other or by earlier scholars, but it is probable that there were interacting influences between them.

Box 6.3 Some features of messes and difficulties

Russell Ackoff first coined the term 'mess' in 1974. He did so in response to the insights of two eminent American philosophers, William James and John Dewey. These philosophers recognised that problems are taken up by, not given to, decision-makers and that problems are extracted from unstructured states of confusion. Ackoff [1, 2] argued, in proposing his notion of mess that:

> What decision-makers deal with, I maintain, are messes not problems. This is hardly illuminating, however, unless I make more explicit what I mean by a mess. A mess is a set of external conditions that produces dissatisfaction. It can be conceptualised as a system of problems in the same sense in which a physical body can be conceptualised as a system of atoms.

From this definition of mess, Ackoff recognised a number of features of messes:

1. A problem or an opportunity is an ultimate element abstracted from a mess. Ultimate elements are necessarily abstractions that cannot be observed
2. Problems, even as abstract mental constructs, do not exist in isolation, although it is possible to isolate them conceptually. The same is true of opportunities. A mess may comprise both problems and opportunities. What is a problem for one person may be an opportunity for another – thus a problem can be an opportunity from another perspective
3. The improvement to a mess – whatever it may be – is not the simple sum of the solutions to the problems or opportunities that are or can be extracted from it. No mess can be solved by solving each of its component problems/opportunities independently of the others because no mess can be decomposed into independent components
4. The attempt to deal with a system of problems and opportunities as a system – synthetically, as a whole – is an essential skill of a systems practitioner

In contrast, difficulties are:
simple situations that can be improved by extracting one problem from them and solving it. These are called difficulties and they are seen as exceptions rather than the norm in terms of decisions that are needed in environmental, organisational and other information-related contexts.
(Following Ackoff [1, 2])

It is certainly known from our 30 plus years of teaching Systems at The Open University, UK, that the mess/difficulty distinction has great utility for most students; an account is given in Chapman's publication 'System Failure' [8]. In many ways the practice response seems clear – recognise these situations for what they are! But is this a trap awaiting the unwary?

6.2   The Trap of Reification

It seems to me that policy makers and others could make significant progress in addressing seemingly intractable issues such as climate change or the global water crisis if only more people were aware of the understandings that the distinction mess/difficulty or wicked/tame problem evoke. Important though this might be, would it be enough? And, after all, these terms were coined over 40 years ago! With these reservations in mind I want to delve further and point to a trap we can find ourselves in if we reify these terms without being aware of the unintended consequences that reifying can produce.[14] But first I will explain what I mean by reification. To do so I have to revisit the type of relational thinking that I introduced in Chapter 3 in my example of walking as a form of practice (Fig. 3.7).

6.2.1   Our Inescapable Relational Dynamic with 'Our World'

Let me start by asking you to think of a typical young child – parent interaction, in which the child's name is repeated until the child learns its own name and begins to associate this name with an identity. The same practice goes on with naming all sorts of things – a child will be corrected if they have it wrong from the perspective of the parent or whoever is in the conversation. In other words we come into living in language in a manner that suggests all of these entities, or things, exist independently of our own acts of distinction. What is more, we perpetuate this type of practice in our daily life when we invent new terms, generate categories or devise new theories. What has become lost in this short-cut that we have adopted as a manner of living is the realisation that we have some choice, some agency in how we engage with situations. Put another way, situations do not have characteristics that pre-determine how we should engage with them. As Dave Snowden [33] points out:

> Humans have a strong tendency to classification using existing frameworks; this is a form of pattern matching which also leads to stereotyping (racism being only one of many examples). [My approach] focuses on sense making. Individually, the term "sense making" describes the complex process by which a person makes sense of the situations they find themselves in and acknowledges the many subtle influences of perceptions, biases, goals, identities, and memories. Organisational sense making involves all of this at the collective level, describing how groups of people develop shared meanings that make sense to the group [14, p. 5].

The process that Snowden describes as 'sense-making' is, for me, a key element of juggling the E-ball.

---

[14] A particular motivation for me arises from reflecting on how distinctions or neologisms such as 'wicked problems' are taught as say part of an MBA. If these distinctions are in the curriculum at all the learner can probably learn and list the features of a 'wicked' or 'tame' problem but this, as I argue in Chapter 5, is not the same as 'knowing a wicked situation' because they are not exposed to the same experiences that Rittel and Webber had that motivated them to coin the term.

6.2.2 Making Distinctions and Living with Them

Although we humans probably couldn't operate if we didn't describe, classify and categorise, I argue that somewhere along the way we have lost sight of the implications of doing this. There are two practices that are poorly understood yet important for reflecting on what we do when we do what we do. The first, neologising, as I have outlined, involves using or coining new words or expressions (someone who does this is a neologist and the result of this practice is a neologism). The second practice is reification. Etienne Wenger [37, p. 58] draws attention to some of the implications of reification in his work on communities of practice (CoPs) (see also [5]). He describes reification as the practice of 'making into a thing' or in other words creating an object in language.[15] This is something we do all the time. Although widespread, the implications of this practice are not well understood. It has particular implications when we make or treat an abstraction, such as justice or learning, as a concrete material thing (as exemplified with the common statues of a blindfolded woman who is justice or attributing a number to learning as described in Chapter 1). Wenger [37] says: 'we project our meanings into the world [through living in language] and then we perceive them as existing in the world, as having a reality of their own' (p. 58). He goes on to use the abstract concept of reification to refer to "the process of giving form to our experience by producing objects that congeal this experience into 'thingness'" and points out that he is introducing reification 'into the discourse because he wants to create a new distinction to serve as a point of focus around which to organise [his] discussion' (p. 58).[16] As I have already pointed out Wenger's notion of 'creating a new distinction' is important as it gives rise to what we call experience.

Avoiding the trap of reification involves moving from a first-order logic to a second-order logic. To do so involves reflecting on practice itself (i.e., the practice of practice).

6.2.3 Reflecting on the Practice of Practice

When reflecting on his own professional experience of engaging with complex situations, Donald Schön [29], author of Educating the Reflective Practitioner (1987) had this to say:

---

[15]As I read it this sentence has an awkward construction that I realise could be made easier to read if I said 'the practice of making something into a thing'. But if I did that I would be falling into the very trap I am trying to escape – of granting thingness an independent existence.

[16]In other words Wenger in his coining of 'reification' is creating a neologism; he exemplifies the dynamic he is trying to draw attention to – i.e., he is inviting us to see the term 'reification' in a new way – as having a new, underlying logic, which, when we accept it, brings with the acceptance new distinctions that become 'useable' as part of our tradition of understanding.

In the swampy lowland, messy, confusing problems defy technical solution. The irony of this situation is that the problems of the high ground tend to be relatively unimportant to individuals or society at large, however great their technical interest may be, while in the swamp lie the problems of greatest human concern. The practitioner must choose. Shall he [sic] remain on the high ground where he can solve relatively unimportant problems according to prevailing standards of rigor, or shall he descend into the swamp of important problems? (p. 28)

Schön also argued that:

all professional practitioners experience a version of the dilemma of rigor and relevance and they respond to it in one of several ways. Some of them choose the swampy lowland, deliberately immersing themselves in confusing but critically important situations. When they are asked to describe their methods of inquiry they speak of experience, trial and error, intuition or muddling through. When teachers, social workers, or planners operate in this vein, they tend to be afflicted with a nagging sense of inferiority in relation to those who present themselves as models of technical rigor. When physicists or engineers do so, they tend to be troubled by the discrepancy between the technical rigor of the 'hard' zones of their practice and apparent sloppiness of the 'soft' ones. People tend to feel the dilemma of rigor or relevance with particular intensity when they reach the age of about 45. At this point they ask themselves: Am I going to continue to do the thing I was trained for, on which I base my claims to technical rigor and academic respectability? Or am I going to work on the problems – ill formed, vague, and messy – that I have discovered to be real around here? And depending on how people make this choice, their lives unfold differently [30, p. 28].

In a reflective piece on social work practice, Clark [10] notes how difficult it is to 'think about thinking because this asks professionals who are naturally oriented to actively solving client and social problems to become unusually reflective and self-critical'. He goes on to observe that 'along with philosophical astuteness, asking external questions[17] demands patience, enthusiasm, humility, and risk-taking because such queries are often unwelcome and dismissed as irrelevant or obstructionist – accusations particularly inimical for professionals' (p. 38).

Donald Schön[18] has also concerned himself with what I regard as the dynamics of juggling the E-ball. He suggests [28, p. 7] that "when we identify something as an instance of a concept already given we do nothing to modify our conceptual scheme, we simply order experienced things in terms of it." He goes on to say (p. 8) that "once having resolved a problematic area of experience, once having found a way of looking at (and therefore dealing with) a situation which was at first novel and puzzling, our impulse to stick with it is overwhelmingly powerful. We have adapted to it, and through it". Our concept-forming apparatus operates under a categorical imperative of "let well enough alone". This serves us well most of the time but of course it can constrain innovation and change.

---

[17] Following philosopher Ralph Carnap, Clark [10] identifies 'external' questions as involving inquiry about problems external to any language or symbol system… questions that ask about the ultimate purpose of a profession's existence (p. 38) – not a perspective I particularly share though I agree with Clark's general point.

[18] The Displacement of Concepts (1963) and then re-issued as Invention and the Evolution of Ideas (1967) [28].

An unintended consequence of the processes or practices of neologising and reifying is that we can become trapped in particular ways of engaging with situations, of juggling the E-ball. It is interesting that Schön, Ackoff, Rittel and Webber all had professional backgrounds connected with social planning. It is not surprising, therefore, that they made similar distinctions when describing, or accounting for, their experiences in the messy business of planning. What these planners had in common was that they recognised that if the situation is engaged with as a difficulty, there will be an outcome that will be different than if a situation is engaged with as a mess. They also agreed that the traditional 'problem solving' methods, which are often associated with fields such as operations (or in the UK, operational) research (OR), or 'scientific management,' become useable only after the most important decisions have already been made. In other words, a difficulty is first abstracted from the mess and then the difficulty is treated using a traditional 'problem-solving approach'. Reacting to the limitations of the traditional approach Thompson and Warburton [34] once sensibly set out to find out what was wrong with the Himalayas acknowledging that the problem was to know what the problem was! From this perspective situations such as climate change adaptation and the "global water crisis" are the new Himalayas![19]

6.2.4   Some Implications Arising from Neologising and Reifying

The field of Operations Research (see Fig. 2.3) provides an interesting case study in changing ways of juggling the E-ball. In recent times special editions of OR journals have been devoted to 'problem structuring methods' (PSMs).[20] The editorial to a special issue on PSMs described their concerns as:

> Problem structuring methods (PSMs) are a collection of participatory modelling approaches that aim to support a diverse collection of actors in addressing a problematic situation of shared concern. The situation is normally characterised by high levels of complexity and uncertainty, where differing perspectives, conflicting priorities, and prominent intangibles are the norm rather than the exception. Typically, the most challenging element in addressing these common managerial situations is the framing and definition of the critical issues that constitute the problem, as well as understanding the systemic relationships between these issues. PSMs provide analytical assistance through 'on-the-hoof' modelling, which are used to foster dialogue, reflection and learning about the critical issues, in order to reach shared understanding and joint agreements regarding these key issues. [32]

This special edition reflects a changing perspective within the OR community and practices that encourage dialogue, reflection and learning as part of practice are to be

---

[19] How does 'reifying' differ from being stuck in a concept with a premise, attitude, or an argument that prevents you from considering something else? From my perspective it differs in the sense that the product of reification is that it creates ontologies.

[20] The Journal of the Operational Research Society (2006) 57 [32], was devoted to 'Problem structuring methods: new directions in a problematic world'.

welcomed. The perspective of those subscribing to PSM, may be characterised with the approaches associated with traditional OR in the early 1980s which included [26]:

- Problems and opportunities are formulated in terms of a single objective that can be optimised. Trade-offs are made by reducing variables to a common scale
- Doing OR has overwhelming data demands, which leads to problems of distortion, data availability and data credibility
- Subjected to demands of science (scientisation), assumed to be depoliticised and that consensus exists
- People are treated as passive objects
- Assumes a single decision maker with abstract objectives from which concrete actions can be deduced for implementation through a hierarchical chain of command
- Attempts to abolish future uncertainty and pre-take future decisions

One way of interpreting the traditional OR approach is practitioners choosing to stay on the high ground, of treating the 'real-world' situations with which many practitioners engage, as made up of difficulties to be solved rather than messes to be improved. In contrast more recent approaches do not seek optimisation as the only outcome, are prepared to combine qualitative and quantitative methods, may involve people as active subjects and accepts uncertainty, aiming to keep options open for later resolution [26, pp. 1–20]. Yet even the term PSM reifies a conception that problems exist as 'real states' that can be structured.

Similarities can be found between the developments within OR and the following observation attributed to Richard Dawkins [24]:

> If I hold a rock, but want it to change, to be over there, I can simply throw it. Knowing the weight of the rock, the speed at which it leaves my hand, and a few other variables, I can reliably predict both the path and the landing place of a rock. But what happens if I substitute a [live] bird? Knowing the weight of a bird and the speed of launch tells me nothing really about where the bird will land. No matter how much analysis I do in developing the launch plan … the bird will follow the path it chooses and land where it wants.

Plsek, a change consultant based in the USA, used the 'rock-bird' story in an address to a UK National Health Service (NHS) Conference entitled: 'Why Won't the NHS Do As It Is Told?' The UK NHS was then the world's third biggest employer after the Chinese Red Army and Indian Railways. Understandably many people involved in the NHS experience it as complex. In his presentation Plsek evoked different metaphors as a means for the audience to make new distinctions. He contrasted the machine metaphor (as characterised by traditional OR, scientific management and the rock governed by Newtonian physics) with an alternative metaphor of complex adaptive systems (CAS) as exemplified by the bird in the rock-bird story.

I admit to being constantly amazed that the distinctions entailed in the rock-bird anecdote are so profound for many audiences. To me, immersed as I am in my own tradition of understanding, this is so obvious! But clearly it is not for others, and is perhaps reflective of how profoundly the modernist conceptions of mechanism and linear causality are embedded in our society. Whilst recognising that for many the

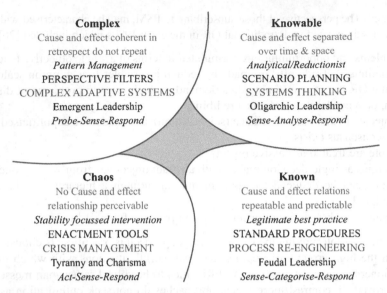

Complex
Cause and effect coherent in
retrospect do not repeat
Pattern Management
PERSPECTIVE FILTERS
COMPLEX ADAPTIVE SYSTEMS
Emergent Leadership
Probe-Sense-Respond

Knowable
Cause and effect separated
over time & space
Analytical/Reductionist
SCENARIO PLANNING
SYSTEMS THINKING
Oligarchic Leadership
Sense-Analyse-Respond

Chaos
No Cause and effect
relationship perceivable
Stability focussed intervention
ENACTMENT TOOLS
CRISIS MANAGEMENT
Tyranny and Charisma
Act-Sense-Respond

Known
Cause and effect relations
repeatable and predictable
Legitimate best practice
STANDARD PROCEDURES
PROCESS RE-ENGINEERING
Feudal Leadership
Sense-Categorise-Respond

Fig. 6.1 The Cynefin framework for sense making in relation to situations [14]

CAS neologism creates new distinctions, I fear that its users and proponents too frequently fall into the trap of reification that I seek to avoid. As a neologism it combines three concepts to produce a fourth – complex, adaptive, system and CAS. My own preference is to treat the three concepts which combine separately, or alternatively to frame my use of CAS through the organising question: what might I learn if I were to engage with this situation as if it were a CAS?

In recent times Dave Snowden has elaborated a form of sense making practice that distinguishes four domains; the known, the knowable, chaos and complex (Fig. 6.1).

This model, the Cynefin framework, contains many of the elements that I have been concerned with when I consider the E-ball:

> [the framework] is a sense making model designed to allow boundaries to emerge through the multiple discourses of the decision making group. It is not a categorisation model in which the four domains are treated as four quadrants in a two by two matrix. None of the domains described is better than or desirable over any other in any particular context; there are no implied value axes.... The model has five domains, four of which are named, and a fifth central area, which is the domain of disorder. The right-hand domains are those of order, and the left-hand domains those of un-order. [14, p. 4].

Whilst I do not agree with all of the descriptors placed in the four quadrants in Fig. 6.1, I do find the approach to practice of interest in respect to juggling the E-ball; what is of most interest is how such a framework is used as part of praxis. Snowden illuminates this by referring to game playing:

> we can look at the Cynefin model and its use in decision making in a more concrete way by using it to consider the way we all learn to make decisions – by playing games. It is well known that children's games reflect the tasks adults face because by and large, play fills

the function of rehearsing adult life.[21] Children's games tend to start in known space, move through knowable space as they grow, and end up in complex space. Games in chaotic space use shock, surprise and opportunity to open up unexpected and unmanageable potentialities, and are often played at social occasions where "mixing things up" is desired.

From my perspective Snowden's practice constitutes a hybrid of influences drawing on phenomenology, complexity theory, systems thinking, narrative theory and other influences that are adapted in context; as such he would seem an adept juggler of the E-ball, though I have not personally experienced him in action. His practice reflects a confluence of lineages in that it draws heavily on phenomenology, as does Checkland's [9, pp. 273–279] applied systems approach, often labelled as 'interpretivist' (see Fig. 2.3).

To conclude this section I want to consider what Ackoff (or Rittel, Webber or Schön) was doing in terms of a practitioner juggling the E-ball. I will answer this question by grounding my answer in our everyday use of language. In everyday speech if we describe our experience of a situation and say 'it is a mess', or 'it is really complex', or 'I find it hard to understand it all', you will notice I have used the word 'it' each time.[22] The word 'it' suggests the existence of something, an entity, a 'real-world' situation with which I have engaged. This structure of the language ties us into a linguistic trap – the naming of an 'it' that is independent of my act of distinction. Getting out of this trap means finding a language that avoids the implication that there is a pre-existing 'it' waiting to be noticed. As someone once said every noun obscures a verb!

Illustration 6.1

---

[21] This is a common assumption that has to do with the implicit requirement for a "purpose" which is quite different from "play is that activity which has no predetermined outcome" [20]. Maybe the difference is between "game" which has rules and an end point, and "play" which is open?

[22] People may also say "what a mess" and "I've got myself in a mess" but despite the absence of an explicit 'it' the practical result is the same.

The same could be said of thinking that there is a NHS which *is* either a Newtonian machine (i.e., a thrown rock) or a CAS. I can get out of this trap by claiming a mess or a difficulty arises in the distinctions that a practitioner makes in a particular situation. If this is the case, a mess or a difficulty is not a property of the situation but arises as a distinction made by a systems practitioner – someone aware of the conceptual distinctions between seeing a mess and experiencing a difficulty and being aware that this is a choice to be made – in the process of engaging with a particular 'real-world' situation (see Fig. 3.5). It could be said that we bring forth the situation.

If, on the other hand, a politician, or a manager were to say, 'Oh, I know what the problem is – we just have to do X and that will fix things – then that person would be implicitly seeing the situation as a difficulty.'[23] As I will show in Reading 4, the use of scientific explanations all too often have the same effect. We can imagine scenarios where a politician or manager was unaware of the distinctions between mess and difficulty (as many decision-makers seem to be). This leads to the search for quick-fixes, rather than engaging with situations systemically (see Seddon [31]). I have represented the implications of this dynamic in Fig. 6.2. You will notice several trajectories, through the metaphor of 'the chosen path', which are distinguished by awareness or not.[24] If someone is aware of the operation

Fig. 6.2 A metaphorical account of the choices and trajectories that arise from the dynamics of making distinctions and then reifying them as objects. The choice of a reification without awareness puts you on a fixed path (the walled path)

---

[23] And if they were to do this they would probably be acting in an emotion of certainty – the key point being that changing the 'framing' changes the underlying emotional dynamics and thus manner of engagement with situations.

[24] By trajectory I mean a particular manner of unfolding the co-evolutionary dynamic discussed in Chapter 1.

of the dynamic I am describing then they have a choice to pursue any trajectory. If they are not aware then what happens is that they conserve a manner of living that limits choices without awareness that that is what they do.

6.3 Exemplifying Juggling the E-Ball

In my experience one of the main ways that change occurs in human social systems is when, through travel, we encounter others (situations, people etc.) who challenge our assumptions. The result is a fresh perspective from which to engage with our day to day situations. It is in this spirit that I introduce the next reading – it has the potential to take you into new experiential territory and may offer insights into your own circumstances and your own practice. I cannot imagine that many readers of this book will have been to Nepal (if you have so much the better) but the place and subject matter of this reading is not my primary concern. As with earlier readings I am inviting you to read this with an awareness of the question: *what is it that the authors claimed they did when they did what they did?* As you read focus your attention on your understanding of juggling both the E and B-balls and how, if at all, the authors exemplify these practices for you.

This reading describes a case study which illustrates the links between 'problem structuring', multiple epistemologies (ways of knowing about the situation) across different scales (i.e., subsystems, nested in systems, nested in a supra-system), assessment of what to do and remediation (i.e., action to change things for the better). The case study initially concerned cystic echinococcosis, a parasitic disease of people associated with a gastro-intestinal tapeworm of dogs. Since the tapeworm usually cycles between canids (dogs, wolves etc.) and other vertebrate animals, the parasite is linked to food safety through animal slaughtering techniques, which in turn are related to changes in the characteristics of animal production and household practices. The researchers found out that these mainly biological issues could not be dealt with without addressing the socio-economic and cultural aspects of the situation, that is, what they called 'the eco-social narratives' which people (including scientists) used to structure their daily lives. A 10 year series of research projects in Nepal demonstrated that conventional science could provide explanations but had a mixed record at achieving solutions. Effective solutions were arrived at only after local stakeholders were engaged in the problem structuring process and the governance (i.e. institutional and management) structures were also examined and changed. The authors conclude that assessment (placing values on scientific measurements) and remediation (acting on those values) requires both citizen engagement and what they describe as 'a nested, complex systemic epistemic stance'.[25] They claim an important role for the creation of culturally acceptable narratives as the main means to cement or synthesise the different elements of their practice.

---

[25] This is not my language – my interpretation of what they mean could be simplified to doing what the systems practitioner as juggler does, as I argue in this book.

Reading 4

Agro-urban ecosystem health assessment in Kathmandu, Nepal: a multi-scale, multiperspective synthesis [36].[26]

David Waltner-Toews and Cynthia Neudoerffer

Introduction

Urban agriculture has become an increasingly important source of food and income for rapidly growing populations in almost every large city in the southern hemisphere. In many situations, these agricultural activities in the midst of dense urban sprawl have arisen when rural peoples have migrated to the city and set up enterprises doing what they know best. The context for these urban farms, however, is utterly different from that within which agricultural practices evolved. Along the banks of the Bishnumati River in downtown Kathmandu, for instance, gardening and animal slaughtering practices imported from the countryside had, by the early 1990s, created an environmentally devastated landscape.

The public health, environmental and eco-social consequences of such urban agriculture are both immense and poorly studied. The complementary problem in the northern hemisphere – the rapid expansion of urban settlements into intensively farmed landscapes – is embedded in a similar problematic situation. The temptation is high to believe that local, technical assessments and engineering solutions are adequate to the task; that belief has led – and will continue to lead – to the waste of a great deal of good science. This paper presents a case study and a general argument for a multi-criteria, multi-scale, participatory and narrative-based synthesis and management.

Echinococcus granulosis is one of several tiny tapeworms of dogs [canids], essentially worldwide in distribution, which infects livestock and people. In canids, which acquire the parasite from eating infective cysts, this parasite is of little consequence. In people and livestock, who acquire the infection through exposure to dog faeces, the parasite is expressed clinically as hydatid disease, a slowly growing parasitic tumour. Depending on where these cysts reside, they may have major or minor clinical consequences in these other species. Dogs are re-infected when cysts are excised from livestock at slaughter and are cast away. With very few exceptions, for instance where people may be buried in shallow graves accessible to canids, humans are usually a dead end host. There are few good treatments other than surgery which, in many parts of the world, is a high risk undertaking. In the late 1980s and early

---

[26] This is an edited extract of the paper available at http://www.maweb.org/documents/bridging/papers/waltnertoews.david.pdf (Accessed 25 May 2009).

(continued)

Reading 4 (continued)

1990s, this appeared to be particularly true in Nepal, where some 20% of surgical patients with hydatid cysts died [4].

Phase 1: Epidemiological approaches

Beginning in the 1990s [researchers] initiated a series of epidemiological studies to determine rates of disease in animals and people, and identify risk factors which could – in theory – be manipulated to prevent the disease.[27] These risk factors related to human-dog interactions in the community, and open-air slaughtering along the banks of the Bishnumati River in Wards 19 and 20 in Kathmandu. Our work was based on the premise that many countries, ranging from Iceland and New Zealand to Chile and Cyprus, have successfully undertaken aggressive and intensive control programs based, or accompanied by, similar research programs [13]. These programs entailed both strict dog control measures and modernisation and securing of slaughtering facilities.

The science which informed this work was normal, in the Kuhnian sense, that is, based on accepted epidemiological ways of thinking.[28] Echinococcosis, that is, infection with the parasite regardless of species, is usually described in terms of its basic life-cycle between canids and other species (Fig. 6.3). Despite rhetoric about "webs of causation", many epidemiological studies reduce their models to the common denominator of their statistical tools, and produce, at most, complicated models of disease causation [18]. Most are basically simplistic, such as that depicted in the linear causal model, i.e., dogs eating offal lead to people being exposed to infested feces so they get sick. These were the models that informed our early work in Nepal, which allowed us to identify risk factors for infection in people and animals, devise public health statements, and have no impact whatsoever on outcomes.

By the mid-1990s, we had gathered an impressive amount of information on infection rates in people and animals, dog behaviour and risk factors for acquiring infection [4]. Nevertheless, little had changed in the communities with whom we were working. Those solutions commonly promoted in other eradication programs – mass killing of stray dogs, restriction and strict control of dog ownership, building of secure, modern slaughterhouses – seemed unlikely to succeed in Nepal. Slaughtering was still done in the open air along the riverbank, amid piles of offal and manure through which dogs, pigs and children wandered at will. A survey of community members at the end of the project listed water quality, health, and waste generation and disposal, particularly from animal slaughtering, as being their most important, on-going concerns.

---

[27] This study, and other similar ones, can be found described at www.nesh.ca, in the "projects" section.

[28] Thomas Kuhn [19] referred to 'normal' and 'post-normal' science.

(continued)

Reading 4 (continued)

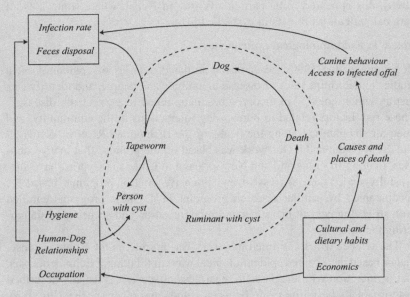

Fig. 6.3 The life cycle of *Echinococcus granulosus* (inside dotted line) (© NESH, 2003. Reproduced with permission)

By the end of the first set of epidemiological studies, we proposed a more complex model which we termed an ecosystem health model. This model, which was systemic, but assumed a single "correct" perspective, informed the beginnings of another set of studies. The model itself, one of several presented to professionals and scientific experts by one of the researchers, was useful in stimulating new ways of thinking. Several of the proposed strategies to "fix" the situation – such as composting and biogas generation – were in fact adopted by some of the more affluent slaughterhouse owners in the community. The model, however, omitted key elements of the situation – such as socio-economic, political and caste status, gender, and livelihoods. Since the overall ecosocial community was in fact an emergent property of how local citizens went about their daily tasks, and since these citizens were not engaged in the processes of problem formulation and solution-seeking, these new models had minimal impact.

Phase 2: Eco-systems approaches

In 1998 [an NGO][29] joined with researchers from Guelph and a variety of community-based stakeholder groups to carry out a project on ecosystem

---

[29] Social Action for Grassroots Unity and Networking (SAGUN).

(continued)

Reading 4 (continued)
***

approaches to health in the two urban wards of Kathmandu. Thus the focus of activity shifted from a specific parasite, the research team expanded to include the community members themselves, and the methods expanded to incorporate a variety of participatory and qualitative tools.

Details of this study are reported elsewhere [23, 27]. The new project encompassed a wide range of investigative methods, reflecting both the ambitious goals of the researchers, and an emerging consensus among scholars in this field that methodological pluralism must be central to any new science for sustainability [22]. We used conventional, quantitative scientific methods including epidemiological surveys, water quality monitoring and a variety of health assessments. These were complemented with more qualitative tools, drawn from Participatory Action Research (PAR) and related fields such as participatory urban appraisal, gender analysis, semi-structured surveys, focus group discussions, appreciative inquiry and stakeholder analysis.

Community researchers, hired and trained by the research team, and members of the local community, were key facilitators in such processes. This was to ensure the development of local capacity for participatory action and research through generation of awareness among people. Various stakeholder groups in the community developed action plans based on group narratives and priorities; these were implemented to varying degrees. However, there was a sense that the collective narrative of the community was not being adequately understood or addressed, and that the multiple perspectives and methods left a sense of fragmentation.

Near the end of the project, the work was re-assessed using AMESH, an Adaptive Methodology for Ecosystem Sustainability and Health, first developed in the context of similar projects in Peru and Kenya [22, 35]; (Fig. 6.4).

AMESH brings together critically reflective public participation with insights from self-organising, holarchic, open systems theories [16].[30] It calls for methodological pluralism and multi-scalar participation of stakeholders. Beginning with a problematic situation and a "given" history, AMESH then engages all legitimate stakeholders to identify key issues and their policy and decision-making contexts; from this emerge narratives, which are then structured into systemic descriptions. These, finally, are used by decision-makers to choose a course of action, identify indicators, and implement correct

***

[30] Self-organising is described elsewhere in the book – a holarchy, 'in the terminology of Arthur Koestler, is a hierarchy of holons – where a holon is both a part and a whole. The term was coined in Koestler's [17] book The Ghost in the Machine. The term, spelled holoarchy, is also used extensively by American philosopher and writer Ken Wilber. The "nested" nature of holons, where one holon can be considered as part of another, is similar to the term Panarchy as used by Adaptive Management theorists Lance Gunderson and C.S. Holling' (Source: http://en.wikipedia.org/wiki/Holarchy). An open system usually refers to a system open to the free flow of energy and matter.

(continued)

Reading 4 (continued)

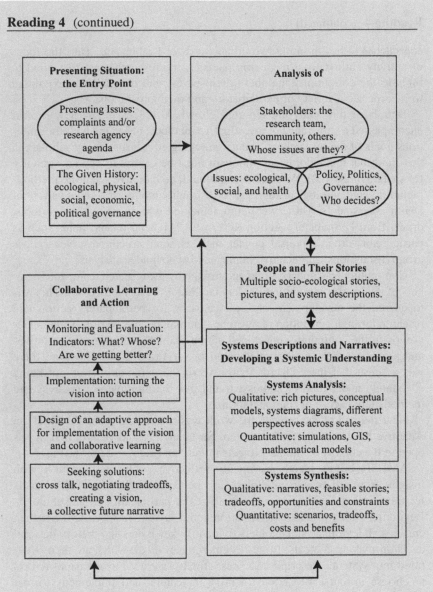

Fig. 6.4 An adaptive methodology for ecosystem sustainability and health (AMESH)
(© NESH, 2003. Reproduced with permission)

actions. The methodology, which we have now applied in Nepal, Kenya, Peru and Canada is iterative and self-correcting, that is, adaptive.

In this paper, the focus is on only one aspect of this complex process – the changing models of reality we had to incorporate into our activities, and the critical point at which everything changed.

(continued)

Reading 4 (continued)

Systems descriptions: one scale, many perspectives

After the epidemiological studies, a full review of the situation, followed by a community-based workshop, led to the new initiative, the "Participatory Action Research Urban Eco-system Health Project", with a major leadership role taken on by the NGO. What emerged from an intensive program of working with a wide variety of stakeholder groups in the community was a set of 'ecosystem stories' or 'ecosystem narratives', one for each stakeholder group. Describing how each stakeholder group perceived the interactions among themselves, other stakeholder groups, and the local eco-social system, these narratives were translated into a set of influence diagrams (see Fig. 6.5).

Using a technique modified from the work of Thomas Gitau, who had initiated another AMESH project in Kenya [21], these diagrams were able to identify a wide range of interactions within groups, as well as point to areas of potential conflict between groups.

Figure 6.5 is the 'issues and influences' diagram for the butcher stakeholder group. The activities of this group – comprised of wholesalers, retailers and butchers – are related to butchering and selling meat. The ecosystem health issues identified had to do with hygiene, waste management and water quality and quantity. The needs and concerns clearly varied by actor perspective even within this group. [Similar figures were drawn, and explained, for other stakeholder groups but are not presented here.]

For the final workshop of the project, we brought these influence diagrams together in various ways, and presented them back to the community. Figure 6.6 depicts the concerns and perspectives of the various stakeholders of the food and waste system. This enabled the community to identify where there were strongly divergent views requiring negotiation of tradeoffs, future visions, and possible future actions.

The same food and waste system can be seen as a set of expressed needs and how those needs were seen to relate to resource states, which were used as general indicators of ecosystem health.

Although these models were not made explicit until near the end of the project, it was clear that they reflected the mental models used by participants in telling their stories. Furthermore, by drawing on these stories, and through the process of civic engagement, the citizens of these communities completely transformed their neighbourhoods. Small scale slaughterhouses were built and butchers began to compost and recycle; both public and private gardens were planted along the river; public toilets were built; and a program to clean up local water sources was initiated. By 2001, the area had been completely transformed. Most importantly, some of the key actors in the community – the butchers – who generated employment, money and waste, emerged as a potent force for change and renewal.

(continued)

Reading 4 (continued)

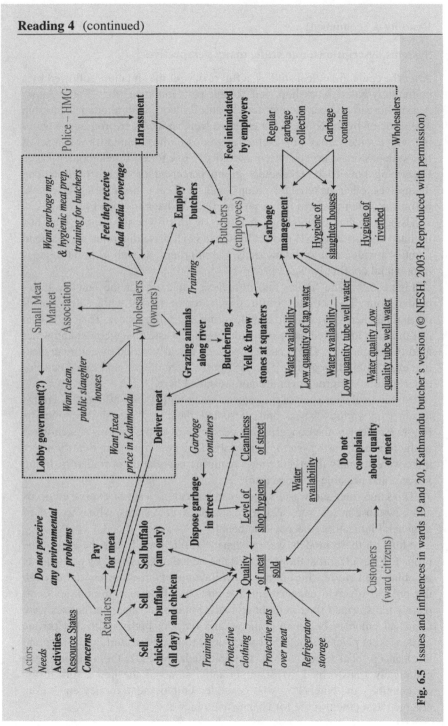

Fig. 6.5 Issues and influences in wards 19 and 20, Kathmandu butcher's version (© NESH, 2003. Reproduced with permission)

Reading 4 (continued)

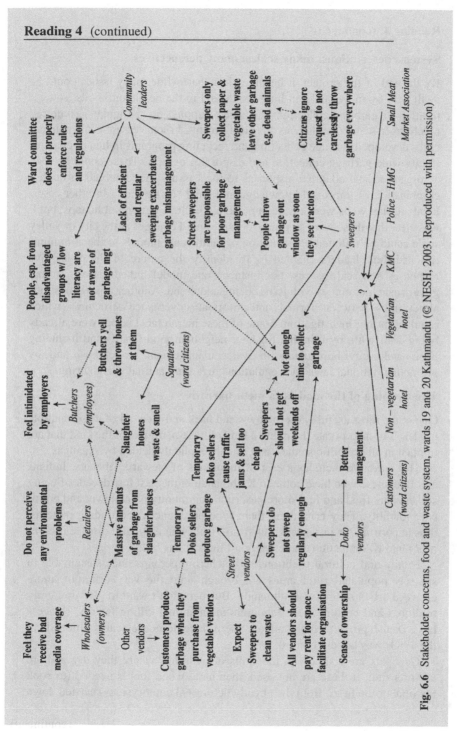

Fig. 6.6 Stakeholder concerns, food and waste system, wards 19 and 20 Kathmandu (© NESH, 2003. Reproduced with permission)

Reading 4 (continued)

Systems descriptions: many scales, many perspectives

By the end of the project, it became clear that, while many issues could be dealt with by individuals, in households and in the neighbourhoods, some – relating to garbage collection and water supplies in particular – required much larger-scale commitments and engagements. In some cases, local volunteer clubs developed garbage collection and recycling programs in lieu of changes in city-wide garbage collection and disposal programs. These could be seen as organisational adaptive responses to local issues that emerged through a combination of improved knowledge and local "ownership". In other cases, local artesian wells were cleaned up and cloth filters put on public taps, but it was clear that these were stop-gap measures until changes at the city or valley scale could be initiated. Presentation of multi-scale models to the communities and their leaders enabled us to identify the nature of these adaptive responses and scale issues. For instance one model linked various formal governance hierarchies with issues being addressed. Another raised questions about links between formal and informal governance structures which required further investigation. Some of these hierarchical issues were already being dealt with, by bringing together neighbourhood (ward) and including Kathmandu city bureaucrats in water management workshops, and by promoting national laws and regulations regarding animal slaughtering.

The meaning of the models: a meta-narrative

One of the most useful ways to now stand back and view all of these multiple models is to incorporate them into a kind of meta-narrative, of the kind that are implicit in all scientific studies, and usually explicitly denied by scientists.

This narrative includes a cast of thousands of Newari, Tibetans, Indians and Bhutanese who have come to Kathmandu in the past few decades fleeing soil erosion, food and fuel shortages, rural community breakdown, and political instability. They bring with them food preferences and trading patterns, thus importing, for instance, goats from Tibet (where hydatid disease is common) into Kathmandu (where we don't think it has been).

Family and cultural traditions of butchering perhaps once sustainable in sparsely populated rural areas are brought into the very different, more crowded urban setting of Kathmandu. Butchers don't want to give up family traditions and become wage labourers in an animal killing factory. Hence a large Danish project which built a modern slaughterhouse near Kathmandu in the 1980s was largely unused.

Cow pats are slapped on the walls of buildings where they dry and are used for fuel. If these are not used, then more wood fuel is needed (to cook the offal or the food) from the already deforested countryside – carried down

(continued)

Reading 4 (continued)

by young girls who are not then getting an education which would enable them to escape these traps. If cow pats are used as fuel, then valuable fertiliser is lost from the countryside. Streets are used for waste disposal, since the government simply cannot afford to maintain a European-style sewage disposal infrastructure.

One way to generate more money to solve this problem is to increase the carpet trade in response to high demands for these from Europe, Germany in particular. This uses vast amounts of water, making less available for public health purposes, and creates serious pollution problems of its own. Because the water system in Kathmandu is ancient and leaky, and because so much water is used to generate income (and money to pay for a better system) by the carpet and tourist businesses, riverbanks are used as public toilets and laundromats.

Groups of dogs – sources of rabies and echinococcosis – serve also as community-watch volunteers for places of worship and meditation. While our University of Guelph field researcher, Dominique Baronet, was in Kathmandu, members of the community where she was working noticed that some of the community dogs died. At the same time, thieves stole some artifacts from a local temple. Their explanation of these events went like this: Canadian woman comes into our neighbourhood, injecting dogs with strange drugs. The dogs, who are our community police, die. The thieves move in. Our version might be that dogs die all the time, but when they have fluorescent collars on, people notice them more; thieves are always on the lookout for portable gods to sell. The events were unconnected. Our version, however, did not determine their behaviour and was irrelevant to resolving the issue, however dearly we might have believed it to be true. Fortunately, Dominique had built up a lot of goodwill in the community and we could continue our work despite the suspicions. The bottom line is that people value their dogs for a whole complex of reasons and getting rid of them – as was done in Cyprus and Iceland – will simply not work.

Rickshaw drivers, taking advantage of increased meat consumption by tourists and increased economic activity of a small upper class, carry meat to market in the morning and tourists to temples in the evening. Streets intended for people on foot are now crowded with families on bicycles and motorcycles, and old vehicles burning fossil fuels, choking the air with pollutants.

The economic, cultural, and family bases of human-dog relations, butchering practices, and their many dependent occupations of small-scale meat transporters and butcher shops throughout the city, cannot simply be altered by decreeing that it should be so. Butchering and food hygiene practices depend not only on knowledge, but on the availability of clean water and affordable fuel and for cooking, thus competing directly with economically powerful activities such as the carpet industry, which use – and waste – huge amounts of water. Even if the dogs could all be treated with an anti-parasitic

(continued)

Reading 4 (continued)

drug, it is clear that the communities involved would still be left with serious public health, economic and environmental problems, many of which appear to be considerably more pressing than this particular parasite. Of all the places these communities could spend what little spare cash they might have, why would they want to spend it on an antiparasitic drug or control program for dogs?

Changing butchering practices seemed to be an essential part of any strategy, but this involves major cultural and economic changes, and not only for butchers. The original program which built the large slaughterhouse assumed that Nepalese people could control the disease (and others) if they behaved like Danes – in fact, if they reconstructed their culture in the model of Denmark. Indeed, much of what is promoted as disease prevention worldwide is based on a science which assumes that its information is objective and globally true. Because this is actually false, the success of our disease control programs depend on the degree to which we could convince the Nepalese to become like Western countries. This explains, in part, why conventional programs to control echinococcosis in New Zealand have been much more successful among settlers of European descent than among the native Maori.

The ultimate effect of conventional public health programs is a narrowing of the cultural base, and a closing off of options for future adaptability to change. They tend, thus, to fly in the face of sustainable development. Just as genetic homogenisation in the populations of plants and animals we use for agriculture is leading inevitably to global epidemics of animal and food-borne diseases, so this cultural smoothing, while solving the disease problems we are focused on, will result in massive public health problems down the road.

Actually, it was even worse than this, because the European and North American models of disease control depend on reducing the complexity of nature to fit the image we have created in our simplistic laboratory models. The implications of this for species extinction, soil erosion, disease epidemics and global climate change we are only now beginning to realise.

Conclusions

The multi-perspective, multi-scale combination of narratives and models might seem overwhelming and perhaps paralysing to someone seeking global quantitative assessments. By starting our system identification from the inside out, based on the priorities of the local stakeholders such as the squatters, street sweepers, businessmen and political leaders, we can begin to understand the meaning of integrated assessments. We cannot of course stop at the "local"; even in our cleaned up wards, we found waste floating into the area from upstream. This is why multi-scale engagements are essential. In on-going debate and adaptation across scales, we can incorporate the insights

(continued)

Reading 4 (continued)

gained from our scientific models, and the concerns of the wider scientific and sustainable development community. This approach to urban agro-ecosystem assessment, then, is not overwhelming, but sensible, reasonable, scientifically sound, and can lead directly to meaningful and convivial changes to the lives of the people with whom we are working.

Waltner-Toews, D., and Neudoerffer, C., 'Agro-urban ecosystem health assessment in Kathmandu, Nepal: a multi-scale, multi-perspective synthesis'.

6.4 Interpreting the Reading

I chose this reading to exemplify how a systems practitioner who is also a researcher juggles the E-ball. In this reading I discern an unfolding evolution in the practice of juggling the E-ball. The sequence described is:[31]

- Phase 1 – engaging as traditional scientists (epidemiology researchers) with a situation framed as a complex biological problem but one which could be solved (and where knowledge and practices from other contexts could be applied – i.e., the knowledge generated was generalisable); the models that informed this early work allowed them to identify risk factors for infection in people and animals, devise public health statements, BUT have no impact whatsoever on outcomes
- Phase 2 – the development of a model of the situation "which was systemic, but assumed a single 'correct' perspective" which 'informed the beginnings of another set of studies. The model itself, one of several presented to professionals and scientific experts by one of the researchers, was useful in stimulating new ways of thinking.' The model, however, 'omitted key elements of the situation – such as socio-economic, political and caste status, gender, and livelihoods. Since the overall ecosocial community was in fact an emergent property of how local citizens went about their daily tasks, and since these citizens were not engaged in the processes of problem formulation and solution-seeking, these new models had minimal impact'
- Phase 3 – In this phase the perspectives of the research team broadened as new, including local, groups joined. The focus of activity also shifted from a specific parasite to a broader framing of the situation (it could be said that the boundaries to their systems of interest changed) and the methods expanded to incorporate a variety of participatory and qualitative tools – described as methodological pluralism. However, despite these innovations, 'there was a sense that the collective narrative of the community was not being adequately understood or addressed, and that the multiple perspectives and methods left a sense of fragmentation'

---

[31]By phase, I mean the phases of juggling practice – not the same phases described in the paper.

- Phase 4 – near the end of phase 3, as good reflective practitioners, and dissatisfied with the effectiveness of what they were doing, they re-assessed their work using AMESH (an Adaptive Methodology for Ecosystem Sustainability and Health). The authors claim that 'AMESH brings together critically reflective public participation with insights from self-organising, holarchic, open systems theories' and 'calls for methodological pluralism and multi-scalar participation of stakeholders.' The authors describe how this methodology is applied and claim experience of application in many contexts. In terms of the E-ball, what is significant is that they have turned towards a designed systemic methodology for engaging which, it is claimed, is flexible and adaptive to context (juggling the C-ball) and involves all 'legitimate stakeholders'
- Phase 5 – From my perspective this phase was marked by the adoption of particular forms of diagramming as a means to engage with the situation. They did this through the 'capturing' of systemic depictions of the situation from the perspectives of different stakeholder groups. These diagrams were able to communicate a wide range of interactions within groups, as well as point to areas of potential conflict between groups. Whilst the exact means of their generation and use are not described fully, this diagramming practice is extremely similar to what the OU has taught systems students for over 30 years as the primary means to engage with complex situations.[32] I use diagrams of this type in my own systems practice and like these authors I often bring these influence and other types of diagrams together in various ways, and present them back to the community. I use the metaphor of 'mirroring back' what we have heard or understood and never make claims that the diagrams represent 'how things are'. They are thus used as a social technology to mediate a conversation from which new understandings, practices and social relations might emerge (Fig. 1.3)

The following table summarises some of the forms of diagramming that we use at The Open University and the systemic insights their use can reveal (Table 6.1).

- Phase 6 – For me this phase is best described (in my language) as the emergent awareness amongst the researchers (arising through their juggling of the B- and E-balls) of what can be gained from the shift towards a more systemic practice. What was particularly telling was how effective the outcomes of Phase 5 were – 'by drawing on these stories, and through the process of civic engagement, the citizens of these communities completely transformed their neighbourhoods. Small scale slaughterhouses were built and butchers began to compost and recycle; both public and private gardens were planted along the river; public toilets were built; and a programme to clean up local water sources was initiated. By 2001, the area had been completely transformed. Most importantly, some of the key actors in the community – the butchers – who generated employment,

---

[32] It is perhaps fair to say that at times in the past diagrams were sometimes taught at the OU as a means to 'represent' or to 'map' systems rather than as a means to engage with situations experienced as complex.

Table 6.1 Some forms of systems diagramming taught to Open University Systems students for engaging with situations of complexity and the systems concepts associated with each (see Table 2.1) [15]

Diagram type	Purpose	System concepts employed or revealed
Systems map	To make a snapshot of elements in a situation at a given moment	• Boundary judgements • Levels – system, sub-system, supra-system • Environment • Elements and their relationships
Influence	To explore patterns of influence in a situation; precursor to dynamic modelling	• Connectivity via influence • System dynamics
Multiple cause	Explore understandings of causality in a situation	• Worldview about causality • Positive and negative feedback
Rich pictures	Unstructured picture of a situation	• Systemic complexity • Reveals mental models and metaphors • Can reveal emotional and political elements of situation
Control model	To explore how control may operate in a situation	• Feedback • Control action • Purpose • Measures of performance

money and waste, emerged as a potent force for change and renewal.' In the language of Fig. 1.3 (see Chapter 1), what was achieved was concerted action amongst multiple stakeholders. Most importantly the circumstances of the people and context were improved. This occurred through a process that involved significant changes to how the researchers juggled the B- and E-balls

As a further reflection I note that the authors engaged in their own processes of neologising – e.g. AMESH, an 'ecosocial community'. If we are to create the circumstances for making new distinctions this is unavoidable but, unfortunately, the practice all too often becomes tied up in 'branding' and attempts to make and hold onto knowledge claims or build institutional capital around a particular group or institute – a perverse product of contemporary academic practices. I also detect in the reading a lack of clarity around the distinctions between situation and system of the type I raised in Chapter 4. This may have some implications for how the AMESH methodology is put into practice by different users, though from this paper it is not possible to say a lot about how AMESH might be enacted.

More importantly though, the breakthrough, in terms of effective and systemic change in the situation, came about when the researchers abandoned the reification of their scientific research results as the 'truth' about the situation. The science was important and necessary but in and of itself was not sufficient to effect systemic improvement. The important step was to realise how their science and subsequent practices could be more effectively contextualised. This is the concern of the next ball, the C-ball, juggled by the systems practitioner.

References

1. Ackoff, R.L. (1974) The systems revolution. Long Range Planning 7, 2–5.
2. Ackoff, R.L. (1974) Redesigning the Future, Wiley: New York.
3. APSC (Australian Public Service Commission). (2007) Tackling Wicked Problems. A Public Policy Perspective. Canberra, Australian Government/Australian Public Service Commission.
4. Baronet D., Waltner-Toews D., Joshi D.D., Craig P.S. (1994) *Echinococcus granulosus* infection in dog populations in Kathmandu, Nepal. Annals of Tropical Medicine & Hygiene 88, 485–492.
5. Blackmore, C. ed. (2010) Social Learning Systems and Communities of Practice. Springer: London.
6. Briggs, L. (2009) Delivering performance and accountability – intersections with 'wicked problems.' The International Society for Systems Sciences annual conference, conference paper. Brisbane, Australia, July 2009.
7. Bunnell, P. (2009) Personal Communication. Systems Ecologist, President of Lifeworks, Vancouver, British Columbia, April 2009.
8. Chapman J (2005) System Failure. DEMOS: London.
9. Checkland, P.B. (1993) Systems thinking, system practice. Wiley: Chichester.
10. Clark, J.J. (2008) Complex approaches to wicked problems: applying Sharon Berlin's analysis of "dichotomous thinking". Social Work Now 39, 38–48.
11. Conklin, J. (2001) Wicked Problems and Social Complexity. CogNexus Institute. http://cognexus.org/wpf/wickedproblems.pdf. Accessed 4 August 2009.
12. Cross, N. (ed.) (1984) Developments in Design Methodology. Wiley: Chichester.
13. Gemmel, M.A., Lawson J.R., Roberts M.G. (1986) Control of echinococcosis/hydatidosis: present status of worldwide progress. Bulletin of the World Health Organization 64, 333–339.
14. IBM (2003) The Cynefin Model. The Cynefin Centre for Organisational Complexity. IBM (UK) Ltd.
15. Ison, R.L. (2008) Systems thinking and practice for action research. In P. Reason and H. Bradbury (Eds.), The Sage Handbook of Action Research Participative Inquiry and Practice (2nd edn) (pp. 139–158). Sage: London.
16. Kay J.J, Regier H., Boyle M., Francis G. (1999) An ecosystem approach to sustainability: addressing the challenge of complexity. Futures 31, 721–742.
17. Koestler, A. (1967) The Ghost in the Machine. MacMillan: New York.
18. Krieger, N. (1994) Epidemiology and the web of causation: has anyone seen the spider? Social Science & Medicine 39, 887–903.
19. Kuhn, T.S. (1962) The Structure of Scientific Revolutions. University of Chicago Press: Chicago, IL.
20. Maturana, H. (1996) Personal Communication. Professor Emeritus, University of Chile, 1996.
21. McDermott J., Gitau T., Waltner-Toews D. (2001) An Integrated Assessment of Agricultural Communities in the Central Highlands of Kenya. Final Report to IDRC. Accessible at the website of the Network for Ecosystem Sustainability and Health (www.nesh.ca). Accessed 3 August 2009.
22. Murray T., Kay J., Waltner-Toews D., Raez-Luna E. (2002) Linking human and ecosystem health on the amazon frontier: An adaptive ecosystem approach. In G. Tabor, M. Pearl, M. Reed, R. Ostfeld, A. Aguirre, J. Patz, C. House (Eds.), Conservation Medicine: Ecological Health in Practice. Oxford University Press: New York.
23. Neudoerffer R.C., Waltner-Toews D., Kay J.J. (2004) AMESH analysis of the urban ecosystem health project, Nepal. In: D. Waltner-Toews, J.J. Kay and N.M. Lister (Eds.), The Ecosystem Approach: Complexity, Uncertainty, and Managing for Sustainability. Columbia University Press: New York.

24. Plsek, P. (2001) Why won't the NHS do as it is told? Plenary Address, NHS Confederation Conference, 6 July 2001.
25. Rittel, H.W.J. and Webber, M. M. (1973) Dilemmas in a general theory of planning. Policy Sciences 4 (2) 155–69.
26. Rosenhead, J. (1989) Introduction: old and new paradigms of analysis. In J. Rosenhead (Ed.), Rational Analysis for a Problematic World: Problem structuring methods for complexity, uncertainty and conflict (pp. 1–20). Wiley: Chichester.
27. SAGUN, NZFHRC and University of Guelph. (2001) Final Report to the International Development Research Centre of the Participatory Action Research on Urban Ecosystem Health in Kathmandu Inner City Neighbourhoods. Accessible through the cybrary at www.nesh.ca.
28. Schön, D. (1967) Invention and the Evolution of Ideas. Tavistock: London.
29. Schön, D.A. (1987) Educating the Reflective Practitioner: Toward a new design for teaching and learning in the professions, Jossey-Bass: San Francisco, CA.
30. Schön, D.A. (1995) The new scholarship requires a new epistemology. Change November/ December, 27–34.
31. Seddon, J. (2008) Systems Thinking in the Public Sector: the failure of the reform regime... and a manifesto for a better way. Triarchy Press: Axminster.
32. Shaw, D., Franco, A. and Westcombe, M. (2006) Problem structuring methods: new directions in a problematic world. Journal of the Operational Research Society 57, 757–758.
33. Snowden, D. (2003) Organising Principles (DRAFT) The Cynefin Centre for Organisational Complexity. IBM.
34. Thompson, M. and Warburton, M. (1985) Decision making under contradictory certainties: how to save the Himalayas when you can't find what is wrong with them. Journal of Applied Systems Analysis 12, 3–34.
35. Waltner-Toews D., Kay J.J., Murray T., Neudoerffer R.C. (2004) Adaptive methodology for ecosystem sustainability and health (AMESH): An Introduction. In: Community Operational Research: Systems Thinking for Community Development. Midgley G. and Ochoa-Arias A.E. (eds). Plenum/Kluwer: New York.
36. Waltner-Toews, D. and Neudoerffer, R.C. (2003) Agro-urban ecosystem health assessment in Kathmandu, Nepal: a multi-scale, multiperspective synthesis. http://www.maweb.org/ documents/bridging/papers/waltnertoews.david.pdf (Accessed 25 May 2009).
37. Wenger, E. (1998) Communities of Practice: Learning, meaning and identity. Cambridge University Press: Cambridge, UK.

Chapter 7
Juggling the C-Ball: Contextualising Systems Approaches

7.1 What Is It to Contextualise?

The word 'contextualise' can be traced back to the Greek 'techne' meaning 'skill or craft' and later to the Latin 'texere' meaning to weave, construct or compose [33]. The other part of the word, "con" means 'with' in the sense of 'with [the rest of] the weave or composition'. In my terms the act of contextualising, of juggling the C-ball, draws on all of these historical meanings. Let me expand a little by explaining what happens when I go swimming, a form of practice, at my local pool. The lanes in my pool are labelled slow, medium, fast, and sometimes, aquaplay. Usually there is at least one lane of each. Over time I have come to contextualise my swimming to a set of circumstances in which I understand that what is fast and what is slow differs with time of day (i.e., who the other swimmers are; whether lap training is happening etc.). I have also come to know that at certain times I can swim in the aquaplay lanes, or that at others I am best to consider myself fast or slow. Because I have flexibility to adapt my swimming to a changing situation I find my practice usually works very well for me ... and presumably those who manage the pool.[1] For me this is an example of juggling the C-ball in my swimming practice.

Thus, just as a juggler might have to compose a performance contextualised to an audience – children or seniors or business people – a systems practitioner has to contextualise their practice to their circumstances including their choice and use of systems concepts, tools, methods etc. In this chapter I explore what is entailed in juggling the C-ball. I will argue that an aware systems practitioner has more choices than the practitioner who is not aware – one aspect of awareness I will draw attention to is use of the distinction between systemic and systematic thinking and action made in Chapter 2 (e.g. Table 2.1).

An aware practitioner, I contend, is able to contextualise a diverse array of systems concepts and methods creating an opportunity for advantageous changes in 'real-world' situations that are systemically desirable and culturally feasible. As the

---

[1]For example if fast and slow were defined by lap times, for example, this would require a whole assessment and monitoring system which would become unwieldy.

R. Ison, *Systems Practice: How to Act in a Climate-Change World*,
DOI 10.1007/978-1-84996-125-7_7, © The Open University 2010.
Published in Association with Springer-Verlag London Limited

authors of Reading 4 (Chapter 6) outlined, the ultimate effect of conventional public health programmes, using traditional approaches to practice, was a narrowing of the cultural base, and a closing of options for future adaptability to change. The traditional approaches, including scientific practice, thus tended to fly in the face of sustainable development. In part, traditional approaches do this because the approach that is taken is generalised as if it were appropriate across contexts, taking on the form of a 'blueprint' or something to be 'rolled out'. The end result can be much like trying to force a square peg into a round hole, i.e., failure or a very uncomfortable or ineffective fit! The authors of Reading 4 advocated a move to what they described as methodological pluralism, which in colloquial terms is a bit like 'picking horses for courses'.[2] This Chapter explores how, through juggling the C-ball, a systems practitioner might avoid a narrowing of possibilities on pathways towards effective action.

Systems concepts and language are useful for talking about just about any subject; this has led some to describe Systems as a meta-discipline or trans-discipline.[3] Fortunately, there is a rich tradition of Systems scholarship upon which to draw to help in meeting the challenges of juggling the C-ball (e.g. Fig. 2.3 and Readings 1–4). The more aware you are of this history and the more it becomes part of your own tradition of understanding – just as in my example of Smilla and the history of the Inuit people – the greater will be your ability to embody particular Systems distinctions in your practice.[4]

My focus is on the thinking that enables you to use relevant tools, techniques and methods in the right context for effecting action. First I describe what I mean by a systems approach and how this relates to purposeful behaviour on the part of the practitioner. Then I distinguish between tools, techniques, method and methodology. Finally, I consider what is involved in contextualising any approach in a given 'real-world' situation. To do this I will ask you to keep in mind a number of questions as you work through the chapter:

- Is it the method, technique or tool or how it is used that is important?
- How are learning and action built in?
- Who is, or could be, involved in the approach?
- What could be said about the politics and practicalities of engaging in a 'real-world' situation?

---

[2] Horses for courses means that what is suitable for one person or situation might be unsuitable for another (Source: http://www.usingenglish.com/reference/idioms/horses+for+courses.html Accessed 2nd August 2009).

[3] The challenge for the systems practitioner is to be able to engage in double learning – learning about the domain in which practice is occurring and learning about the systems approach to the domain as well as juggling the other balls. This is a lot to manage.

[4] The various Systems traditions might equate more to Inuit, Lapp, Anu, or even Tongan, etc. One can be quite proficient with the richness in any one of the branches without being a scholar of the nature of the branches.

7.2   What Are Systems Approaches?

An approach is a way of going about taking action in a 'real-world' situation, as depicted in Fig. 3.5. As outlined earlier, an observer has choices that can be made for engaging with complexity. As I said in Chapter 1, my invitation is to consider choosing systems approaches, as a means to approach the world systemically using systems thinking. Other choices of approach could be made. Think of the everyday ways we use adjectives to describe the word approach. Some that come to mind are a scientific approach; a reductionist approach; an empirical approach; a philosophical approach; an experimental approach; a spiritual approach; a practical approach; a critical approach. You can probably think of more.

Some of these approaches to taking action seem to operate at different levels – there are certainly scientists who see themselves as systems biologists, for example, just as there are many scientists who take a reductionist approach and some, such as Teilhard de Chardin who took a more spiritual approach.[5] Thus Science could also be seen as a meta-discipline and different actions could be taken in either "Systems" or "Science" by an aware practitioner. I have already claimed that both a systemic and a systematic approach can be encompassed within a systems approach by an aware practitioner. Please bear in mind here that I am saying these are choices to be made; I am not commenting on the appropriateness, quality or efficacy of the options for any particular circumstance, nor am I saying they are the only options, nor that they are mutually exclusive.

I argued in Chapter 2 that what constitutes systems thinking and practice arises in social relations but that a key aspect was making a connection with a history of 'doing systems'. I mapped out, from my perspective, some of the Systems lineages in Fig. 2.3. This figure also refers to contemporary systems approaches – these can be understood as approaches to systems practice which appear in current literature and conversations (in many forms; face to face, on-line etc.). My explanation opens up many possibilities for what could be claimed as taking a systems approach. Some might claim that one of the reasons that Systems has not been widely taken up is its lack of key or core concepts. This is a concern but not, in my view, the main issue (see Chapter 13). Instead, I would argue, there exists a great opportunity and need to build and create effective systems practices for our current circumstances.

The absence of dedicated programmes of Systems study is a constraint on existing or potential practitioners, given that the intellectual field is substantial. Thus, whilst the question of choice of systems concepts, methods etc. is in part illuminated by the phrase 'horses for courses', in practice it is much more subtle than this. The metaphor of juggling seems to say much more than this alternative image. It is not just a question of matching a 'horse' – an approach – with a 'course' – or

---

[5] 'French philosopher and Jesuit priest who trained as a palaeontologist and geologist and took part in the discovery of Peking Man.... Teilhard's primary book, The Phenomenon of Man, set forth a sweeping account of the unfolding of the cosmos.'– see http://en.wikipedia.org/wiki/Pierre_Teilhard_de_Chardin (Accessed 16th August 2009).

a 'real-world' situation.[6] This is because taking a systems approach involves addressing the question of purpose. Let me explain what I mean by this.

One of my pet hates is when people say 'you should......' to me – because I experience them as imposing their purpose on to me whenever they use 'should'. This is something we tend to do all the time but attributing purpose to someone else is different to declaring our own purpose and acting according to that in our own actions.[7] The question of purpose is central to what I call aware systems practice and the process of contextualising an approach.

7.3 Purposeful and Purposive Behaviour

It is possible, as observers, to ascribe a purpose to what we or others do based on the actions we see. How particular actions or activities are construed will differ from observer to observer because of their different perspectives, which arise from their traditions of understanding. For example, in Fig. 7.1 the people welding may ascribe their purpose as learning a trade, building a skyscraper or welding joints, whereas an observer may assume they are all just building a skyscraper, or even propose that they are just "taking out their aggressions"! Thus, same actions, different purposes.

Fig. 7.1 An iconic model of how different 'actors' ascribe different purposes to the same action

---

[6] From my perspective the Systems literature about multi-methodology too often falls into this trap (see [23]).

[7] I am not claiming that all human action in its doing is purposeful – in many ways we often do what we do and later attribute purpose to our doing, particularly if an explanation or justification is required.

Within systems thinking, purpose is a contested notion. Historically in the Systems literature two forms of behaviour in relation to purpose have been distinguished. One is purposeful behaviour, which Checkland [4] described as behaviour that is willed – there is thus some sense of voluntary action. The other is purposive behaviour – behaviour to which an observer has attributed purpose. Thus, in the example of the government minister discussed in Chapter 6, if I described her purpose as meeting some political imperative, I would be attributing purpose to her and describing purposive behaviour. I might possibly say her intention was to deflect the issue for political reasons. Of course, if I were to talk with her I might find out this was not the case at all. She might have been acting in a purposeful manner which was not evident to me (in the sense of Fig. 7.1).

Purpose is always attributed to a system by someone. Within systems practice the attribution of purpose can be a creative, learning process. I am reminded of Peter Checkland's [4] story of working to improve prison management and seeing purpose – and thus system – in terms of 'rehabilitating criminals'; 'training criminals'; 'protecting society'; etc. Stafford Beer once said: 'the purpose of a system is what it does'. He may have meant it tongue in cheek, and he may have meant that the "purpose" obviously depends on the description of what the system is doing... but in my view this statement runs the risk of objectifying 'the system'. I would rather employ the notion of purpose as a perspective within the process of inquiry. This leads me to ask: What might I/we learn about the situation if I/we were to think of a prison (for example) as if it were a system to train criminals?

There is also a risk in reducing the notion of purpose to mean an objective or goal that can be achieved, and in some cases optimised. I make this distinction in order to emphasise an important aspect of systemic practice as compared with systematic practice. Namely, systemic practice encourages an approach of exploring or inquiring of a situation: 'What would I learn from attributing purpose to this situation?' Equally, the systemic approach might be posed as the question 'In reflection what purpose do I attribute to my own actions in this situation?'

The purposive – purposeful distinctions have led some to speak of purposive and purposeful systems. The former refer to systems that have an imposed purpose (from outside) and the latter can be seen to be those systems that can articulate as well as seek their own purposes. Some find these distinctions helpful, but from my perspective they have the same limitations as the concept of 'complex adaptive systems'. As terms they arise from a practice of typologising and classifying and are thus prone to objectification or reification as I discussed in Chapter 6. With awareness they sometimes help to create new distinctions and thus understandings. For example, one of the key features attributed to purposeful systems is that the people in them can pursue the same purpose, sometimes called a *what*, in different environments by pursuing different behaviours, sometimes called a *how*.

Note that I have deliberately not used the term goals, because of the current propensity to see goals as quite narrowly defined objectives. Certainly this was the way 'goals' were interpreted in the systems engineering tradition of the 1950s and 1960s and in the traditional Operations Research (OR) paradigm (see Table 7.1). Checkland and his co-workers beginning in the late 1960s reacted against the thinking

Table 7.1 The 'hard' and 'soft' traditions of systems thinking compared (Adapted from Checkland [3])

The hard systematic thinking tradition	The soft systemic thinking tradition
Oriented to goal seeking	Oriented to learning
Assumes the world contains systems that can be engineered	Assumes the world is problematical but can be explored by using system models
Assumes system models to be models of the world (ontologies)	Assumes system models to be intellectual constructs (epistemologies)
Talks the language of 'problem' and 'solutions'	Talks the language of 'issues' and 'accommodations'
Allows the use of powerful techniques	Is available to all stakeholders including professional practitioners
	Keeps in touch with the human content of situations
Assumes that there is a right answer	Does not produce the final answers and accepts that inquiry is never-ending
May lose touch with aspects beyond the logic of the problem situation	Remains aware that there are dimensions of the situation to which linear logic does not apply

in systems engineering and OR at that time and coined the terms 'hard' and 'soft' systems. The terms 'hard' and 'soft' are now widespread in the Systems literature, but as outlined in Chapter 2, have some limitations though the thinking behind the distinctions remains highly relevant.

Like Checkland and colleagues, other systems practitioners have found the thinking associated with goal-oriented behaviour to be unhelpful when dealing with situations understood as messes or 'wicked' (see Chapter 6). This has resulted in a move away from goal-oriented thinking towards thinking in terms of learning. I will also say more about learning in Part III of the book.

7.3.1 Appreciating the Place and Role of Learning and Knowing

Practitioners from many fields, not just Systems, advocate a perspective based on understandings of experiential and action learning. The process is commonly depicted as a cycle of activity of the form that is described in Fig. 7.2. This depiction is one of many manifestations of the experiential learning cycle. If this cycle is completed, it is argued, the purposeful action can be aimed at intended improvements; improvements that is, in the opinion of those who take the action. Those involved in this process learn their way to new understandings of the situation from which decisions about change can be made. Many systems approaches are designed to facilitate cycles of learning of this form. The key shift in understanding that is involved is from seeing 'systems' as having a purpose to seeing purposeful activity as being organised as a system which has embedded within it the act of building, or bringing forth, or modelling systems of interest as epistemological devices.

Appreciating the nature of experiential learning is, in my view, critical to developing good systems practice. That said, I am not convinced that the typical cyclical

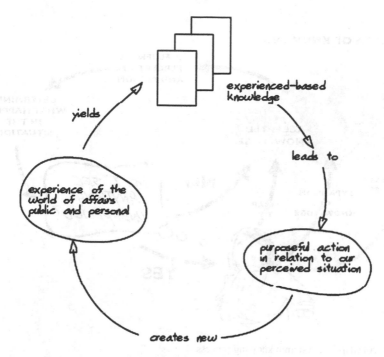

Fig. 7.2 An activity-sequence diagram of the experience-action cycle involving purposeful action [7]

representation of the process is adequate. In many ways my metaphor, or isophor, of the juggler, juggling the B, E, C and M-balls can be seen as an alternative conception. Critical to the juggling process is, I argue, a shift in our ways or patterns of knowing (Fig. 7.3).

Acting in the awareness that bringing forth a system of interest is an epistemological act opens up opportunities to break out of the inner accepted-knowledge reinforcing cycle depicted in Fig. 7.3. Elaborating this purposefully as a 'learning system' is a refinement built upon this awareness. Together they offer opportunities for an expansion of our knowing and for changing the premises on which our knowing is built.

As one develops systems awareness and understanding it is all too easy to recognise situations in which the main players seemingly have no common sense of purpose. Through a systems lens one could claim that there is no agreement on what the system of interest is or what purpose it is seen to have. This seems to be a common situation. For example, there is no shortage of experts, organisations, agencies, governments, and so on engaged in the definition and derivation of values, targets, principles, indicators and standards against which the achievement of the measures of performance of a supposed 'system' might be evaluated, monitored and audited – but little agreement, or even discussion, about purpose. In the UK this can be related to the failures and unintended consequences of policy built around the achievement

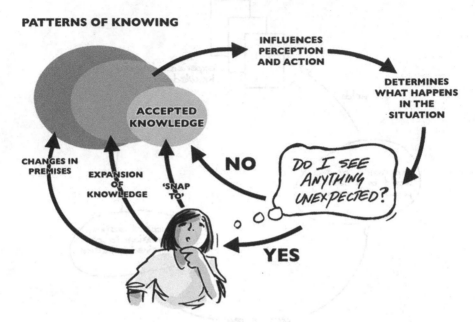

Fig. 7.3 An adaptive, systemic knowing process

of imposed targets as discussed in Chapter 9. In other words many people have a propensity to pursue purposive behaviour that assumes both purpose and measures of performance, rather than engaging stakeholders in a dialogue in which purpose is jointly negotiated. This can have unfortunate consequences.

If a system is conceptualised as a result of the purposeful behaviour of a group of interested observers, it can be said to emerge out of the conversations and actions of those involved. It is these conversations that produce the purpose, and hence the conceptualization of the system. What it is and what its measures of performance are will be determined by the stakeholders involved. This process has many of the characteristics attributed to self-organising systems.

7.3.2 Juggling the C-Ball by Exploring Purpose

Peter Senge suggested the key to finding a strategy that energises and focuses an entire business enterprise without constraining imagination lies in building a deep sense of purposefulness [30]. Thus, when engaging with situations of concern, an initial exploration of purpose – for example using the material in Box 7.1 and Table 7.2 heuristically – can illuminate the question of *what* a system of interest is and move the thinking away from a range of contested *how's* which are not connected

Table 7.2 A checklist of boundary-setting questions for the design of a system (S) of interest (Adapted from Ulrich [38])

Sources of motivation

1. Beneficiary or client: who ought to be/is the beneficiary of the service or system (S) to be designed or improved?
2. Purpose: what ought to be/is the purpose of S?
3. Measure of success: what ought to be/is S's measure of success (or improvement)?

Sources of control

4. Decision maker: who ought to be/is the decision maker (in command of resources necessary to enable S)?
5. Resources: what components of S ought to be/are controlled by the decision maker?
6. Environment: what conditions ought to be/are part of S's environment, i.e. not controlled by S's decision maker and therefore acting as possible constraint?

Sources of expertise

7. Expert (or designer): who ought to be/is involved as designer of S?
8. Expertise: what kind of expertise or relevant knowledge ought to be/is part of the design of S?
9. Guarantor: what ought to be/is providing guarantor attributes of success for S (e.g. technical support, consensus amongst professional experts, stakeholder involvement, political support) and hence what might be/are false guarantor attributes of success (e.g. technical fixes, managerialism, tokenism)?

Sources of legitimation

10. Witnesses: who ought to be/is representing the interests of those affected by but not involved with S, including those stakeholders who cannot speak for themselves (e.g. future generations and non-human nature)?
11. Emancipation: to what degree, and in what way, ought/are the interests of the affected free from the effects of S?
12. Worldview: what should be/is the worldview underlying the creation or maintenance of S? i.e. what visions or underlying meanings of 'improvement' ought to be/are considered, and how ought they be/how are they reconciled?

to a common *what*.[8] Researchers at The Open University have used West Churchman's and Werner Ulrich's notions of purpose to explore how different stakeholders in the contested debate over the release process for genetically modified organisms (GMOs) understand the what/how distinction.[9] This has revealed an unarticulated conflict of

---

[8] It is important to understand how systems thinkers understand the *what, how, why* distinctions. In any given conception of a system of interest *what* refers to the system, *how* to a sub-system and *why* to the supra-system. Of course from another observer's perspective *what* might become *why* and so on. Conceptually the use of these terms involves understanding the systems concept of hierarchy or layered structure and an appreciation that these are not fixed but different ways of looking at (engaging with) a situation.

[9] Exploring purpose also surfaces different boundary judgments that are being made, either explicitly or implicitly. Thus in the debate on GMOs, within the category GMOs I could choose to distinguish two different systems – 'a system of within-species gene manipulation' e.g. traditional plant breeding or 'a system to introduce novel, alien genes into an organism' – transgenics. The lack of differentiation of these two possible ways of seeing GMOs has, in my view, seriously constrained the public understanding of the situation.

purpose amongst key members of some government panels charged with advising government on GMO release.

Similarly one of my colleagues ran a very effective session for one of the committees of the University's governing Council in which they were finally, sometime after being established, able to gain some clarity of purpose. This was achieved through use of a modified form of SSM (soft systems methodology) that invited members of the committee to explore their committee as a system to do P (*what*) by Q (*how*) because of R (*why*).[10]

C. West Churchman listed nine conditions that he considered necessary for assessing the adequacy of a design for a system of interest. The first of these addresses the question of purpose (Box 7.1).[11] Please remember as you read this material that the focus here is on the design of a system and thus the designer, not 'the system' even though the language may sometimes suggest the latter.

Box 7.1 Conditions for Assessing the Adequacy of Design of Any System of Interest

Churchman [9] identified nine conditions for assessing the adequacy of design of any system of interest. He argued that these conditions must be fulfilled for a designed system (S) to demonstrate purposefulness. The conditions are reproduced in summary below (adapted from Churchman [9, p. 43]):

1. S is teleological (or 'purposeful')[12]
2. S has a measure of performance
3. There is a client whose interests are served by S
4. S has teleological components which co-produce the measure of performance of S
5. S has an environment (both social and ecological)
6. S has a decision maker who can produce changes in the measure of performance of S's components and hence changes in the measure of performance of S
7. S has a designer who influences the decision maker
8. The designer aims to maximise S's value to the client
9. There is a built in guarantee that the purpose of S defined by the designer's notion of the measure of performance can be achieved and secured

---

[10] See Checkland and Poulter [6] for an explication of this approach.

[11] I have already made clear my own preferences regarding the ontological status of systems; my preference is to see 'systems' as epistemological devices – thus my use of the language: a system of interest... to someone... which is brought forth (distinguished) as part of practice in a situation. In presenting Churchman's ideas I make no claims or commitments to the study of the evidences of design or purpose in nature, a position I do not find satisfying.

[12] This is a shorthand for 'the designer, or designers, can attribute a purpose to this system of interest', or 'the measure of performance can be taken to be' etc.

Churchman [10, p. 79] later reordered the nine conditions listed in Box 7.1 into three groups of three categories; each group corresponding with a particular social role – client, decision maker, and planner. Werner Ulrich [36], a student of Churchman's, coined two allied categories which he later termed 'role specific concerns' and 'key problems'. In Table 7.2 Ulrich's re-ordered set of questions are presented; they have come to be known as boundary-setting questions. As Ulrich [37] notes:

> we cannot conceive of systems without assuming some kind of systems boundaries. If we are not interested in understanding boundary judgements, i.e. in critical reflection and debate on what are and what ought to be boundaries of the system in question, systems thinking makes no sense; if we are, systems thinking becomes a form of critique.

You will notice that the questions in Table 7.2 are divided into four groups of three; these are:

1. The sources of motivation of those involved – what is the value basis of the design?
2. The sources of control – who has power or authority and on what basis?
3. What are the sources of expertise or know-how and is it adequate?
4. What are the sources of legitimation and its basis?

Two features of the questions in Table 7.2 warrant further elaboration [26]. The three questions associated with each source of influence address parallel issues: the first question in each group (1, 4, 7, and 10) addresses issues of social role; the second question (2, 5, 8, and 11) addresses issues of role-specific concerns; and the third question (3, 6, 9, and 12) relates to key problems associated with roles and role-specific concerns. In more contemporary language, these terms are best associated with 'stakeholders', 'stakes', and 'stakeholdings' respectively.

As indicated in Table 7.2, each question is also asked in two modes, thereby generating 24 questions in total. In critical systems heuristics or CSH (the area of systems thinking developed by Ulrich) all questions need to be asked in a normative, ideal mode (i.e., what 'ought' to be...) as well as in the descriptive mode (what 'is' the situation...). Contrasting the two modes provides the source of critique necessary to make an evaluation.

I have tracked some of the history of the evolution of these boundary setting questions here as it offers insights into how several subsequent systems methodological approaches have been built (e.g., critical systems heuristics; soft systems methodology). Ulrich's set of questions can be used as a device to engage with situations and thus to better contextualise practice. Ulrich's questions in Table 7.2 are thus worthy of consideration as part of bridging practice when juggling the E or C-ball. This history also provides a background to 'systemic inquiry' which I discuss in Part III of this book.

I have found in my own practice that variations on these questions can be used in business and research settings. For example, at the time of writing I am undertaking research into the history of water and river catchment managing in Victoria,

Australia and have been interviewing some key figures. Questions that have proven useful in my interviews include ones like: Who is/ought to be the system's client? That is, whose interests are/ought to be served? [13]

7.4 Tools, Techniques, Method and Methodology

As you engage with systems thinking and practice you will become aware of how different authors refer to systems methodologies, methods, techniques, and tools, as well as systems approaches. Having explained earlier what I mean by a systems approach, I now want to distinguish between methodology, method, technique and tool (Fig. 7.4).

Several authors and practitioners have emphasised the significance of the term 'methodology' rather than methods in relation to Systems. I consider a method as

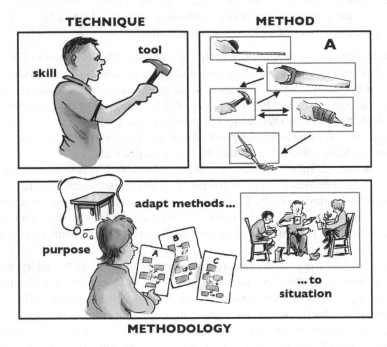

Fig. 7.4 Some distinctions between tool, technique, method and methodology

---

[13] From an aware systems practitioner perspective this question does not assume there is 'a' system to which this question applies, but that different actors will distinguish different systems i.e., make different boundary judgments, and that the role that systems research can fulfil is to surface and explore the implications of the differences as part of a process of change (e.g. as depicted in Figure 1.3).

something that is taken or used as a given, much like following a given recipe in a recipe book. In contrast a methodology can be adapted by a particular user, or users, in a given situation. A methodology in these terms is both the result of, and the process of, inquiry where neither theory nor practice take precedence [3]. For me, methodology involves the conscious braiding together of theory and practice in a given situation [17] – it is thus a context specific enactment. A systems practitioner, aware of a range of systems distinctions (concepts) and having a toolbox of techniques at their disposal (e.g. drawing a systems map) as well as systems methods designed by others, is able to judge what is appropriate for a given context in terms of managing a process.

How a practitioner adapts a method, of course, depends to a large extent on the nature of the role the systems practitioner is invited to play, or chooses to play in a given situation. Are they part of a project group in an organisation, someone with a designated management role, a consultant etc.? When braiding theory with practice, there are always judgements being made: 'Is my action coherent with my theory?' as well as, 'Is my experience in this situation adequately dealt with by the theory?' and, 'Do I have the skills as a practitioner to contribute in this situation?' There are also feelings that often represent a systemic understanding that one is not able to verbalise – 'Does it feel right?'[14]

Unlike the hammer in Fig. 7.4, within systems practice a tool is usually something abstract, such as a diagram, used in carrying out a pursuit, effecting a purpose or facilitating an activity. Technique is concerned with both the skill and ability of doing or achieving something and the manner of its execution, such as drawing a diagram in a prescribed manner. An example of technique in this sense might be as simple as drawing a systems map to a specified set of conventions, or as complex as designing and conducting a workshop that is experienced as effective by the board of a major company.

There is nothing wrong with learning a method and putting it into practice. How the method is put into practice will, however, determine whether an observer could describe it as methodology or method. If a practitioner engages with a method and follows it recipe-like, regardless of the situation, then it remains method. If the method is not regarded as a formula but as a set of 'guidelines to process', and the practitioner takes responsibility for learning from the process, it can become methodology. The transformation of method into methodology is something to strive for in the process of becoming an aware systems practitioner and of course one can draw guidelines from several methods to develop a methodology in any situation.

---

[14]My perspective on methodologies is grounded in my own practice – particularly that of being a systems educator and experiencing how mature age students learn about systems tools, techniques and methods. It is challenging for many students to move beyond the application of method to becoming methodological (not to be confused with "methodical") in their practice. There is also a danger in treating methodologies as reified entities – things in the world – rather than as a practice that arises from what is done in a given situation. My own perspective is that texts that write about 'systems methodologies' too often lead to the sort of reification that I discussed in Chapter 6, thus taking attention away from systemic praxis.

7.5   Contextualising Practice to a Situation

At the beginning of this chapter I posed four questions that I asked you to consider as you worked through it:

1. Is it the method, technique or tool, or how it is used that is important?
2. How are learning and action built in?
3. Who is or could be involved in the approach?
4. What could be said about the politics and practicalities of engaging in a 'real-world' situation?

Like so much in systems practice, there are no definitive answers to these questions other than 'it will depend on the context and your own abilities in that context'. What I hope is clear is that an aware systems practitioner does not force a method on to a context, a 'real-world' situation, to which it is not suited. Posing these four questions and attempting to answer them for yourself would be a good start in juggling the C-ball.

Your ability to contextualise a systems approach, of juggling the C-ball, will be aided if you look and reflect, before you organise your performance! Because most systems practice is carried out in some organisational setting, your ability to contextualise an approach will also be helped if you appreciate that it is not only people who make judgments about what constitutes relevant knowledge, and thus relevant practices, in given contexts. All organisations conserve manners of thinking and acting that have evolved over time, so much so that organisations themselves come to be described as having a culture within which conceptions of what counts as legitimate knowledge are enacted and maintained. These epistemologies, built by individuals and groups at some historical moment, are built into institutional structures and practices. Don Schön [31] cited the example of the typical elementary school that was organised around what he called 'school knowledge' – knowledge contained in the curriculum, the lesson, the module, in the promotion procedures for teachers, the practices of teachers, the organisation of rooms and so on. All of these things enter into the idea of 'school knowledge'.

Some systems practitioners have paid considerable attention to how to understand and manage the process of contextualisation. For example, as part of their SSM practice, Checkland and colleagues recognised that there were always three roles present in the use of SSM in practical situations. These were [6, p. 28]:

1. A person or group who had caused the intervention to happen, someone without whom there would not be an investigation at all – this was the role of 'client'
2. A person or group who were conducting the investigation – this was the role of 'systems practitioner'
3. A person or people who could be named and listed by the practitioner who could be regarded as being concerned about, or affected by the situation and the outcome of the effort to improve the situation – this was the role of 'owner of the issue(s) addressed'

These distinctions led Checkland and Winter [8] to differentiate two modes of doing SSM, what they call SSM(p) – concerned with the process of using SSM to carry out a study in a given situation – and SSM(c) – concerned with the content of the situation of concern (study, investigation etc.). Within the metaphor of the juggler SSM(p) is more akin to juggling the C and M-balls and SSM(c) to juggling the E-ball.

As I have been at pains to emphasise, systems practice does not have to be determined by named methods or methodologies. Nor does it have to be in a professional consultancy situation. In my own work on systemic and adaptive water governance the relatively simple act of developing a systems map of the institutions that a river catchment authority has to deal with (Fig. 7.5) reveals the complexity of the situation as well as providing insights into the different types of knowledge that would be needed to engage in effective river managing in this particular situation. In this example of juggling the C-ball, the technique of systems mapping was adapted to a group-based inquiry process to produce a composite map which captures the understanding of the senior team. I think it is fair to say that they were somewhat shocked at the complexity that this revealed and that they had to grapple with on a daily basis. This was conducted as part of a wider study examining how institutional complexity affects water governance.

Figures 7.5 and 5.8 are both systems maps. In addition to the content, a difference between them is how the technique of systems mapping was used and who was involved. Figure 5.8 involved using systems mapping as a device for sense making in a situation I was experiencing as complex. In contrast Fig. 7.5 was used as a collaborative process with a group of senior managers. Both uses of systems mapping reflect the dynamic depicted in Fig. 1.3, though the answer to the question 'who learns' is different in each context.

To conclude this chapter I wish to introduce another reading (Reading 5) – this one comes from my own systems practice. It introduces material relevant to both the C-ball and the M-ball, the subject of the next chapter. It also introduces and exemplifies aspects of what I mean when I refer to 'systemic inquiry', the focus of Chapter 10 in Part III of this book.

7.6 An Example of Juggling the C-Ball

This reading [16] presents a case study of a systemic inquiry into a knowledge transfer strategy (KTS) by a division of a UK Ministry. My main motivation in introducing this here is that it demonstrates how systems practice can be set up (i.e., contextualised) in circumstances that initially do not seem conducive. On the other hand, it also illustrates the limitations to taking effective action through systemic practices, even when the evidence seems overwhelming that things should change. Two main points arise from the Reading. First that it is possible to 'build' a generalisable form of systems practice as a response to experiences of complexity by initiating a systemic inquiry that fosters the emergence of a learning system.

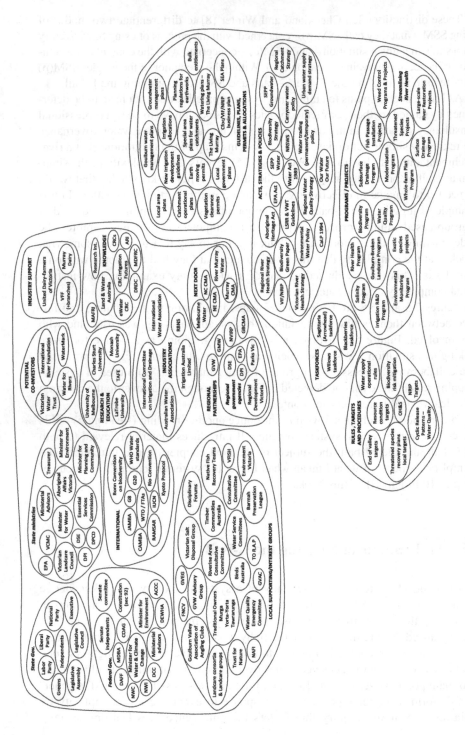

Fig. 7.5 A systems map of the institutions that are involved in river catchment managing from the perspective of senior staff in a Catchment Management Authority in Victoria, Australia

Second, that exploring how metaphors reveal and conceal offers scope for shifting the 'mental furniture' of participants as part of a systemic inquiry.

This inquiry proceeded with a process designed for the circumstances, recognising that there are no blue-prints. A key design aspiration was that those participating might experience a coherence, or congruence, between my espoused theory and theory in use in relation to considering the KTS as if it were a second-order learning system. In this aim it succeeded. The inquiry suggested two sets of considerations for the design of learning systems and a potentially fruitful line of further inquiry.

Reading 5

Some reflections on a knowledge transfer strategy: a systemic inquiry

Ray Ison

1. Systemic inquiry

Systemic inquiry proceeds by enacting a learning process with those who have a stake in a situation experienced as problematic or as presenting an opportunity. The possibility of designing a systemic inquiry is open to anyone who is able to make a connection between a theoretical framework (in this case concerned with systems thinking and practice) a methodological approach and a given situation [24]. For example, Checkland [5] argues that the enactment of Soft Systems Methodology (SSM) is an exemplar of systemic inquiry that results in changing modes of thinking. He argues that: "It is a process in which the thinking (of individuals and groups) is shifted to a different level. It produces 'meta-thinking' – that is, thinking about how you are thinking about the phenomenal world" and 'This mode of thinking rearranges people's mental furniture and enables plausible action-to-improve to be achieved'.

This paper presents a case study of a systemic inquiry initiated in response to a specific experience.

2. The experience

In September 2001 I received an invitation from one of the main organisations associated with agriculture in the UK to attend a one-day 'stakeholder meeting' concerned with their 'knowledge transfer strategy' (KTS). This was a surprise as I had had relatively little to do with UK institutions associated with agriculture since taking up the chair of Systems at The Open University in 1994. Given my research experiences in this area [e.g. [17]] I was intrigued by the invitation and duly accepted.

(continued)

Reading 5 (continued)

The espoused purpose of the KTS was expressed as: 'to encourage improved practice in the agricultural industry towards its sustainable development and to protect the environment from pollution'. The KTS proposed to achieve this purpose by pursuing the following sub-aims: (i) to transfer understanding of environmental issues and natural resource management [from researchers to farmers]; (ii) to put backbone into what we are doing – environmental protection – to be able say what we are aiming for; (iii) to ensure land managers are using the best available knowledge to do their farming; and (iv) to change (farmer/land manager) behaviours.

My experience of the day is best described as being in a conversation that was at least 10 years out of date. This of course says as much about me as it does about those present. During the day I made a number of contributions which were designed to elicit some reflection on the nature of the conversation and particularly the theoretical ideas that I perceived to be operating (whether explicitly or implicitly). I also expressed concern at what I perceived to be the narrow range of 'stakeholders' present. From my perspective farmers and other intended beneficiaries of the KTS were very much underrepresented. I also came to reflect on what the designers of the day had imagined its purpose to be. I wondered how they might have completed the sentence: 'Today can be seen as a system to ....?'

During the day four individuals approached me with a view to following-up some of the points I had made. This resulted in two specific invitations for further conversation and follow-up that form the basis of this paper. The first was from the organisers of the day. The second was from a person central to the development of a farmer-based R&D network based on self-organising groups in the south-west of England [35]. In the following section I outline a negotiated response to the first invitation. I also report some considerations for the design of learning systems that have emerged from this experience which I connect to my own tradition of understanding as embodied in my own research practice [17].

3. Responding to an invitation

3.1. Design considerations

Following the initial meeting I received an invitation to make a presentation of my ideas to some of the London-based head-office staff responsible for formulating and delivering the KTS. My response was to propose an alternative to a standard presentation. My purpose was to attempt to create the circumstances that had the possibility of initiating a systemic inquiry. That is, I wanted to avoid going to London to tell people what I thought they should do! From my perspective, to have done so would have fallen into the same trap that I was critical of in the KTS i.e., engaging in the linear, 'transfer of technology' mode

(continued)

Reading 5 (continued)

---

of research practice (see [11, 14, 27] for an explication of these ideas). Instead I proposed a one day process in which I spent time interviewing (listening) to some of the key managers of the KTS, the outcomes of which I would then mirror-back, along with my reflections on the initial stakeholder meeting.

Kersten [19] and Kersten and Ison [20] report on the role of relationship building and listening in the design of R&D based on dialogue rather than debate. The process of 'mirroring-back' is described in some detail in Webber [39]. The central feature is that it is a dialogic process in which those aspects of the researcher's experience of the interviews which most take their attention are held up for consideration by participants as a basis for triggering discussion and learning. (A contrasting position would be to use the output of the interviews as a basis for presenting the facts of the situation).

My proposal was accepted. Subsequently five 20-minute interviews were conducted followed forty minutes later by a joint meeting of researcher and interviewees (November 9, 2001). Spray diagramming was the main technique I used to record and make sense of the interviews [25]. In both the initial stakeholder meeting and during the interviews I paid particular attention to the metaphors in use that for me revealed and concealed particular theoretical positions [22]. The joint session lasted almost two hours. The format proposed and followed for this session was:

(i) to mirror back some of the outcomes from listening to multiple perspectives on the KTS from some key stakeholders;
(ii) 'mirroring back' some of the metaphors I had heard (from listening and reading project documents);
(iii) Exploring some theoretical and practical implications;
(iv) questions and discussion.

To aid my own learning about the process design those involved in interviews were invited some weeks later to provide feedback. The questions and feedback are outlined in Section 4.

3.2. Revealing and concealing metaphors in the KTS

Metaphors provide both a way to understand our understandings and how language is used. Our ordinary conceptual system, in terms of which we think and act, is metaphorical in nature. Paying attention to metaphors-in-use is one means by which we can reflect on our own traditions of understanding. Our models of understanding grow out of traditions, where a tradition is a network of prejudices that provide possible answers and strategies for action. The word prejudices may be literally understood as a pre-understanding, so another way of defining tradition could be as a network of pre-understandings. Traditions are not only ways to see and act but a way to conceal [27].

(continued)

Reading 5 (continued)

Traditions in cultures embed what has, over time, been judged to be useful practice. The risk is that a tradition can become a blind spot when it evolves into practice without any manner of critical reflection being connected to it. The effects of blind spots can be observed at the level of the individual, the group, a organisation, the nation or culture and in the metaphors and discourses in which we are immersed. Experience suggests that often we cannot see what the problem is because we cannot identify our own blind spots. It is only when we attempt to step out of the situation (reflect) that we can begin to see it from another perspective or from another level.

Metaphors also both reveal and conceal but because we live in language it is sometimes difficult to reflect on our metaphors-in-use. The strategy of mirroring-back particular metaphors or metaphor clusters thus holds open the possibility for reflection and learning. For example, as outlined by McClintock [22] the metaphor countryside-as-a-tapestry reveals the experience of countryside as a visually pleasing pattern, of local character and diversity and of what is lost when landscapes are dominated by mono-cultures. However the metaphor conceals the smell, danger, noise and activity of people making a living in the countryside. By exploring metaphors we are able to make part of our language use 'picturable' and thus rationally visible, publicly discussible and debatable as well as a psychological instrument which can be a practical resource 'with which and through which we can think and act' [34].

In Table 7.3 some of the metaphors elicited from the initial meeting, the written project material and the five interviews are clustered into three groupings. Within each some of their revealing and concealing aspects are suggested. This is not exhaustive as it is not the role of the researcher to classify and name the revealing and concealing aspects in a process aimed at triggering reflection and learning. The clusters are indicative of a small sample of the many metaphors that could have been reported. Within the spirit of 'mirroring back' and in terms of making connections with my own traditions of understanding these have been selected. The first cluster relates to why I experienced the conversation at the initial meeting as 10 years out of date.

The 'communication as signal transfer' cluster reflects traditions that have become blind spots or have been subjugated, not only in agriculture but other sectors of the community. As outlined by Fell and Russell [11] this is a legacy of the use by Heinz von Foerster of 'information' to replace 'signal transfer' when writing up the proceedings of the Macy conferences in the 1950s. Communication as information transfer is based on the mathematical model of Shannon and Weaver [32]. This in turn has been incorporated in the technology transfer and its associated "diffusion of innovations" models. Ison [14] outlines how Everett Rogers, in his preface to the third edition of "Diffusion of Innovations", acknowledges that "many diffusion scholars have conceptualised

(continued)

Reading 5 (continued)

Table 7.3 Metaphor clusters associated with the knowledge transfer strategy

Metaphors	Reveals	Conceals
1. Communication as signal transfer	*Shannon and Weaver's theoretical model in action*	*The biological basis of human communication*
Examples	'Key messages as deliverable'	
	'Farmers as knowledge users'	
	'Advice as target-able or deliverable'	
	'Information as relay-able'	
	'Advice as understandable or knowable objectively'	
	'Knowledge as transferable'	
	'A knowledge transfer strategy as deliverable'	
2. Information as deterministic	*Diffusion of innovation theory*	*The nature of networks, relationships and co-learning processes*
Example	'Barriers to uptake of advice and research messages'	
3. Advice as changing behaviour	*The imperative is to change someone else*	*The ethics of practice (i.e. giving advice)*
Examples	'Farming industry as able to be influenced'	
	'Regulation as command and control'	
	'KTS as delivering public goods'	
	'KTS as an economic argument'	
4. KTS as role clarifying	*Alternative possibilities for practice and for power relationships*	*How and by whom roles will be clarified*
Examples	'Advisers as service providers'	
	'Advice provision as able to be pictured'	
	'Farmers as champions'	
	'Regulation as self-organising (helping themselves)'	
	'Thinking outside the box'	
	'Farmers (or land managers?) as environmental improvers'	

the diffusion process as one-way persuasion" and that "most past diffusion studies have been based upon a linear model of communication defined as the process by which messages are transferred from a source to a receiver."

First-order communication is based on simple feedback (as in a thermostat) but should not be confused with human communication, which has a biological basis. Second-order communication is understood from a theory of cognition that encompasses language, emotion, perception and behaviour. Amongst human beings this gives rise to new properties in the communicating partners who each have different experiential histories. Second-order communication reveals the limitations of the 'knowledge/knowing as commodity' metaphor and also reveals the extent emotioning and power have been ignored in considerations of most KT strategies.

(continued)

Reading 5 (continued)

Exploring these ideas enables recognition that the following claim is made from a first-order communication perspective:

...that is what we are coming to – a melding of computers and communications to produce knowledge.... If that pool of information, of knowledge is over there, over here we have the users, the seekers of knowledge, the needful of information. (The fact that some of them do not yet realise that they need this information or knowledge is not germane to the issue. There is a lot of education needed to show the people what is available.

The reference to 'education needed to show the people' sounds like a euphemism for the next metaphor cluster (Table 7.3), that of 'advice as changing behaviour'. It was acknowledged during the conversation that changing landholder's behaviour was the major aim of the strategy, but in reflection all those present acknowledged that despite their awareness of environmental issues they had not really changed their own behaviour. A second-order explanation of communication posits that information arises within (from the Latin *in formare*, formed within) and that knowledge is not something "we have" but "the knowledge of the other is my gift... which arises in interpersonal relations" [21]; and that experience arises in the act of making a distinction, it is not something external to us. These explanations based on the biology of cognition suggest that "all knowing is doing" which arises in daily life.

Within all dominant discourses there is always resistance (following Foucault [12]); not surprisingly we can all be the repositories of seemingly paradoxical notions and thus bring forth alternative metaphors. The fourth metaphor cluster (KTS as role clarifying, Table 7.3) contains some metaphors that I considered as evidence of questioning the dominant discourse. These represented some sites of resistance to the more common metaphors found in the first two clusters (and the overall name of the strategy). For example, 'regulation as self organisation' was clearly an alternative to that of 'regulation as command and control'.

3.3. Exploring some practical implications

Based on my own learning from the interviews I suggested the following opportunities and threats were worthy of discussion and consideration for the KTS in the light of the espoused purpose. Opportunities included:

- to move towards a facilitated model of behaviour change which is local and contextualised (for example a key value driving those present was that they were responsible for implementing EU legislation such as the Water Framework Directive. From their perspective responsibility for implementation was an imperative leaving no scope for systemic, learning-based approaches. From my perspective they had fallen into the trap of conflating

(continued)

Reading 5 (continued)

what with how, i.e. the imperatives of the Directive were now law so the *what* was established, but *how* it was implemented in local contexts was very much open. This was something that I sensed had not occurred to those present)

- the Division responsible for KTS becomes a pilot for (organisational) culture change within the revamped Ministry (this was a choice available to some divisions in recognition of a need to think smarter and work in different ways, particularly following recent controversies in UK agriculture)
- avoiding infractions (it became apparent in the interviews that the whole of England was in imminent danger of being declared a 'nitrate vulnerable zone' (NVZ). The imperative for the civil servants, who saw this as an opportunity, was to protect their Minister in ways that gave scope for 'innovative' action)
- some budgets available (i.e. resources were available but further release of money required treasury approval. It transpired that meeting treasury requirements, both real and perceived, was the key design variable for all policy initiatives)
- realisation that the traditional approach to KT has not worked well in the past (on reflection all those present admitted that past KTSs had not worked but that the reasons why were often lost from institutional memory because of staff transfers and lack of continuity of focus)

Some of the threats included that:

- the strategy (KTS) is swamped in a plethora of initiatives in the Civil Service (in the light of BSE and foot and mouth and an inquiry into the future of rural areas this was a valid concern)
- there is a risk of over-selling the KTS strategy (i.e. making promises that could not be met – from my perspective this seemed a real possibility)
- the KTS is perceived as involving losing control (I think this was one of the main concerns of the civil servants – perhaps not as individuals but in terms of civil service culture and the likely reaction of superiors)
- the KTS is perceived as costing to much
- a new chief scientist is about to be appointed (also a possible opportunity)
- the public good arguments are not won with Treasury (if their funding or agreement is required)
- criteria for success are not conceptualised appropriately (this too seemed possible from my perspective)

The final part of the session invited those present to consider how they might use the inquiry results and data at their disposal in the design of one or many learning system(s) to achieve the espoused purpose of the KTS.

(continued)

Reading 5   (continued)

3.4. KTS as the design of learning systems

As I engaged with the KTS I realised that the design considerations we have used at The Open University (OU) to evolve a pedagogy (a learning strategy) for Systems course development had features which might be used in a KTS imagined as 'a R&D strategy to design learning systems'. The pedagogy we have evolved at the OU has the features described in Table 7.4 [14, 15].

My conviction that these eight considerations had something to offer was reinforced in the second conversation that followed the first meeting. This was conducted with the principal of TACT consulting who had played a major role in initiating and overseeing the development of a farmer-based

Table 7.4 Two sets of design considerations for the design of learning systems

Eight design features of Systems courses at the Open University	Ten design considerations for the SWARD project including some key initial starting conditions
1. Ground concepts and action as much as possible in the student's own experience	1. A perceived issue or need which had local identity
2. Learn from case studies of failure	2. Active listening to stakeholder perceptions of the issue/need
3. Develop diagramming (and other modelling) skills as a means for students to engage with and learn about complexity	3. Good staff – in this case young, motivated and proactive women
4. Take responsibility as authors (or researchers) for what we say and do (epistemological awareness)	4. No, or very limited forms of, control
5. Recognise that learning involves an interplay between our emotional and rational selves	5. Proper resourcing particularly in the early stages
6. Develop skills in iterating – seeing learning as arising from processes that are not deterministic	6. A minimum number of initial group leaders who acted as 'key attractors'
7. Introduce other systems concepts, tools, methods, and methodological approaches so as to develop skills in 'formulating systems of interest… for purposeful action' (an example would be my exploration of metaphors for this inquiry)	7. Scope for self-organisation around particular enthusiasms
8. Use verbs not nouns! (i.e. verbs denote relationships and activity and are key to the process of activity modelling which is one of the main features of SSM)	8. An appropriately experienced participant conceptualiser
	9. Some small 'carrots' for participants at the beginning
	10. A supportive local press creating a positive publicity network

(continued)

Reading 5 (continued)

R&D (or learning) network (the SWARD project) in Cornwall and Devon. An account of the project suggested many successes but also concerns about evaluation and scaling up. The reflection, which our conversation enabled, suggested ten design features including key initial starting conditions for the project (Table 7.4). These features emerged in our conversation because it was not clear whether scaling up the KTS meant expansion or starting again in a new context.

From my perspective (and thus from my own understandings) a potential way forward for the KTS could have been its conceptualisation as a systemic inquiry that attempts to make transparent understandings and practices of those involved (e.g. by exploring what metaphors reveal and conceal) and possibly, rearranging stakeholders' mental furniture. It could be argued that a potentially useful starting point would be to consider the two sets of design criteria described in Table 7.4. This would involve conceptualising the KTS as if it were a learning or researching system.

Not surprisingly, given my own history, my design for this inquiry had many features similar to the design features suggested for second-order R&D [28]. Russell and Ison [28] describe second-order R&D as practice which seeks to avoid being either subjective (particular to the individual) or objective (independent of the individual) because the objects of our actions and perceptions are not independent of the very actions/perceptions that we make. From this perspective problems and solutions are both generated in the conversations that take place between the key stakeholders and do not arise, or exist, outside of such engagements. Second-order R&D is built on the understanding that human beings determine the world that they experience.

4. Feedback and reflections on process issues

My purpose in inviting feedback was primarily to aid my own reflection and learning, particularly in terms of process design. It was not designed to establish cause and effect in terms of outcomes. In inviting feedback I posed a number of questions; these concerned what, if any change our joint activity might have triggered in (i) the KTS and (ii) in the understandings or practices of the participants (the KTS proponents).[15] I also asked them about their experience of the 'inquiry process' and an open-ended question inviting any other feedback.

---

[15] Triggered is used here as a term as an attempt to avoid the more usual application of linear cause-effect thinking; my understanding of 'learning systems' of the form created through this inquiry is that they are not deterministic but create the circumstances for emergence. I also invited in feedback an accounting for any changes that had happened – in asking this question my aim was to provide an opportunity to tell a story, which may, or may not involve a story about causation on the part of the responder.

(continued)

Reading 5 (continued)

From the feedback it was clear that 'thinking on KT has shifted... towards a more participatory model.' A claim was made that 'this was happening before the [initial] event, but the language may not have caught up. The concept of stakeholder participation and ownership have certainly come to life as a result of discussions at [the first meeting], and afterwards.' This feedback in itself is evidence of linguistic shift – these terms were not in evidence in either written or spoken form prior to my entry into the situation. From my perspective it remained an open question as to whether thinking (i.e. understanding) had genuinely shifted or whether this was merely evidence of a semantic shift that had no concurrent change in practices of the form depicted in Fig. 1.3.

For some the tension between holding onto central control and allowing local stakeholding to develop had come into focus: 'the difficult issue is still how to ensure some degree of uptake of Government agenda, whilst still allowing real ownership and decision-making by land owners at local level.' For another respondent: 'the need for local issues to be resolved locally rather than centrally seems to have gelled, but with need for central guidance. Not sure whether funding fits the same bill? Use of local facilitators seems to be the way forward. And maybe slightly different models will be appropriate in different areas. I think these developments have been brought about by continued discussion.'

In response to the question: How has your own thinking/action about 'Knowledge Transfer' changed since [the first meeting]? one respondent reported that: 'I have become more convinced of the need for local solutions (within some central guidance), and a need to provide a local context so that farmers, landowners and others understand the state of their local environment and more clearly how their actions affect the environment, and how changes to practices could lead to improvements. We need to be honest about the costs to businesses.' However another respondent said: 'So the one thing that bothers me is that we have not sought opinion or buy in from farmers etc. – the very people we want to influence. We can interact with bodies like the NFU [National Farmers Union] and CLA [Country Landowners Association] – but this does not provide direct feedback. And we need to do this in the local context as well.'

There were contradictions in the feedback; another respondent said: 'my perception is that KTS has developed gradually but is not that far removed from where I imagined it might be. The positive factors have been the input from other stakeholders and some agreement on the issues, ownership or part ownership of the problems, and possible offers of help.'

In one area it seemed little had changed – that of the commitments to particular metaphors about human communication as evidenced by one respondent who said: 'we have discussed brigading messages in some way

(continued)

Reading 5 (continued)

such that the key organisations are seen to be in agreement – this would reduce confusion and conflict (at least in the subject areas and options for improvement that we can agree). This will need more discussion and interaction among the stakeholders. We do have some examples of multi-badging publications already.' And another: 'and all assuming we can agree a joined up message on some key activities on the farm.'

In relation to their experience of the inquiry approach one respondent said: "[the] interview technique was interesting – certainly appeared as 'practice what you preach'." But this same respondent harboured concerns of a cause effect nature: 'not absolutely convinced conclusions would not have been the same without interviews – we were a receptive audience and RLI probably had a reasonable idea of where the strategy was going from [the first meeting] and our subsequent group discussions on his follow-up meeting.' Another respondent recognised the value of the personal interviews: 'a one to one is useful to identify key issues and key concerns.'

My main personal reflections are that more time (an extra hour) would have been desirable for joint discussion, some of which I would have allocated to synthesising some of the interview data for the following session (as it was it was particularly rushed). I would also have liked more time working interactively through the metaphors that were elicited. Under similar circumstances I might in future try to negotiate a more explicit 'social contract' prior to undertaking such a task (other than travel expenses no fees were paid – and it is often the commissioning for payment that establishes de-facto the social contract). I was pleased that what I did was experienced as coherent with my espoused theoretical position.

5. General conclusions

This small inquiry reflects how pervasive particular metaphors (and thus theories-in-use) are in institutions responsible for environmental and agricultural policy development and natural resource management. Regardless of whether they are changing, their pervasiveness is a cause for concern. It is a concern because language and our underlying conceptions both constrain and make possible the choices that are made. Exploring what particular metaphors reveal and conceal enables a dialogue to begin about our taken-for-granted traditions of understanding. This is a starting point for triggering change in our 'mental furniture'. But clearly this takes time.

In this reflection I present a mode of practice with the potential to trigger reflections on metaphors-in-use in a manner that is coherent with the theory that is espoused. I contend that the lack of coherence between espoused theory and theory-in-use acts as a major constraint to researching with people and the translation of learning theory into practice [18]. I have also tried to

(continued)

Reading 5 (continued)

convey the idea that by thinking about my experience and responding to the invitation in a particular way a generalisable model of practice (systemic inquiry) is demonstrated. Further development and refinement is warranted.

What has also emerged from this inquiry is the articulation of two sets of criteria for the design of learning systems, which despite differing provenance, have features in common. Together with those articulated by Russell and Ison [28] for second-order R&D this suggests a potentially fruitful line of further research inquiry.

Source: Ray Ison (2002) 'Farming and Rural Systems Research and Extension', Proceedings from the Fifth IFSA European Symposium, Florence, April. [16]

Illustration 7.1

7.6.1   Responses to the Four Organising Questions

In respect of the experiences I report in Reading 5, the answers that I give to the four questions I posed earlier are:

Q1. Is it the method or how it is used that is important?

R1. From my perspective it is the praxis that is important – not the method. Unfortunately the civil servants in the situation I encountered were not engaged in praxis – they were responding to a set of pressures – to Treasury, the Secretary of the Department, the high turnover of staff and thus loss of 'organisational' memory, the felt need to be seen to be doing something, in a manner that they had done before – they were thus acting out a familiar routine because it was part of their repertoire, even though they knew it would not work.

Q2. How are learning and action built in?

R2. My response in the situation that presented itself was to build an invitation upon an invitation. This opened up a reflective opportunity, something that is quite rare it seems in organisational life. Other systems practitioners may have offered a method and, possibly, intentionally or not, implied that the method had capabilities to solve the issues of concern. Thus, whenever a tool, technique, method or methodology is brought in, then it is wise to be aware that the underlying concern might be with delivering control of, or certainty in, the situation. This delivers an underlying emotional dynamic that can preclude learning and sustained action because it involves an abrogation of responsibility. In these situations systems practice remains a silent practice. Juggling the C-ball involves an awareness of how opportunities for learning and taking responsibility might be opened up. Acting effectively in these spaces, however, is likely to take time as my inquiry shows.

Q3. Who is, or could be involved in the approach?

R3. On reflection I might have paid more attention to identifying who in the Ministry might have been capable of supporting the five key stakeholders on a learning journey and inviting them to be part of the process (I have subsequently done this in other situations with positive outcomes). Equally it might have been worth trying to understand who, if anyone, had the power to hold open a reflective space in which doing something different could be tried. With further involvement, it might have been possible to create the circumstances for a joint workshop between some of the civil servants and farmers, so that they could experience how things might be done differently.

Q4. What could be said about the politics and practicalities of engaging in a 'real-world' situation?

R4. If I had been free to continue my engagement with the group then an opportunity for a co-inquiry or co-research could have presented itself.[16] Had I proposed this action it is highly possible that it might have happened.

I imagine you might still be left with a number of unanswered questions as a result of engaging with Reading 5. In Part III, I will be elaborating further on systemic inquiry of the form introduced here; perhaps some of your questions will be answered then?

7.6.2 Implications for Practice

My Google search on 'knowledge transfer strategy' in 2009 returned almost two million hits. When I drop off 'strategy' in my search it goes up to 16 million hits. 'Knowledge transfer' also has its own Wikipedia site. So clearly the concept and some associated practices are widespread and not just restricted to agriculture and natural resources management. In my own experience the phrase has 'taken off' in Universities, business, medicine and those concerned with the 'adoption' of any form of new knowledge or practices. Unfortunately as experiences in the agricultural domain have shown those who promote it, build it into policies or are employed to do it, rarely understand what it is they do when they do what they do! All too often the practice that emerges is that of trying to change someone else's behaviour without their genuine participation in the process. In my terms they drop the B-ball – particularly that pertaining to an ethics of practice.

I return to this topic in Chapter 11. However, at this stage I invite you to reflect upon the following claims in relation to this reading. The claims I make are that:

---

[16] My existing commitments precluded my offering this as a strategy. This is also a limitation of my engagement as I did not have someone with me as part of the process who may have been able to take such a strategy forward.

1. Systems practice can be contextualised to unfolding, changing circumstances
2. The practitioner can actively contribute to making the circumstances conducive to their systems practice e.g. inviting engagement through an interviewing and inquiry process rather than merely presenting a seminar
3. Responding to, and creating invitations, because of the underlying emotional dynamics, creates a different relational and thus practice dynamic to that of practice understood as an 'intervention' – whether as internal change management or as an 'external consultant' (I have called this the politics of invitation and intervention)[17]
4. Systems practice does not have to rely on established methods or 'methodologies' – in my case I make connections with second-order cybernetic understandings of language and communication and the work of Donald Schön on metaphor and organisational change
5. Systemic change does not rely only on changes in understandings, as there can be deeply embedded structural impediments to the development and embedding of systemic understandings and practices, e.g. Treasury demands and budgetary procedures, high staff turnover and other institutional arrangements

In finishing this section on the C-ball, and following on from the experiences I describe in Reading 5, I want to point out that sometimes there are circumstances or situations where attempts at juggling the C-ball lead to the realisation that the opportunities for systemic improvement are minimal or non-existent. This sometimes leads to a crisis of ethics or an emotion of despondency. John Seddon exemplifies what I mean when he asks 'Who is to blame?' in respect of the failure of public sector reform in the UK under what he calls the 'Blair regime'. He says:

> Public sector managers cannot be blamed for doing as they are told; specifiers cannot be blamed for doing their job, for creating specifications as directed by ministers; ministers justify themselves by pointing to consultations carried out and the advice of consultants, who in turn will point the finger at those charged with implementation.... It is a regime that ensures that no one is responsible. It needs radical reform. The present regime is systemically incapable. It can't learn.... The public-sector reform that is most needed is the one that is never talked about – that of the regime itself, the vast pyramid, hundreds of thousands strong, of people engaged in regulating, specifying, inspecting, instructing and coercing others doing the work to comply with their edicts. [29, pp. 192–193][18]

---

[17]See High et al. [13]; an invitation is not an invitation if one is not willing to accept no for an answer – if one is upset that an invitation is not accepted then what was issued as an invitation was, more often than not, a demand or an attempt at coercion disguised as an invitation. An invitation and an intervention thus have very different underlying emotional dynamics. Intervene is derived from the Latin 'venire', and has the sense of 'coming between'. 'Advice' has a similar root [33]. On the other hand 'invitation' can be traced back to the Sanskrit 'vita' meaning loved [1].

[18]Briggs in her 2009 [2] presentation points to the fundamental tensions between the vertical accountabilities in the Westminster system of Cabinet Government, with its underlying accountability of individual Ministers to Parliament, where differences are ultimately resolved in the Cabinet or by the Prime Minister *and* the horizontal responsibilities in whole of government approaches, where differences are expected to be resolved within and between agencies and, potentially, other stakeholders. Thus the current (UK, Australian) governance situation can be usefully understood as a structure determined system that does what it is structured to do, and no more. But clearly it is no longer adequate for contemporary circumstances.

The remaining chapters of the book will provide further answers to the four questions I posed and answered above. Answers to these questions will also depend on the role a systems practitioner is playing, e.g. inquirer, problem solver, facilitator, investigator, consultant, expert. It will also depend on whether your engagement with a situation comes about through an 'intervention' or an 'invitation'. They are qualitatively different modes of engaging because the underlying emotions are different. Whatever the role, they will involve managing, juggling the M-ball, which is the subject of the next chapter.

References

1. Barnhart, R. (2001) Chambers Dictionary of Etymology. Chambers: USA.
2. Briggs, L. (2009) Delivering performance and accountability – intersections with 'wicked problems.' The International Society for Systems Sciences annual conference, conference paper. Brisbane, Australia, July 2009. See http://www.apsc.gov.au/media/briggs150709.htm (Accessed 14th August 2009).
3. Checkland, P.B. (1985) From optimizing to learning: a development of systems thinking for the 1990s. Journal of the Operational Research Society 36, 757–767.
4. Checkland, P.B. (1993) Systems Thinking, System Practice. Wiley: Chichester.
5. Checkland, P.B. (2001) Presentation to a joint meeting of UKSS/OUSys. OU Systems Society Newsletter, February.
6. Checkland, P.B. and Poulter, J. (2006) Learning for Action: a Short Definitive Account of Soft Systems Methodology and its Use for Practitioners, Teachers and Students. Wiley: Chichester.
7. Checkland, P.B. and Scholes, J. (1990) Soft Systems Methodology in Action. Wiley: New York.
8. Checkland, P.B. and Winter, M. (2006) Process and content: two ways of using SSM. Journal of the Operational Research Society 57, 1435–1441.
9. Churchman, C.W. (1971) The Design of Inquiring Systems: basic concepts of systems and organisations. Basic Books: New York.
10. Churchman, C.W. (1979) The Systems Approach and its Enemies. Basic Books: New York.
11. Fell, L. and Russell, D. (2000) The human quest for understanding and agreement. In R.L. Ison and D.B. Russell (Eds.), Agricultural Extension and Rural Development: Breaking out of Traditions (pp. 32–51). Cambridge University Press: Cambrige, UK.
12. Foucault, M. (1972) The Archaeology of Knowledge and the Discourse on Language. Pantheon: New York.
13. High, C., Ison, R., Blackmore, C. and Nemes, G. (2008) Starting off right: Reframing participation though stakeholder analysis and the politics of invitation. Proc. Working Group 13 'The OECD's New Rural Paradigm', XII World Congress of Rural Sociology, Seoul, Korea.
14. Ison, R.L. (2000a) Technology: transforming grazier experience. In R.L. Ison and D.B. Russell (Eds.), Agricultural Extension and Rural Development: Breaking out of Traditions (pp. 52–76). Cambridge University Press: Cambridge, UK.
15. Ison, R.L. (2000b) Systems approaches to managing sustainable development: experiences from developing supported open-learning. Invited Paper: Ecological Sustainability Theme, Proc. 16th Symposium of the International Farming Systems Association and 4th Latin American Farming Systems Research & Extension Symposium, 27–29 November 2000, Centro de Extension Universidad Catolica, Santiago Chile.
16. Ison, R.L. (2002) Some reflections on a knowledge transfer strategy: a systemic inquiry. Fifth European Farming Systems Research Conference, Conference Paper, Florence, April 2002.
17. Ison, R.L. & Russell, D.B. (Eds.) (2000) Agricultural Extension and Rural Development: Breaking Out of Traditions (pp. 239). Cambridge University Press: Cambridge, UK.
18. Ison, R.L., High, C., Blackmore, C.P. and Cerf, M. (2000) Theoretical frameworks for learning-based approaches to change in industrialised-country agricultures. In LEARN Group M. Cerf,

D. Gibbon, B. Hubert, R.L. Ison, J. Jiggins, M. Paine, J. Proost and N. Roling (Eds.), Cow up a Tree. Knowing and Learning for Change in Agriculture. Case Studies from Industrialised Countries (pp. 31–54). Paris: INRA (Institut National de la Recherche Agronomique) Editions.

19. Kersten, S. (2000) From debate about degradation to dialogue about vegetation management in western New South Wales, Australia. In LEARN Group M. Cerf, D. Gibbon, B. Hubert, R.L. Ison, J. Jiggins, M. Paine, J. Proost and N. Roling (Eds.), Cow up a Tree. Knowing and Learning for Change in Agriculture. Case Studies from Industrialised Countries (pp. 191–204). Paris: INRA (Institut National de la Recherche Agronomique) Editions.

20. Kersten, S. and Ison, R.L. (1998) Listening, interpretative cycles and dialogue: Process design for collaborative research and development. The Journal of Agricultural Education & Extension 5, 163–178.

21. Maturana, H. (1988) Reality: The search for objectivity or the quest for a compelling argument. Irish Journal of Psychology 9, 25–82.

22. McClintock, D. (2000) Considering metaphors of countryside in the United Kingdom. In LEARN Group M. Cerf, D. Gibbon, B. Hubert, R.L. Ison, J. Jiggins, M. Paine, J. Proost and N. Roling (Eds.), Cow up a Tree. Knowing and Learning for Change in Agriculture. Case Studies from Industrialised Countries (pp. 241–251). Paris: INRA (Institut National de la Recherche Agronomique) Editions.

23. Mingers, J. and Gill, A. (Eds.) (1997) Multimethodology: The Theory and Practice of Combining Management Science Methodologies. Wiley: Chichester.

24. Open University (2000a) Managing complexity. A systems approach. (T306). The Open University: Milton Keynes.

25. Open University (2000b) Systems Thinking and Practice: Diagramming. (T552). The Open University: Milton Keynes.

26. Reynolds, M. (2006) Critical appraisal in environmental decision making. Book 4. T863 Environmental Decision Making: a systems approach. The Open University: Milton Keynes.

27. Russell, D.B. and Ison, R.L. (2000a) The research-development relationship in rural communities: an opportunity for contextual science. In R.L. Ison and D.B. Russell (Eds.), Agricultural Extension and Rural Development: Breaking out of Traditions (pp. 10–31). Cambridge University Press: Cambridge, UK.

28. Russell, D.B. and Ison, R.L. (2000b) Designing R&D systems for mutual benefit. In Ison, R.L. & Russell, D.B. (Eds.), Agricultural Extension and Rural Development: Breaking out of Traditions (pp. 208–218). Cambridge University Press: Cambridge, UK.

29. Seddon, J. (2008) Systems Thinking in the Public Sector: the failure of the reform regime… and a manifesto for a better way. Triarchy Press: Axminster.

30. Senge, P. (1998) A brief walk into the future: speculations about post-industrial organisations. The Systems Thinker 9, 1–5.

31. Schön, D.A. (1995) The new scholarship requires a new epistemology. Change November/ December, 27–34.

32. Shannon, C. and Weaver, W. (1949) The Mathematical Theory of Communication. University of Illinois Press: Urban, IL.

33. Shipley, J.T. (1984) The origins of English words: a discursive dictionary of Indo-European roots. John Hopkins University Press: Baltimore, MD.

34. Shotter, J. (1993) Conversational Realities: constructing life through language. Sage: London.

35. Thomson, David (2002) Principal, TACT Consulting, Bath UK. Personal communication.

36. Ulrich, W. (1983) Critical Heuristics of Social Planning: a new approach to practical philosophy. Stuttgart: Chichester/Haupt: Wiley (paperback edition).

37. Ulrich, W. (1996) A Primer to Critical Systems Heuristics for Action Researchers. Centre for Systems Studies, University of Hull, Hull, UK, 31 March 1996, 58pp. (ISBN 0-85958-872-6)

38. Ulrich, W. (2000) Reflective practice in the civil society: the contribution of critically systemic thinking. Reflective Practice 1(2): 247–268.

39. Webber, L. (2000) Co-researching: braiding theory and practice for research with people. In R.L. Ison and D.B. Russell (Eds.), Agricultural Extension and Rural Development: Breaking out of Traditions (pp. 161–188). Cambridge University Press: Cambridge, UK.

Chapter 8
Juggling the M-Ball: Managing Overall Performance in a Situation

8.1 Perspectives on Managing

My focus in this chapter is on the M-ball being juggled by a systems practitioner. As I outlined in Chapter 4 the M-ball is concerned with juggling as an overall performance, managing both the juggling and the desired change in the world. Another way to describe this is as co-managing self and situation. As the term managing is often used to describe the process by which a practitioner engages with a 'real-world' situation, it can be considered as a special form of engagement. In this chapter I hope to enable you to appreciate the diversity of activities that might constitute managing. Managing also introduces the idea of change over time, in the situation, the approach and the practitioner – of adapting one's performance.

I see the M-ball as where responsibility for creating the pattern of an overall performance resides. So it is not just juggling per se but juggling in context and with others (Fig. 8.1). Thus managing, for me, involves managing the ongoing juggler-context relationship, a co-evolutionary dynamic of the form described in Chapter 1. This involves looking out to the situation as well as looking inwards to the dynamics of juggling the B-ball.

It seems that many of the situations I encounter, whether personally or through the media, might well be considered as wicked, messes or complex or uncertain situations. Having these experiences on a regular basis I was attracted to the research of Mark Winter [29, 30] who asked the question "Why not think of 'managing' in more generic terms?" Following his lead I have illustrated this in the form depicted in Fig. 8.2. Winter's research and practice resonated with me as earlier I had carried out a simple exercise of considering some of the verbs that I associated with the word managing. I came up with: understanding, surviving, seeing, visioning, allocating, optimising, communicating, commanding, controlling, helping, defending, leading, supporting, backing, enabling, coping, informing, modelling, facilitating, empowering, encouraging, delegating.[1] In order to make more sense of my list of verbs I identified three categories: (i) getting by, (ii) getting on top of, (iii) creating

---

[1] I generated this list with my colleague Rosalind Armson in the mid-1990s; I make no claim that this list is definitive or that this is the only way to categorise my list.

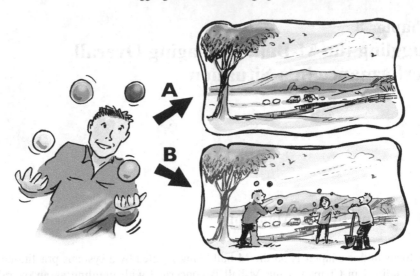

Fig. 8.1 Juggling with awareness opens up different choices for practice (A or B), including how to build collaborative performances with others (B)

Fig. 8.2 Perspectives on managing – an image to engage with the question "Why not think of 'managing' in more generic terms?"

space for.[2] Undoubtedly if you were to do the same exercise your list and categories would be different. My motivation for generating this list and categories was similar to Winter's. I wanted to understand managing in all its manifestations and particularly how these practices (as exemplified by the verbs) might become embodied in a

---

[2] This is a good example of how, even in my own practice, it is hard to escape the practice of classifying or typologising – though I would claim that I do so with awareness of what I do when I do what I do!

particular manager. I was also reacting to what I considered as a narrowing of consideration and learning under the rubric of 'management education'.[3]

In doing this simple task I was also drawing on an underlying logic that has been incorporated into soft systems methodology by Checkland and colleagues. I worked backwards (compared to normal SSM use) from a set of activities (denoted by verbs), generating a set of higher order concepts or categories (i.e., getting by, getting on top of and creating space for) and implicitly recognising a possible system of interest. Why do I call this a system of interest? Well, because it has the potential to enhance my learning and thus action, in the situation I am in by engaging with it systemically. I specified 'a system to understand managing by recognising a range of activities in order to appreciate what capabilities might be needed for juggling the M-ball'.

A key concept in systems thinking is that of transformation (Table 2.1); unfortunately the operational dynamics of the transformation process is generally not well understood [5, p. 47]. For example in the system description I have just developed the transformation, that which the system does, is the following:

managing not understood → managing understood

Thus, in conceptual terms I have 'formulated' a description of a system of interest comprising three sub-systems – getting by, getting on top of and creating space for, which through its enactment or operation has the potential to transform my understanding of managing.

When I first did this exercise I did not make all of my underlying thinking explicit as I have attempted here. Much of this way of thinking has become part of what I do – part of my own tradition of understanding. I do remember though that the transformation I have articulated above happened. I became aware that:

- 'Getting by' could be seen as those activities devoted to personal maintenance, activities that might be exacerbated or lost with different life-work balances
- 'Getting on top' of could be seen in one of two ways – a resort to command and control procedures associated with hierarchical management or enhancing my juggling so that my performance was always better adapted to context (these are very different choices)
- 'Creating space for' could be seen as delegation (as a generic set of activities) but for me, more importantly, could be seen as creating the circumstances for emergence and novelty (innovation) through self-organisation and removing barriers to others being responsible (i.e., creating response-able circumstances)

If I step back out of the specifics of this inquiry for a moment you might see that it is of the form: What is the system to which doing X is the answer (the how)? The systemic exploration of this question enables new systems of interest to be formulated, which can be used by those in the situation to arrive at new understandings on which to base their actions.

---

[3] This is the same issue taken up by Henry Mintzberg [16].

As a result of this mini-systemic inquiry, when I now think of a manager, I think of anyone in any context who is engaged in taking purposeful action. The purposefulness arises whenever we act to open a reflective space, as I have tried to exemplify when thinking about what managing might be. The emergent property of this process is enhanced awareness of what we do when we do what we do! This can be done in all situations with any phenomenon or issue of concern and may or may not involve systems thinking. In this regard it seems to make sense to me to envisage a role for the systems practitioner as someone engaged in managing in everyday situations which are increasingly characterised by surprise or messiness.[4] I do not have a new professional management elite in mind – though this could also exist – but more a citizenry enabled with systems thinking and practice.

8.1.1 Transforming the Underlying Emotions of Managing

Henry Mintzberg, in his critique of management education, particularly MBAs, argues that "we need balanced, dedicated people who practise a style of managing that can be called 'engaging'". He goes on to claim that the 'development of such managers will require another approach to management education, likewise engaging, that encourages practising managers to learn from their own experience. In other words, we need to build the craft and art of managing into management education and thereby bring these back into the practice of managing' (p. 1). I agree wholeheartedly with this critique! However, I find it intriguing that in the index to Mintzberg's book there is no listing for 'emotion' or 'ecology'. It would seem that even in a profound critique such as Mintzberg's, and thus in management education discourse, it is still not completely legitimate to talk about emotions or the ecological basis on which consumption, and thus business, rests! This is despite the widespread success of Daniel Goleman's book on 'emotional intelligence' [9] and his more recent one on 'ecological intelligence' [10].

Part of my thesis in this book is that we can no longer leave emotions and ecology out of our concerns and by this I mean all considerations of what we do when we do what we do! So how might understandings of managing be transformed in ways that are more sensitive to underlying emotional and ecological dynamics?

As I outlined earlier, I am aware, for example, of the millions of people around the world who now engage in local and family history research. It seems many live

---

[4] I am aware that my language is not really adequate here – I trust that by now you will have become at least partially aware that because we are human beings, then it follows that my perspective is that the practitioner – situation relationship is best understood as a relational dynamic – they bring forth each other. So whenever I use the word situation I am using a short hand description for a practitioner (observer) – situation relationship that mutually construct each other.

with a passion for explaining who they are and where they have come from. I would describe all of these people as small 'r' researchers. So for me researchers are not just confined to laboratories and universities. We can make a choice to see ourselves as small 'r' researchers; such a framing for managing is a way of breaking down a commitment to certainty. A shift such as this, and this is certainly not the only shift imaginable, transforms our underlying attitude or predisposition and thus brings our living onto a different trajectory. Small 'r' research in the sense I use it is willed and reflexive action, done for a purpose, though the purpose may not be clear initially and involve a mix of emotion and intellect. Others may call this reflexive inquiry, or appreciative inquiry.[5]

In my own practice I have refined my small 'r' position to focus more on what I call systemic inquiry – something that I exemplified in Reading 5 (Chapter 7) and my relatively simple example above. I will say much more about this in Part III of the book. Mark Winter, in his own inquiries into managing, has chosen to draw on the work of Geoffrey Vickers [28], to cast the act of managing in terms of a process of relationship maintaining (Fig. 8.3). As depicted, relationships are maintained through conversation which as it unfolds creates a matrix of relationships. When the conversation breaks down, whether through choice or other factors, then those 'nodes' drop out. To paraphrase Vickers, one's standards of fact and value emerge out of this network. I consider this a very useful conception, though, as outlined in Chapter 1, I would want to add additional relationships, that of humans to the biosphere, to other species and to future generations.[6]

Another choice that can be made to transform the underlying dynamics of managing is to think of it in terms of learning. Some of my colleagues, for example, see enactment of systems thinking through the isophor of a juggler as understandable in terms of enactment of an experiential learning cycle (of the form depicted in Fig. 7.2). The key here is that in considering practice it is experience that provides the initial starting conditions. The history of David Kolb's articulation of the experiential learning cycle can be traced back to the work of Kurt Lewin shortly after the Second World War. Lewin is generally recognised as the originator of the notion of action research. When we connect with this history, it is possible to recognise the experiential learning model as a model for action research as well. The idea of the systems practitioner as action researcher is a powerful one in my experience (e.g. [4, 14]). I will say more about this in Part III (Chapter 11) of the book.

---

[5] These are parallel and potentially synergistic traditions to my own [6, 20].

[6] Since the global financial crisis there have been many commentators who argue that what has been lost is 'trust'. This is a claim that is hard to disagree with, but much harder to think of in terms of how trust is built and sustained. I would argue that trust is an emergent property of the process of relationship building and maintaining. So trust *per se* cannot be managed; it is about managing the quality of relationships.

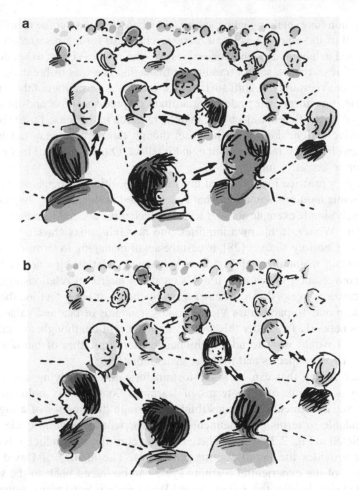

Fig. 8.3 An unfolding network of conversation and relationships. 'Managing' involves maintaining a network of asynchronous relationships in the context of an ever-changing flux of events and ideas. As any manager engages in one conversation, others are engaged in different conversations. As individuals participate in different conversations a coherent network of conversations results (Expanded from Winter [29, p. 67 and p. 83])[7]

These ideas about managing that take a small 'r', or learning, turn also provide a conceptual framework to imagine what the life-long learner might be, the idealised person that has become so popular in recent discourse in education circles.[8]

---

[7] Both Figs. 8.2 and 8.3 take as their inspiration work done by Mark Winter and reported in his Ph.D. thesis [29].

[8] For example see http://en.wikipedia.org/wiki/Lifelong_learning (accessed 15th August 2009).

8.2 Managing with Systemic Awareness

A systemic approach involves using systems thinking to construct ways of knowing, and thus acting, as part of an inquiry process. This can be done in any type of situation but my main focus is on those situations which we experience in particular ways that lead us to call them complex, messy, wicked, uncertain etc. Through systems practice, the aspiration is that we can generate fresh and insightful explanations which trigger new ways of taking purposeful action that is systemically desirable. To have arrived at this way of acting requires a systemic sensibility – to be aware of the distinctions between systemic and systematic approaches (as discussed in Chapter 2). This in turn means being willing to choose and contextualise depending on your mode of engaging with a situation.

What do I mean by choose? All too often when faced with this question the listener tends to provide a response in either/or terms, i.e. I need to choose systemic, not systematic! This is not really a choice in a systemic sense because the act of choosing one is a negation of the other. It has an intrinsic linear logic which is described as a dualism. Unfortunately dualisms abound in our daily living, e.g. objective/subjective, right/wrong etc. The alternative, and this is a choice that can be made, is to regard a pair such as systemic/systematic as a duality, a totality which together go to make up a whole, or a unity.

Based on my experience, a choice made to pursue only the systematic route is inherently conservative and likely to result in first-order change: doing the same thing more effectively or optimally [15]. This has its place. However, the systemic route opens up the possibility of second-order change, change that changes the 'whole situation'.[9,10] In Table 8.1, I summarise some of the features I have come to recognise as being associated with systemic and systematic thinking and action (remembering that systemic/systematic comprise a duality, not an either/or, and that the features I describe suffer the limitations of all generalisations abstracted from context).

Systematic and systemic thinking and practice together build a powerful repertoire for juggling the M-ball. It is important therefore to realise that I am not in any way trying to set up the idea that systemic is good, systematic is bad. In the hands of an aware practitioner they are not in opposition. My perspective, when managing or engaging in messy situations, is that it is usually more appropriate to approach the

---

[9] In this section I am reminded of the book by Scott M. Peck [21]. For me it acts as a metaphor for my concerns and how to think about systems approaches in an evaluative sense. Thus the systemic is the road less travelled – but making the choice takes one to different places. Traditional evaluation approaches do not, in my experience, do justice to the implications of making and pursuing these choices.

[10] The same choices face complexity theorists or thinkers. They may choose to see complexity existing in the world or they may see complexity thinking as providing the means to formulate an epistemological device, a way of knowing, that is capable of generating new explanations about the world, as opposed to descriptions of it.

Table 8.1 A summary of the characteristics that distinguish systemic thinking and action and systematic thinking and action

Systemic thinking	Systematic thinking
Properties of the whole are said to emerge from their parts, e.g. the wetness of water cannot be understood in terms of hydrogen and oxygen	The whole can be understood by considering just the parts through linear cause-effect mechanisms
Boundaries of systems are determined by the perspectives of those who participate in formulating them. The result is a system of interest	Systems exist as concrete entities; there is a one to one correspondence between the description and the described phenomenon
Individuals hold partial perspectives of the whole situation; when combined, these provide multiple partial perspectives	Perspective is not important
Systems are characterised by feedback; may be negative, i.e. compensatory or balancing, or positive, i.e. exaggerating or reinforcing	Systems are comprised of chains of cause-effect relationships
Systems cannot be understood by analysis of the component parts. The properties of the parts are not intrinsic properties, but can be understood only within the context of the larger whole through studying the interconnections	A situation can be understood by step-by-step analysis followed by evaluation and repetition of the original analysis
Concentrates on basic principles of organisation	Concentrates on basic building blocks
Systems are seen as nested within other systems – they are multi-layered and both intersect and interconnect to form networks	Systems are hierarchically organised
Is contextual in approach	Is analytical in approach
Concerned mainly with process	Concerned mainly with entities and properties
The properties of the whole system are destroyed when the system is dissected, either physically or theoretically, into isolated elements	The system can be reconstructed after studying the components

Systemic Action	Systematic Action
The espoused role and the action of the decision-maker is very much part of an interacting ecology of systems. How the researcher perceives the situation is critical to the system being studied. The role is that of participant-conceptualiser	The espoused role of the decision-maker is that of participant-observer. In practice, however, the decision maker claims to be objective and thus remains 'outside' the system being studied
Ethics are perceived as being multi-faced as are the perceptions of systems themselves. What might be good from one perspective might be bad at another. Responsibility replaces objectivity	Ethics and values are not addressed as a central theme. They are not integrated into the change process; the researcher takes an objective stance
It is the specification of a system of interest and the interaction of the system with its context (its environment) that is the main focus of exploration and change	The system being studied is seen as inherently distinct from its environment. It may be spoken of in open-system terms but intervention is performed as though it were a closed system
Perception and action are based on experience in the world, especially on the experience of patterns that connect entities and the meaning generated by viewing events in their contexts	Perception and action are based on a belief in a 'real world', a world of discrete entities that have meaning in, and of themselves
There is an attempt to stand back and explore the traditions of understanding in which the practitioner is immersed.	Traditions of understanding may not be questioned although the method of analysis may be evaluated.

Illustration 8.1

task systemically. In other words, systemic thinking provides an expanded context for systematic thinking and action (Fig. 2.1). Thus my ideal, aware, systems practitioner is one who is able to distinguish between systemic and systematic thinking and is able to embody these distinctions in practice. This has implications for the initial starting conditions for any form of purposeful action – i.e. to start out systemically or systematically?

Of course, this is an 'ideal model' and day-to-day experience is different from this. No person can expect to become the ideal overnight. It requires active engagement in a process of experiential learning. And even an aware practitioner will at times respond to the situation without considering the systemic/systematic alternatives.

8.3   Skill Sets for Managing Systemically

I now want to describe some of the possibilities for juggling the M-ball that I see as being available in the repertoire of an aware systems practitioner able to connect with the history of systems thinking (in which I include many of the so-called theories of complexity).

In a survey where employers were asked to indicate what they most desire in new employees they said [23]: 'graduates must understand that the world is not linear – They need the ability to manage ambiguity and connectivity and to be comfortable with provisionality – making decisions when you don't really know what is going to happen, e.g. with e-commerce. They must also be comfortable with emergence'. These employers said that they considered the traditional skill set didn't go far enough if graduates wanted to be employable internationally. What was missing, they claimed, were 'complexity skills'. Employers assessed graduate communication skills, disciplinary knowledge, teamwork, information technology and interpersonal skills as ok, leadership as adequate but understanding of the nature of globalisation, working with cross-cultural sensitivity and sensitivity to different ethical positions as poor.

Robertson's [23] findings make a strong case for graduates with systemic awareness able to appreciate the nature of globalisation and work with cross-cultural and ethical sensitivity – this, I would claim, requires recognition and adeptness in juggling the different balls I have been describing. The same applies in the public sector. For example, Geoff Mulgan [18] identified seven factors that increased the relevance of systems thinking to policy making and to the functions of government. These were:

1. The ubiquity of information flows, especially within government itself
2. Pressure on social policy to be more holistic
3. The growing importance of the environment, especially climate change
4. Connectedness of systems brings new vulnerabilities
5. Globalisation and the ways in which this integrates previously discrete systems
6. Need for ability to cope with ambiguity and non-linearity
7. Planning and rational strategy often lead to unintended consequences

He concluded that out of all these factors has come a 'common understanding that we live in a world of complexity, of non-linear phenomena, chaotic processes, a world not easily captured by common sense, a world in which positive feedback can play a hugely important role as well as the more familiar negative feedback that we learn in the first term of economics.' He also recognised that 'so far remarkably little use has been made of systems thinking or of the more recent work on complexity' and that in part this is 'to do with the huge sunk investment in other disciplines, particularly economics' (see also [3]). The trends and imperatives recognised by Mulgan have, if anything, become more pronounced since 2001.

In 2009 the Australian Public Service Commissioner made the point that new skills in problem framing and boundary setting were needed so as to [2]:

- Generate fresh thinking on intractable problems
- Work across organisational and disciplinary boundaries
- Make effective decisions in situations characterised by high levels of uncertainty
- Be able to tolerate rapid change in the way problems are defined
- Engage stakeholders as joint decision-makers (not just providers or recipients of services)

She argued that (i) not all public servants will need to work this way all the time (some may not be affected at all) but many will be confronted by ambiguous and complex problems at some point, and (ii) it is important for senior levels of the public service to exercise the kinds of leadership that these problems require.[11]

So what might managing be in these contexts? Patricia Shaw [25] refers to consultants who operate from a complexity perspective, in contrast to what she describes as a traditional perspective (Table 8.2).

---

[11] I would have said situations rather than problems!

These business and public sector examples point to the contexts in which systemic managing will be required in future. On the other hand, if these understandings are absent in situations you find yourself in (as in the traditional situation in Table 8.2), then it is highly possible the situation will not be conducive to

Table 8.2 Contrasting perspectives that consultants may hold in undertaking interventions in an organisational setting [25]

The consultant with a traditional perspective	The consultant with a complexity perspective
Designs and implements an educational strategy to realise planned change intended to improve the organisation's position in its environment	Stimulates conditions of bounded instability in which the organisation co-evolves with its environment through self-organisation
Understands organisational change in terms of temporary transitional instability between system-wide stable states	Understands change dynamics as unfolding in the ongoing tension between stability and instability in which islands of order arise and dissolve
Contracts to deliver a pre-determined objective or outcome	Contracts for a step-by-step process of joint learning into an evolving future
Sees large scale project plans and political and ideological control strategies as useful only in circumstances closer to certainty and agreement	Dissuades managers from using inappropriate forms of control to manage the anxieties raised when operating far from certainty and agreement
Chooses an effective marginal or boundary position from which to diagnose the state of the system as a whole	Becomes an active agent in the life of the organisation, by participating in its shadow and legitimate systems to engage in complex learning processes
Tries to create an intended change in people's shared beliefs, values and attitudes	Seeks to stimulate and provoke conditions in which people's mental models are continuously revised in the course of interaction
Focuses on global, whole system change whether that of groups, individuals or organisations	Focuses on feedback loops operating at a local level through which activity may be escalated up to system-wide outcomes
Designs and facilitates off-site meetings to develop strategies and plans and build teams	Intervenes in the ongoing conversational life in organisations in which people co-create and evolve their action-in-contexts or contexts-in-action
Collects data on generic system variables through surveys, interviews and other instruments to feedback the legitimate system	Invites an exploration of the relationship between the system's formal agenda (what the legitimate system says it knows) and the multitude of informal narratives by which the organisation is working (what the shadow system knows). These feedback loops generate their own outcomes
Emphasises the need for alignment and consensus around clear directions.	Amplifies existing sources of difference, friction and contention, so that complex learning might occur, provided people's anxiety in the face of such learning is well contained.

your systems practice. This could be challenging – but awareness opens up possibilities.

At the beginning of this chapter on juggling the M-ball, I explored the three categories I used for making sense of a list of verbs associated with managing (i.e. 'getting by', 'getting on top of' and 'creating the space for'). When I did this work I was not aware of the set of three distinctions made by Gerard Fairtlough [7] as the 'three ways of getting things done'. He refers to hierarchy as the most common and recognises the hegemony of hierarchy in our organisational practices.[12] His second category is heterarchy – multiple rule with a balance of powers rather than a single rule through hierarchy (e.g. a group of partners in a law firm). His third category is 'responsible autonomy' in which an individual or group has autonomy to decide what to do but is accountable for the outcome of the decision (p. 24). Within my categories 'creating the space for' is the liberating and encompassing systemic category that more closely aligns with the skill set that is desired in business and the public sector. The same applies to Fairtlough's category of 'responsible autonomy' – they are I think much the same. Because I associate this category with the question: How can I create the space for emergence? I now want to address this question in relation to the question of purposefulness and self-organisation as part of the skill set for juggling the M-ball.

8.4   Clarifying Purposefulness in Managing

Research conducted by Ralph Stacey [26] showed how business managers often behave in a way contrary to espoused policies and expectations. Rather than adhering to conventions of long-term planning and accepted orthodoxies and procedures, in practice they often tend to make a succession of unrelated, adaptive responses to changing situations as the need arises. This is often, and rather disparagingly, labelled muddle-through or crisis management but can result in adaptive action and organisation.

I use these outcomes from Stacey's research to make clear that when I speak about purposeful behaviour I am not equating it with behaviour normally associated with blueprint planning or other forms of purely rational planning. Purposeful behaviour is willed behaviour and this may be triggered by personal circumstances which, on reflection, we regard as being rational or emotional behaviour.[13] There are widespread and conflicting understandings and assumptions

---

[12] He argues that hierarchy becomes hegemonic when everyone accepts that it is normal, that that is how things are!

[13] I use this expression though I understand being rational as another form of emotion.

about the nature of human behaviour, particularly within science and economics. Within economics and the messy issue of tax policy, the following exemplifies what I mean:

> ...all neoclassical [economic] analysis [is] based on an erroneous model of human behaviour that assumes the choices we make are always carefully calculated to maximise our material wellbeing. For the past 20 or 30 years, behavioural economists have been pointing out to conventional economists all the flaws in their assumption that people are always rational, but this seems to have had zero impact on the happy analysis of the tax economists. Assuming people always behave in the manner economists regard as rational makes it relatively simple to predict their behaviour in particular circumstances (and to do it using mathematical equations which economists think is really cool). Only problem is, since people rarely behave the way neoclassical economists assume they do, the predictions these economists make are frequently way off beam. Funny that. [8]

In a study of nine companies, Stacey showed how attempts to overcome ambiguity through planning failed completely except over short time periods. Furthermore, at least seven of the companies made significant shifts in how they operated despite the failure of their attempts to predict and plan. All the changes emerged unexpectedly and unintentionally. As Stacey observed: 'The changes occurred, not because we were planning, but because we were learning in a manner provoked by the very ambiguity and conflict we were trying to remove.' Managers, he argued, have to strike an appropriate balance between too much and too little control. They have to balance two tendencies within their organisations, programmes or projects. Too much control and blueprint-based planning leads to an inability to respond to change or to an unexpected eventual ossification. Too little control leads to fragmentation and disintegration. Success, it is argued, lies somewhere between these extremes.

A key point from Stacey's research is that too much control or attempts to intervene according to any pre-conceived view and necessarily partial view, or blueprint plan, stunts the process of self-organisation. Change and adaptation in human institutions occur through social interaction. Apparent fixes can inhibit the emergence of organisation and relationships that are most appropriate to any particular situation, such that solutions arrived at in this way are likely to be short-lived. It is in this sense that I see creating the space for spontaneous behaviour and emergent phenomena as a key element in managing for self-organisation.

Stacey's perspective is not a strategy for avoiding planning. It allows space for creative conflict, negotiation, interaction and learning wherein assumptions may be dashed but the seeds of new perspectives and formulations may be nurtured. Which seeds eventually develop and emerge depends on politics and negotiation and on the skills of those promoting or inhibiting the new perspectives. Systemic approaches in the hands of skilled and aware practitioners contribute to the surfacing of adaptive and innovative actions.

8.5   Managing for Emergence and Self-Organisation

One way of knowing that you are developing your systemic awareness is when you react with alarm at hearing the phrase 'roll-out'! It is a good indicator in my experience of a likely failing in managing, mainly because those concerned will have metaphorically dropped the C-ball. John Seddon provides an excellent example of what happens as a precursor to 'roll out' of policies and practices that all too often subsequently fail [24, p. 24]:

> The [UK] government's CBL [Choice-based letting] scheme took an idea that was developed in Holland, modified it to fit with existing practice and, as a result, removed its essential value. This is a classic example of copying without knowledge, rather than seeking first to understand the thinking and principles behind the original design. Inadequate research has subsequently been used to support the scheme; it is though the research priority is to find support for the policy rather than learn about what works.

This is an example of failure to engage in context sensitive design i.e., to privilege systematic practice over systemic. All too often, even with pilot testing, a 'roll-out' whether of a policy, a practice or a method fails to be meaningfully enacted; this is a failure of practice, the practice of creating a context sensitive performance.[14] Setting out to manage for emergence, or to create the circumstances for self-organisation can avoid the limitations of the 'roll-out' phenomenon. But even here we have a lot to learn, including whether it is possible to purposefully create conducive conditions for emergence or self organisation.

The example of Linux is now a well known model of software development based on self-organising dynamics. The mode of development of Linux is contrasted with the approach formerly employed by Microsoft and many other firms, described in Box 8.1 as "the old 'closed shop' model of commercial software producers". This is contrasted with the idea of altruistic programmers, working together across the Net on freely distributed code that is open for everyone's perusal and tinkering so that the result is 'more powerful and reliable software' than Microsoft's. Eric Raymond's [22] commentary on a leaked document explores this argument (see Box 8.1 also):

> To put it slightly differently: Linux can win if services are open and protocols are simple, transparent. Microsoft can only win if services are closed and protocols are complex, opaque. To put it even more bluntly: 'commodity' services and protocols are good things for customers; they promote competition and choice. Therefore, for Microsoft to win, the customer must lose. The most interesting revelation in this memo is how close to explicitly stating this logic Microsoft is willing to come. (From website: http://sagan.earthspace.net/esr/writings/cathedral-bazaar/)

It is possible to say that Linux emerged through a form of self-organisation. Self-organisation is the phenomenon associated with a system distinguished by an

---

[14] For example the introduction of SSM, a systems methodology, into the guidelines for government procurement and implementation of information systems failed because it became a blueprint, in which what could have been a context sensitive methodology became reduced to a technique [11].

observer, which is able to construct and change its own behaviour or internal organisation. Computer simulations have shown, for example, that the behaviour of a flock of flying birds can be replicated by a few simple rules, which, if changed, results in the emergence of new patterns of simulated flock behaviour. Of course self organisation in human activities has more dimensions of organisation than the flight of a flock of birds. Self-organisation is also sometimes considered as the acquisition of variety by a system or the progressive emergence of novelty when removal of constraint or control releases capacity for autonomous action. These features seem to be present in the Linux case. My colleague John Naughton [19], for example, who writes regularly for the Observer newspaper about technology, considers the Internet a wonderful example of self-organisation (Box 8.1).

Box 8.1 Open Versus Closed Systems of Innovation – an Example of Self-Organisation?

There is a saying in the computer business that 'only the paranoid survive'. The man who has taken it most to heart is Microsoft's Boss of Bosses, Bill Gates. Although he is the richest man alive and his company has a stranglehold on the world's computer screens, Gates is forever looking over his shoulder, trying to spot the newcomer who will wipe him out.

One can understand his anxiety. The pace of change in the computing industry is such that if you blink you might not spot the threat. Gates himself blinked spectacularly in 1994, when Netscape was founded. He failed to appreciate the looming significance of the Internet and Netscape had captured a huge slice of the web-browser market before he woke up.

From that moment onwards, Microsoft's corporate ingenuity was devoted to finding ways of crushing Netscape. Its crass attempts to do so eventually stung the US Department of Justice into launching the anti-trust suit which is currently being decided in an American court. But while the eyes of the media are on the trial, those of the Net community have been focused elsewhere – on a leaked Microsoft internal memorandum which is far more revealing than anything released in court. For it shows that Gates & Co have finally realised where the Next Big Threat is coming from. And it's nothing to do with Netscape – or browsers. They're yesterday's battlegrounds.

The leaked memo is now all over the Net. It was written by a Microsoft engineer called Vinod Valloppillil last August (1998), but is universally known as the 'Halloween Memo' because it was leaked last weekend (November 1998). Its purpose is to explain to Microsoft bosses the nature and extent of the threat posed by a free operating system called Linux and the 'Open Source' software development community that built it.

(continued)

Box 8.1 (continued)

To appreciate the memo's significance, you need to remember that Microsoft dominates the world market in operating systems – the complex programmes which transform computers from paperweights into machines which can do useful work. The Windows operating system is the jewel in Gates's crown and anything that threatens it threatens his company's dominance. Microsoft's long-term strategy is to move us all on to a version of it called Windows NT (for 'new technology'). But NT is in trouble. The release date for the next version has been postponed so often that it has had to be renamed 'Windows 2000'. And as NT flounders, the world's attention has increasingly focused on a rival operating system called Linux which offers many of the same facilities as NT, is incredibly stable and reliable – and is free. Anyone can download it, free gratis, from the Net.

Linux is free because it was developed collectively across the Net by skilled programmers working in the Open Source tradition which created the Internet and which holds that software should be freely accessible to the community. The name comes from the fact that 'source code' is computer-speak for the original version of a programme – as distinct from the version you buy and install on your computer. If you have the source code you can do what you like with it – alter it, damage it, improve it, whatever. Linux is powerful and stable because it was created by clever people working collaboratively on the source code and because it's been tested to destruction by more programmers than Microsoft could ever muster. The Halloween Memo warns Gates that Linux and its ilk pose a serious threat to Microsoft. It argues that Open Source software is now as good as, if not better than, commercial alternatives, concedes that 'the ability of the OSS process to collect and harness the collective IQ of thousands of individuals across the Internet is simply amazing', and concludes that Linux is too diffuse a target to be destroyed by the tactics which have hitherto vaporised Microsoft's commercial rivals. The people who built Linux cannot be driven out of business, because they're not 'in' business. Henceforth, Microsoft will be fighting not another company, but an idea.

The Halloween Memo provides a chilling glimpse into the Darth Vader mindset of Microsoft. The reason Linux is so powerful, reasons Valloppillil, is that its basic building blocks – its technical protocols – are free, openly distributed and not owned by anyone. The only way to kill it therefore is for Microsoft to capture the protocols by pretending to adopt them and then 'extending' them in ways that effectively make them proprietary. The new (Microsoft) revisions will – surprise, surprise! – be incompatible with the 'free' versions. Gates calls this process 'embrace and extend'. In reality it's 'copy and corrupt'.

The coming battle, then, will be between two philosophies – closed shop versus Open Source, commercial paranoia versus altruism and trust. The outcome is already predictable. Microsoft's difficulties with Windows NT show that some software is now too complex for even the richest, smartest company.

(continued)

Instead of trying to suborn Linux, what Gates should do is release the NT code and let the collective IQ of the Net fix it for him. He won't do it, of course, which is why his company has just peaked. If you have Microsoft shares, prepare to sell them now.

Naughton, J., 'Internet–It's Free and it works. No wonder Bill gates Hates it', The Observer, Sunday 8th November 1998, Copyright Guardian News and Media Ltd 1998.

The key aspects that emerge from the Linux case study relevant to managing the M-ball are that in the right context (i.e. in a world with internet connections) and with some relatively simple rules it is possible to create the conditions for cooperative self-organisation from which a new product, Linux, has emerged.[15] The same is possible, although not always easy, in organisational and project settings (e.g. through the types of practices undertaken by a consultant with a complexity perspective as outlined in Table 8.2).

In my own work on the governance and management of river catchments we have seen it as appropriate to understand sustainable and regenerated water catchments as the emergent property of social processes and not the technical property of an ecosystem [13, 17, 27]. That is, desirable water catchment properties arise out of interaction (stakeholders engaging in issue formulation and monitoring, negotiation, conflict resolution, learning, agreement, creating and maintaining public goods, concertation of action) among multiple, inter-dependent, stakeholders in the water catchment. We describe this overall set of interactions when it occurs in a complex natural resource arena as social learning. I will say more about this in Part III.[16] In taking this perspective we have made a choice to perceive 'ecosystems' as bounded by the conceptualisations and judgements of humans, as are agreements to what constitutes an improvement. This contrasts with the mainstream position which has come to see ecosystems as having an existence of their own totally independent of human engagement.

8.6 A Case Study: Aspects of Juggling the M-Ball

Additional challenges to be faced in managing the M-ball include allowing for the emergence of new insights from the use of systems methods in their entirety (i.e. as conceptualised by their developers) as opposed to just picking and using parts of them. Based on experience I would argue for attempting them in their entirety first and until they begin to feel familiar or embodied. Another challenge, as I have

---

[15] However, this is not magic and involves a lot of work by many people and excludes as many (such as those without internet access) as it potentially includes.

[16] See also Blackmore [1] whose book deals more with multi-stakeholder processes, communities of practice and social learning.

mentioned already, is understanding that all practice is in a context or environment and that what also needs to be managed is the practice-context relationships. The final Reading in Part II (Reading 6) exemplifies a systems practitioner juggling the M-ball; he addresses in part the two challenges I have just raised. He does this through reflecting on 25 years of systems practice primarily, but not exclusively, within the Systems Dynamics (SD) tradition in the business and government sectors. Importantly his reflection exemplifies how juggling the M-ball extends beyond a simple focus on systems methods or technique, combining all the elements of the B, E, C and M-balls. In particular he explores the relationships between his systems practice and generic aspects of 'project management' which is a major issue I will address in Part III of this book. The overwhelming lesson, he claims, is that the quality of the work is secondary to the manner in which projects are managed within organisations.

Reading 6

Reflections on SD Practice [12]

Tim Haslett

Introduction

This paper reflects on a body of work commissioned by commercial clients. While a number of academic papers have been published as a result, this work has essentially been pragmatic and driven by the fundamental consideration of the usability of the models that were developed. The process of making models usable is deeply embedded in the political and organizational processes of the client. This means that reflection on SD [Systems Dynamics] practice is about processes much broader than the process of model building itself.

The clients

The client group has been wide and diverse. The projects have included modelling the capabilities of the Joint Strike Fighter, the use of golf courses, restrictions to blood donor groups, the national superannuation system, the heroin trade, ambulance service demand, patient flows in hospitals, privatized garbage collection, the viability of WorkCover insurers, a call centre, oil and delivery transport systems, case management in the courts and an accident repair centre.

SD is inherently difficult to understand and the explanations more difficult

The theoretical foundations of System Dynamics are at once the greatest strength and the greatest weakness for our discipline. The strength is that we have very sure foundation upon which to build and proceed with our work. In this, it is always appropriate that the practitioner should acknowledge the

(continued)

Reading 6 (continued)

seminal work of Jay Forrester. We are also fortunate that the body of literature, particularly that produced in the System Dynamics Review, has helped to define the methodology and the field of endeavor.

The downside of this however is that the body of theory that is needed to understand the discipline is relatively complex. For instance, the concept of feedback is commonly interpreted as meaning what people receive on their performance appraisals. Engineers understand feedback and control systems but when dealing with other professional groups this assumption cannot be taken for granted. When this idea is complicated with the addition that feedback is either reinforcing or balancing and that the systemic impact of both of these types of systems need to be considered, clients begin to view the consultant with the degree of suspicion reserved for door-to-door vacuum salesmen and snake charmers. While the feedback concept is relatively easy to explain to an interested audience, the implications of its importance are not as easy.

The great power to be able to analyze systems in terms of positive and negative feedback systems rests in the fact that the results of the analysis are often counterintuitive and provide insights that have not previously been possible. The difficulty is that the results of SD analysis can often be at odds with the thinking of managers and decision-makers whose mental processes are essentially linear. Linear thinking is characterized by an assumption that an intervention in a system will have a chain reaction effect uncomplicated by a feedback from unintended consequences.

Our failure to help a client understand these and other principles of SD modelling can have a number of important consequences both at the beginning and the end of the project. A recent presentation to a senior parliamentarian set out the preliminary stages of a large project. As the presentation progresses, signs of glazed eyes and a decreasing attention span became increasingly obvious. The explanation was clearly not getting through and as a consequence of this one senior decision maker's confidence in the technique was seriously weakened. Within a fortnight, the project had been cancelled. Attributing causation, particular when the causal structures are not well known, is always dangerous. However, it is likely that our failure to communicate clearly what we were able to do was a contributing factor into the cancellation of the project.

The problem of counter-intuitive outcomes

The more frustrating aspect of the failure to help the client to understand what SD will deliver is when the consequences of the lack of understanding become obvious at the end of the project. Clients can be swept up by enthusiasm for a technique that has the potential to make the implications of policy decisions clear, but this is often supported by the hope that it will provide a justification for current policy. As the project progresses and the causal structures become clearer, it becomes clear that counterintuitive outcomes are not

(continued)

Reading 6 (continued)

going to provide support for current policy. Often the project will be completed, only to be shelved.

A major motor services company that had developed an Accident Repair Centre (ARC) designed to act like a production line had encountered problems with bottlenecks developing within the system. The problem was multifaceted and involved the manner in which the work was scheduled and the availability of spare parts. The spare parts problem led to cars being offloaded from the production line to the extent that the whole system became grid locked. The client required a simulation model to see how the problem could be solved. The complexity of the model is an indication of the complexity of the system that had been developed. It was little wonder that it was unmanageable. The most noticeable complexity was the complicated sets of conditions that had to be met before cars could move from one station to another.

[...]

The results of modelling can best be captured in the lines of the immortal poet Hillaire Belloc:

> The Chief Defect of Henry King
> Was chewing little bits of String.
> At last he swallowed some which tied
> Itself in ugly Knots inside.

> Physicians of the Utmost Fame
> Were called at once; but when they came
> They answered, as they took their Fees,
> There is no Cure for this Disease.
> Henry will very soon be dead."

The design of the ARC which had cost millions of dollars to build and was to revolutionize car repair in Victoria [Australia] was so fundamentally flawed that it was never going to work properly. This was not what the client had expected. They had hoped that the problem would be solved, not proved to be insoluble. Within a short time, the client closed the ARC and began building a new one.

The tendency of SD modelling to produce counterintuitive outcomes produces significant difficulties for some clients. Given that many organizations do not have an inbuilt capability for Systems Thinking and consequently an understanding of the impact of feedback, it is highly likely that some of the results of modelling exercises will run contrary to the accepted wisdom within the organization.

Some years ago work with a provider of major emergency ambulance services wished to understand the drivers of demand for emergency ambulance services. The funding model was that ambulances were purchased when

(continued)

Reading 6 (continued)

demand went above current capacity and the projected utilization of the new vehicles demonstrated the demand would be sustained.

The purchase of the new ambulance had the immediate effect that the ambulance service was under pressure to use the increased capacity. This was often achieved by using expensive and well-equipped emergency ambulances for tasks such as ferrying patients between hospitals, such as between acute and sub-acute facilities to ensure utilization of the vehicle. The consequences of this were that the specialized vehicles took out some of the workload of less specialized vehicles that then had to look for some way to utilize their excess capacity. This meant there was a knock on effect down through the ambulance services with everybody looking for extra work. This was often achieved by providing a much wider range of services to people in the communities. Anecdotal evidence existed of emergency ambulances being used to take patients to appointments at outpatients units. Very soon the demand for these services increased so that it was necessary to purchase new ambulances in a variety of categories where demand had been stimulated. These new ambulances then went out looking for work (Fig. 8.4).

The conclusion was that one of the key drivers of this demand for emergency services is the provision of an increasing range of services from the emergency service itself. Demand was endogenously driven. This may not surprise the experienced modeller but it certainly came as a shock for the service and provided very little help in their budget submissions.

The first difficulty in this situation is that the model has produced a counterintuitive outcome for the client. This outcome was based on the well-

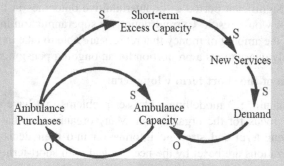

Fig. 8.4 Endogenous ambulance demand. An 'S' indicates that the causal variable has the Same effect on the dependent variable. As New Services go up then demand for those services also goes up. An 'O' indicates that the causal variable has the Opposite effect on the dependent variable. If Demand goes up then Ambulance Capacity goes down

(continued)

Reading 6 (continued)

understood principle that systems generate their own behavior and it was to be expected that the client would have had a major impact on the dynamics of the system. What emerges is the question: "So what do we do now?" The situation was clearly not impossible but was going to require a major policy and behavioral changes for the client and other major stakeholders. It's very easy for these situations to fall into the "too high basket".

Sometimes the opposite happens. A recent project involved in the modelling of the impact of compulsory superannuation in Australia. The work that was conducted in the ATO [Australian Tax Office] Superannuation Branch was very successful for a number of reasons. The first was the ability of the client group to accept the possibility of the counterintuitive. One of the critical dynamics for the introduction of compulsory superannuation was that it takes 30 to 40 years before people are retiring with maximum benefits. For significant periods of time, people would be leaving the workforce with relatively small amounts invested in superannuation. The question at issue was: "What will these people do in this situation?"

The consensus was that people would make additional payments into superannuation to increase their final benefit. However, it was agreed that there was no empirical evidence to support what seemed to be a fairly logical conclusion. The possibility of this assumption being wrong led to a significant investment in market research. The results of research justified the investment. Not only was it extremely unlikely that people would top up their superannuation, it was extremely likely they would take their superannuation as a lump sum and spend it on a caravan or trip around the world and then go on the pension. As one of the main thrusts of the superannuation policy was to make people less dependent upon government funded pensions, it looked likely that a significant period of counterintuitive outcomes in the superannuation sector would result. Current policy for superannuation in Australian now restricts the amount of money that retirees are able to take in a lump sum and effectively quarantines a proportion for an ongoing pension.

The problem of the short term v long term

By its very nature, SD modelling addresses problems that have a long-term and strategic focus for the organization. Many organizations have difficulty integrating long-term and strategic information into their decision-making. Many organizations are beset by the need to deal with short-term key performance indicators and the immediate effects of a turbulent environment. Some organizations find themselves constantly in "semi-crisis" mode. This is particularly true of organizations where resources are stretched to meet the demand of what appears to be an unpredictable environment. Health services

(continued)

Reading 6 (continued)

appear to operate in this mode for a good proportion of their time. A recent modelling assignment from a major hospital was commissioned as a result of a number of the senior administrators completing SD programs at Monash. The discussions to develop a long-term modelling capability to predict and manage demand for hospital beds was frequently interrupted by quarter-hourly updates on the status of individual patients and a frequent absence of the manager to head off the latest crisis in bed allocation. Unfortunately, stamping out spot fires does not stop a major bushfire. While the immediate and practical reality is that the spot fires need to be fought, the difficulty is that dealing with the causes of the spot fires needs to be done at the same time as trying to manage them. This dilemma is most acute in organizations that become "addicted to crisis". Managers become adept at dealing with the day-to-day crises of the organization. Often their efforts are nothing short of heroic and frequently lifesaving. At a personal level, the hero manager has nothing to gain from improving the structures that provide opportunities for managerial heroism.

A senior medical administrator accounted with great pride how she had spent an entire Saturday searching one of our major cities for a compatible blood donor for a dying child. She finally located the child's aunt and brought her in to make a donation that saved the child's life. It is impossible to question the efficacy or importance of what was done in that situation. But when the administrator announced with pride that she was the only person who had the power and authority to do this, two questions inevitably arises: "But what if you weren't there? and What systems do you have in place to deal with that?"

It is true that structures determined system behavior. It is equally true that the heroic efforts of individuals can overcome the influence of counter-productive structures in the short term. It is the heroism of individuals that often stands in the way of long-term strategic thinking. Structural change is time consuming, costly and risky. Often SD modelling indicates the need for significant structural change but when the immediate problems demand attention, managers rarely have time to make the necessary change.

A recent assignment in a major Victorian Hospital was able to deal with this problem in a different way. The hospital was building a short-stay facility and wished to know whether the plans that had been drawn up would meet the performance criteria that have been set by the government. The critical dynamic in this case was that between the operating theatres, which generate patients, and the post-anaesthetic care units (PACU) where they recover. The new facility with its improved procedures was going to make significant improvements on patient throughput. While it is possible to bring about significant improvements in the processes surrounding the operating theatres, it is not possible to get patients to improve their post-anaesthetic recovery times. The question was one of operating theatre capacity with PACU capability.

(continued)

Reading 6 (continued)

If PACU reaches capacity, the hospital staff had no options for moving patients. They can only leave PACU when they are fully recovered and walking. Once PACU is full and there is no recovery space available, the operating theatres must stop. Failure to keep the operating theatres working efficiently has a highly detrimental effect on KPIs (Key Performance Indicators). This meant that PACU capacity was a key leverage point in the system.

The model simulations indicated that the PACU capacity in the original plans would not be sufficient to cope with all the services that will be taken to the new facility. A decision was made to limit the procedures that would be carried out in the new facility and bring the output of the operating theatres within the PACU capacity.

This application of SD modelling was interesting for a number of reasons. It tested the assumptions of the design of a facility in a way that allowed the managers to short-circuit potentially problematic situations before they developed. It also combined architectural plans and SD modelling, certainly for the first time in the Victorian health system.

Internal sponsorship and the rich and famous

SD modelling projects are generally initiated by someone within the organization who already has a working knowledge of the capabilities of the technique. In many cases, this has been a student who has studied at Monash University. It is very important from the perspective of a consultant to have a very realistic view of the organizational status of the project initiator. The key is to understand whether their budgetary discretion encompasses the cost of the project. If this is not the case, then there needs to be a series of presentations to convince someone higher in the organization to approve the project. It is important to have a clear understanding of the roles of the people now potentially involved in the project. The person who approves the budget now becomes the project sponsor within the organization and this role requires careful management on the part of a consultant. The same applies to the person who originally introduced the project to the organization who is likely to become the internal project manager.

This is where the concept of "rich and famous" becomes important. In every consulting assignment, someone must become rich and famous. It is important to have a very clear understanding of who those people will be. It is the project sponsor, the approver of budgets, who must become pre-eminently famous. The consultant becomes rich but never famous. To achieve their fame, the project sponsor must be in a position to demonstrate the progress, efficacy or importance of the project. They must be brought up to speed about the nature the SD modelling and provided with enough information to be able to speak formally and informally about the benefits of the project. In doing,

(continued)

Reading 6 (continued)

this they become famous. The consultant takes no role in representing the project in the organization. The emphasis on using the project sponsor and project manager to represent the project within the organization means that the emphasis is constantly on relating the progress and outcomes of the project to organizational objectives. It also enables the project manager and project sponsor to take ownership of the project.

The communication processes surrounding the project are as important as the project itself and structures need to be put in place to ensure that they work well. Stafford Beer's "Viable Systems Model" provides an excellent template for structuring not only the project, but also the communication processes that must support it (Fig. 8.5).

The project team is Beer's System 1. This is where the work of building causal diagrams and modelling is done. This group will clearly contain the project manager and if possible some members of the organization who have demonstrated enthusiasm for developing SD mapping and modelling skills. It is from this group that a replacement for the project manager must be found

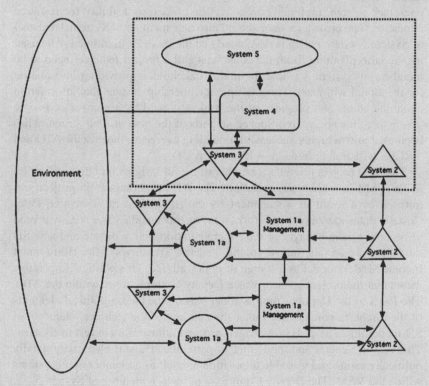

Fig. 8.5 Stafford Beer's Viable Systems Model

(continued)

if the project manager moves to a new role. Developing requisite variety at all levels of the project serves as a protection against the loss of expertise as a result of staff movement.

The project manager will be effectively a one person System 2 responsible for coordinating System 1. The project manager and the project sponsor will be members of System 3 which is responsible for maintaining the stability of the project, interpreting policy decisions from system five, allocating resources to System 1 and carrying out audits of progress. System 4 is the intelligence gathering function that will consist of the various stakeholders who have input to the modelling process. System 5 has a policy development role and will ultimately be the user of the models.

Once the VSM has been established within the client organization, it is necessary to establish the membership of the five systems. The critical aspect of the membership of the systems is that membership should be multiple and overlapping. The project manager will need to be a member of System 5. Their expertise and knowledge of SD modelling will allow them to act as mentor to the modelling acolytes. They will also need to be a member of a least one other system, preferably System 3 that has responsibility for resource allocation. The project sponsor should also be a member of System 3 and also of System 5 where policy is developed and the model will ultimately be used for scenario planning. Both the consultant and a project manager need to be members of System 4 where the major stakeholders providing information for the model will meet. This multiple membership ensures that information about the status and progress of the project is held by a number of people giving them the ability to influence all parts of the system. This structure has been used in two highly successful modelling exercises, those at the ATO and ARCBS (Australian Red Cross Blood Service).

The ATO project provides a number of useful insights into the application of the VSM in a client organization. The ATO superannuation project was initiated as a result of a statement by the [then] Federal Treasurer, Peter Costello that superannuation reform was "on the agenda". The ATO was keen to become a major player in the superannuation policy debate and saw SD modelling as an alternative to the Federal Treasury's RIM (Retirement Income Model) model as a source of policy advice. This political imperative meant that the project sponsor was a Deputy Commissioner within the ATO. The focus of the Deputy Commissioner was on the macro-political elements of the Canberra bureaucracy rather than on any of the technical elements of SD modelling. The project manager was an enthusiastic convert to Systems Thinking in general and modelling in particular. He was also exceptionally politically astute and was able to position himself as a member of all systems within the VSM. The Deputy Commissioner was a member of System 5.

(continued)

Reading 6 (continued)

System 4 was known as the High Level Modelling Team and consisted of stakeholders from ATO offices across Australia. In addition to this wide exposure that the project had through System 4, the ATO also ran a series of workshops on Systems Thinking and SD modelling techniques. It was from the participants in these workshops that the members of the System 5 modelling team were selected.

In addition to this significant level of organizational support, the modelling team also had a dedicated modelling room in the ATO offices. As the causal diagrams for the superannuation scheme were developed, they were displayed on the walls of the modelling room. ATO staff were free to come and look at the work and a number of presentations to external bodies were conducted in the modelling room. This enabled members of the modelling team to become moderately famous. The project leader became rather more famous and made presentations on the work to other departments including one to the Assistant Treasurer. The project sponsor presented the work to a number of external industry bodies and later left the ATO to work for a major industry organization. Fame of course comes at a price. The project ended when Treasury offered all of System 5 and the project leader jobs in Treasury. All accepted. The move of these key players out of ATO Superannuation branch effectively removed any policy capability from the branch.

The counterintuitive outcome of the ATO's desire to develop its policy capability in superannuation was that its success led all the key players to move on to policy roles within Treasury. The deputy commissioner had moved to work with a superannuation industry body. Everyone was famous and the consultant was slightly rich.[17]

The pragmatics of SD modelling

As an academic and a teacher of SD, it is possible, and indeed necessary, to involve oneself in the more arcane aspects of SD modelling. However, the concerns of the academic community are rarely reflected in the concerns of the business and government sectors. Their concerns focus around outcomes, answers and action. The technical complexity of SD modelling opens it up to the accusation of being "academic", which is one of the more pejorative terms used in the business world. This outcomes, answers and action focus was shown in an exercise modelling the patient flow in a subacute hospital facility. The project was stopped after the development of the causal diagrams and the early development of the simulation model. The project team believed that they now under-

---

[17] But an unintended consequences was that the capability to do SD modelling was lost – those who moved to Treasury soon discovered that SD would never displace the extant modelling approaches already embedded there. Hence all SD practice capacity was soon lost.

(continued)

Reading 6 (continued)

stood the problem sufficiently to be able to take effective action. The change program that arose from the modelling exercise reduced the patient stay time in the facility from 35 days to 17 with no change in level of patient care.[18]

This experience neatly defines the role that our work plays in organizations. At best, we can be a catalyst for change by providing information for the decision-makers who must ultimately take responsibility for the change. We can provide little information about the way to initiate, implement or evaluate the change. The criteria for our work must be that we are able to provide information that allows responsible managers to make informed choices about the changes in their organization.

Haslett, T., (2007) 'Reflections on SD Practice, Systemic Development: Local solutions in a global environment. James Sheffield (ed.), ISCE Publishing.

In choosing this reflection by Tim Haslett as a reading, I am perhaps open to criticism for ending my account of the four balls juggled by the systems practitioner on a sombre note. However, if we are to employ systems approaches as a way of acting in a climate-change world, then Tim's account points to where future attention and action is required. One of my reasons for selecting this reading was that his account highlights how inimical institutional settings are to the establishment and conservation of systems thinking and practice. A particular constraint is the projectified world we inhabit! Without changes in these institutional settings it is unlikely that systems thinking will be able to make the contribution that it could to the quality of our co-evolutionary future.

In Part III, I examine four major constraints to the emergence of systems practice into the mainstream. I then look at three conducive meta-framings for systems practice: systemic inquiry, my own antidote to living in a projectified world in which, historically, certainty has been privileged above uncertainty, systemic action research and systemic intervention.

References

1. Blackmore, C.P. ed. (2009) Social Learning Systems and Communities of Practice. Springer: London.
2. Briggs, L. (2009) Delivering performance and accountability – intersections with 'wicked problems.' The International Society for Systems Sciences annual conference, conference paper. Brisbane, Australia, July 2009.
3. Chapman, J. (2002) System Failure. Why governments must learn to think differently. Demos: London.

---

[18] Tim's experience is not unique in this regard – within the SD community of practitioners qualitative SD practice, based on conceptual modelling rather than conceptual + quantitative modelling is now a recognised form of SD practice; as with Donella Meadows and other SD practitioners 'systems' in this reading are referred to in ways that grant them an ontological status.

4. Checkland, P.B. (1999) Soft Systems Methodology: A 30 Year Retrospective. Wiley: Chichester.
5. Checkland, P. B. and Poulter, J. (2006) Learning for Action: a Short Definitive Account of Soft Systems Methodology and its Use for Practitioners, Teachers and Students. Wiley: Chichester.
6. Cooperrider, D.L. (1998) What is appreciative inquiry? In S.A. Hammond and C. Royal (Eds.), Lessons from the Field. Applying Appreciative Inquiry. Berrett-Koehler: San Francisco, CA.
7. Fairtlough, G. (2007) The Three Ways of Getting Things Done. Hierarchy, Heterarchy and Responsible Autonomy. Triarchy Press: Axminster.
8. Gittins, R. (2009) Tax inquiry reveals economists at their most clueless. The Age, Monday June 29, 2009.
9. Goleman, Daniel (1997) Emotional Intelligence. Why it can matter more than IQ. Bantam Books: New York.
10. Goleman, D. (2009) Ecological Intelligence: How Knowing the Hidden Impacts of What We Buy Can Change Everything. Broadway Books: New York.
11. Government Centre for Information Systems (GCTA) (1993) Applying Soft Systems Methodology to an SSADM Feasibility Study. CCTA: Millbank.
12. Haslett, T. (2007) Reflections on SD Practice. Proceedings of the 13th ANZSYS Conference, plenary paper. Auckland, New Zealand, 2–5 December 2007.
13. Ison R.L., Röling, N. and Watson, D. (2007) Challenges to science and society in the sustainable management and use of water: investigating the role of social learning. Environmental Science & Policy 10 (6) 499–511.
14. Ison, R.L. (2008) Systems thinking and practice for action research. In P. Reason and H. Bradbury (Eds.), The Sage Handbook of Action Research Participative Inquiry and Practice (2nd edn) (pp. 139–158). Sage Publications: London.
15. Ison, R.L. and Russell, D.B. eds (2000) Agricultural Extension and Rural Development: Breaking Out of Traditions. Cambridge University Press: Cambridge, UK. 239p.
16. Mintzberg, H. (2004) Managers Not MBAs. A hard look at the soft practice of managing and management development. Prentice Hall: London.
17. Morris, R.M., Roggero, P.P. Steyaert, P., van Slobbe, E. and Watson, D. (2004) Ecological constraints and management practices in social learning for integrated catchment management and sustainable use of water. SLIM Thematic Paper 1, (available at http://slim.open.ac.uk).
18. Mulgan, G. (2001) Systems Thinking and the Practice of Government. Systemist 23, 22–28.
19. Naughton, J. (1998) Internet – It's Free and it Works. No Wonder Bill Gates Hates it. The Observer, 8 November 1998. Accessible on the website at http://molly.open.ac.uk/personal-pages/pubs/981108.htm
20. Oliver, C. (2005) Reflexive Inquiry. A framework for consultancy practice. Karnac Books: London.
21. Peck, M.S. (1978) The Road Less Travelled. Simon and Schuster: New York.
22. Raymond, E. http://sagan.earthspace.net/esr/writings/cathedral-bazaar/
23. Robertson, D. (1998) What employers really, really want. Presentation to the Society for Research into Higher Education.
24. Seddon, J. (2007) Systems Thinking in the Public Sector: the failure of the reform regime... and a manifesto for a better way. Triarchy Press: Axminster.
25. Shaw, P. (1996) Intervening in the shadow systems of organisations: consulting from a complexity perspective. Complexity and Management Papers No 4, Complexity and Management Centre, University of Hertfordshire.
26. Stacey, R. (1993) Strategy as order emerging from chaos. Long Range Planning 26, 10–17.
27. Steyaert, P. and Jiggins, J. (2007) Governance of complex environmental situations through social learning: a synthesis of SLIM's lessons for research, policy and practice. Environmental Science & Policy 10 (6), 575–586.
28. Vickers, G. (1978) Responsibility – its Sources and Limits. Intersystems Publications: Seaside, CA.
29. Winter, M.C. (2002) Management, Managing and Managing Projects: Towards an extended soft systems methodology, PhD Thesis, University of Lancaster, UK.
30. Winter, M.C. and Szczepanek, T. (2009) Images of Projects. Gower: London.

Part III
Systemic Practices

Chapter 9
Four Settings That Constrain Systems Practice

9.1 Juggling Practice and Context

The systems practitioner as juggler introduced in Part II was created as an ideal type. One of my main aspirations for the juggler isophor was that through its use I could convey a sense of possibility for engaging in, or further refining your own, systems practice. Underpinning my conception of the isophor is the view that anyone can engage in systems practice and that it is, in general, a competence worth having, not only at a personal level but at a societal level as well. At the end of Part II, I broached the subject of whether it was enough to develop one's juggling (systems practice) in isolation from the contemporary contexts in which systems practice seems needed, yet is not flourishing? Hence, in Part III I first want to explore four pervasive institutional settings which I consider inimical to the flourishing of systems practice and which are also unhelpful in preparing us for our living in a climate changing world (Chapter 9). In each of the remaining three chapters of Part III (Chapters 10–12) I will introduce a way of framing systems practice with potential to address one or more of the inimical settings described in this chapter. I contend that each of these three framings for systems practice has relevance in our current circumstances.[1] My 'framing' choices are not the only viable options for fostering future systems practice but highlight some of the different issues at stake.[2]

When I wrote about a systems practitioner juggling the M-ball in Chapter 8, I described it as involving managing the ongoing juggler – context relationship which I understand as a co-evolutionary dynamic (Chapter 1).[3] In my experience

---

[1] I find it difficult to find the right language to express what I mean at this point – I choose 'framings' in the sense that each of the three I have chosen can be understood as incipient social technologies.

[2] For example I could have chosen 'systemic development' as explicated by Bawden [4]. One reason I did not is that material based on the so-called 'Hawkesbury' tradition has been included in Blackmore [7] which, like this book, also doubles as a set book for the OU course 'Managing systemic change: inquiry, action and interaction' (code TU812). Another option could have been the Imagine Methodology (see [6]).

[3] A juggler-context relationship could be understood as exemplifying the structural coupling of an organism with its milieu or environment.

R. Ison, *Systems Practice: How to Act in a Climate-Change World*,
DOI 10.1007/978-1-84996-125-7_9, © The Open University 2010.
Published in Association with Springer-Verlag London Limited

systems practice which only focuses on methods, tools and techniques is ultimately limited in effectiveness. This is particularly so at this historical moment because the environment, or context, is generally not conducive to enacting systems practice. For example, in Chapter 1 I referred to the widespread desire for certainty which, in my view, is inimical to the development and expansion of systems practice capacity. The desire for certainty is both an attitude or an emotion but also a demand that has been built into our institutional arrangements through devices like 'deliverables', 'key performance measures' or 'goals'. Thus, to be truly effective in one's systems practice it may mean that changes have to be made in both practice and context. This is not easy – but it is not something that has to be done alone. Many people now recognise the limitations of our current modes of being and doing. Appreciating this opens up two opportunities: (i) to spot those contexts which favour the initiation and development of systems practice and (ii) networking with others to bring about changes in context so that more conducive conditions for effective systems practice are created.

In my own professional life I have done my fair share of attempting to spot favourable contexts for usefully developing and deploying systems practice.[4] Some have been more successful than others. My overall experience is, however, that working with others to change the context for future systems practice, particularly some of the institutional settings, is urgently needed. In this chapter the four contemporary settings that constrain the emergence of systems practice that I want to address are:

1. The pervasive target mentality that has arisen in many countries and contexts
2. Living in a 'projectified world'
3. 'Situation framing' failure
4. An apartheid of the emotions

My argument put simply is that the proliferation of *targets* and *the project* as social technologies (or institutional arrangements) undermines our collective ability to engage with uncertainty and manage our own co-evolutionary dynamic with the biosphere and with each other. This is exacerbated by an insitutionalised failure to realise that we have choices that can be made as to how to *frame situations*. And that the framing choices we make, or do not make, have consequences. In doing what we do we are also constrained by the institutionalisation of an intellectual apartheid in which *appreciation and understanding of the emotions* is cut off from practical action and daily discourse. In this chapter I outline the basis for my claims and mount arguments about the need for 'antidotes'.[5]

In Chapter 10, I exemplify my own attempt, assisted by others, to invent a new social technology, systemic inquiry, as an antidote to targets and the projectified world. In Chapter 11, I introduce systemic action research, an approach to research practice in which I and others incorporate understandings of human emotioning. In keeping with the spirit of this book of introducing other voices, Chapter 12 is devoted to 'systemic intervention' an approach to practice that makes it clearer how

---

[4] As an example see Ison and Armson [19].

[5] The word antidote means to 'give as a remedy' – it is often linked to health matters as in 'an antidote for the poison'.

different systems methods and techniques might be successfully combined as part of systems practice. All three approaches demand awareness of the choices that can be made in relation to framing of situations.

9.2 Managing Systemic Failure – the Travesty of Targets

There are seemingly no shortage of situations in which failing to think and act systemically leads to breakdown or some form of 'failure' [14, 15], either of policy or practice. Several commentators have, for example, described the GFC (Global Financial Crisis) as exemplifying systemic failure. By its very nature there are many factors that give rise to any case of systemic failure. However, when choosing to describe situations as if they were systemic failures there has been a tendency, in recent years, for Government Ministers, bureaucrats, CEOs etc. to use the term 'systemic failure' as a means to abrogate responsibility. Whilst the claims these people make may have some validity it is, from my perspective, unacceptable for such claims to be made without some accompanying appreciation of how the systemic failure came to happen! By so doing these commentators fail to recognise that they are a participant and the opportunity (or part of the opportunity) to shift the dynamic may reside in their jurisdiction, hence there is an ethical responsibility, i.e. a higher calling to respond! When the explanation 'systemic failure' is not accompanied by a commitment and openness to inquiry into the circumstances of the 'failure' then an abrogation of responsibility occurs. Responsibility is denied when those involved are not open to learning and change.[6] Otherwise the claim to systemic failure is no better than an attempt to explain something away through a sort of magic!

In my first example I want to address the failings of what some have called the 'targets culture', a culture that has become endemic in the British New Labour government as well as widespread in other areas of government and corporate life. I have come to understand this situation as exemplifying the privileging of systematic approaches over systemic, sometimes at considerable social cost. Take for example the case described by Simon Caulkin [8] in Reading 7.[7]

---

[6]Russell Ackoff [1] argues that we cannot learn from doing anything right. In exploring why organisations fail to adopt systems thinking he points to a general phenomenon – that of failing to embrace and learn from mistakes. He cites August Busch III, then CEO of Anheuser Busch Companies, who told his assembled vice presidents, "if you didn't make a serious mistake last year you probably didn't do your job because you didn't try anything new. There is nothing wrong in making a mistake, but if you ever make the same mistake twice you probably won't be here the next year". Of particular concern is organisations and individuals who transfer responsibility for their mistakes to others thus avoiding learning.

[7]This reading concerns situations connected with the UK's National Health Service (NHS), a very large and complex organisation. MRSA, or *methicillin-resistant Staphylococcus aureus*, is a bacterium responsible for difficult-to-treat infections in humans (see http://en.wikipedia.org/wiki/Methicillin-resistant_Staphylococcus_aureus. Baby P, (also known as "Child A" and "Baby Peter") was a 17-month old boy who died in London in 2007 'after suffering more than 50 injuries over an 8-month period, during which he was repeatedly seen by social services' (Source: http://en.wikipedia.org/wiki/Death_of_Baby_P).

Reading 7

This isn't an abstract problem. Targets can kill.

Simon Caulkin[8]

MRSA, Baby P, now Stafford hospital. The Health Commission's finding last week that pursuing targets to the detriment of patient care may have caused the deaths of 400 people at Stafford between 2005 and 2008 simply confirms what we already know. Put abstractly, targets distort judgment, disenfranchise professionals and wreck morale. Put concretely, in services where lives are at stake – as in the NHS or child protection – targets kill.

There is no need for an inquiry into the conduct of managers of Mid Staffordshire NHS Foundation Trust, as promised by Alan Johnson, the health secretary, because contrary to official pronouncements, it is exceptional only in the degree and gravity of its consequences. How much more evidence do we need?

Stafford may be an extreme case; but even where targets don't kill, they have similarly destructive effects right across the public sector. Targets make organisations stupid. Because they are a simplistic response to a complex issue, they have unintended and unwelcome consequences – often, as with MRSA or Stafford, that something essential but unspecified doesn't get done. So every target generates others to counter the perverse results of the first one. But then the system becomes unmanageable. The day the Stafford story broke last week, the Daily Telegraph ran the headline: "Whitehall targets damaged us, says Met chief", under which Sir Paul Stephenson complained that the targets regime produced a police culture in which everything was a priority.

Target-driven organisations are institutionally witless because they face the wrong way: towards ministers and target-setters, not customers or citizens. Accusing them of neglecting customers to focus on targets, as a report on Network Rail did just two weeks ago, is like berating cats for eating small birds. That's what they do. Just as inevitable is the spawning of ballooning bureaucracies to track performance and report it to inspectorates that administer what feels to teachers, doctors and social workers increasingly like a reign of fear.

If people experience services run on these lines as fragmented, bureaucratic and impersonal, that's not surprising, since that's what they are set up to be. Paul Hodgkin, the Sheffield GP who created NHS feedback website Patient Opinion (www.patientopinion.org.uk) notes that the health service has

---

[8]http://www.guardian.co.uk/business/2009/mar/22/policy

(continued)

been engineered to deliver abstract meta-goals such as four-hour waiting times in A&E and halving MRSA – which it does, sort of – but not individual care, which is what people actually experience. Consequently, even when targets are met, citizens detect no improvement. Hence the desperate and depressing ministerial calls for, in effect, new targets to make NHS staff show compassion and teachers teach interesting lessons.

Hodgkin is right: the system is back to front. Instead of force-fitting services to arbitrary targets (how comforting is hitting the MRSA target to the 50% who will still get it?), the place to start is determining what people want and then redesigning the work to meet it.

Local councils, police units and housing associations that have had the courage to ignore official guidance and adopt such a course routinely produce results that make a mockery of official targets – benefits calculated and paid in a week rather than two months, planning decisions delivered in 28 days, all housing repairs done when people want them. Counterintuitively, improving services in this way makes them cheaper, since it removes many centrally imposed activities that people don't want. Sadly, however, the potential benefits are rarely reaped in full because of the continuing need to tick bureaucratic boxes; and in the current climate of fear, chief executives are loath to boast of success built on a philosophy running directly counter to Whitehall orthodoxy.

The current target-, computer- and inspection-dominated regime for public services is inflexible, wasteful and harmful. But don't take my word for it: in the current issue of Academy of Management Perspectives, a heavyweight US journal, four professors charge that the benefits of goal-setting (i.e. targets) are greatly over-sold and the side-effects equally underestimated. Goal-setting gone wild, say the professors, contributed both to Enron and the present sub-prime disasters. Instead of being dispensed over the counter, targets should be treated "as a prescription-strength medication that requires careful dosing, consideration of harmful side effects, and close supervision".

They even propose a health warning: "Goals may cause systematic problems in organisations due to narrowed focus, increased risk-taking, unethical behaviour, inhibited learning, decreased co-operation, and decreased intrinsic motivation." As a glance at Stafford hospital would tell them, that's not the half of it.

Caulkin, S., 'This isn't an abstract problem. Targets can kill'. The Observer, Sunday 22 March 2009, Copyright Guardian News & Media Ltd 2009.

Simon Caulkin has been one of the leading critics of the UK government's target's mentality. From this article it is easy to appreciate the distorting effects of the woolly thinking associated with imposing common targets across diverse contexts and the ill-informed use of 'goal-oriented' thinking. This example is an archetypical, though shameful, case of dropping all of the balls of concern to a systems practitioner as juggler! It is in such situations where systemic inquiry can find a place – as an institutionalised form of practice – particularly as a replacement for, or complement to regulation, policy prescriptions (blueprints) and targets. As I outline in Chapter 10 traditional policy instruments are, in a climate changing world, increasingly blunt or totally inappropriate instruments because, once formulated, they generally:

1. Fail to be institutionalised in an adaptive manner that is open to revision as the situation evolves
2. Are easy to develop but much more difficult and expensive to monitor and police (i.e. the effectiveness of many regulations is often not known until after some form of breakdown in the situation where the regulations were designed to operate)[9]
3. Preclude context sensitive local design and the establishment of more effective measures of performance of a policy or practice in relation to a situation or issue of concern

How to institutionalise adaptive practice is a key but by no means simple issue because of the propensity for enacting institutions in manners that tend to simplify, or at least focus, their reward feedback on whatever is easy to "measure fairly and consistently" whether or not that is relevant to the work! Together these factors militate against the fostering and development of capability for systemic and adaptive governance.

Once engaged in a systemic inquiry process other factors begin to reveal themselves. Are, for example, the failures described in Reading 7 symptomatic of an even bigger issue? Goals or targets imposed in the way that Caulkin critiques, whether knowingly or not, perpetuate a command and control mentality. In many ways Caulkin's article epitomises the failure of the pervasive hierarchical model of getting things done in our organisational life as described by Gerard Fairtlough [11] – see also Chapter 8. With these distinctions in mind what is it that New Labour has sought to control? The espoused aim was to control the phenomena to which the targets were aimed, e.g. hospital waiting lists. However was this really the case?

Some other commentators have seen the 'targets mentality' from another perspective, claiming that:[10]

---

[9] And of course the monitoring and policing becomes a self conserving praxis... which blinds people to the indicators that mean something!

[10] Source: The ultimate turnaround from Labour, the dying Government. By abandoning targets, Labour is admitting the depth of its failure, says Philip Johnston [22].

Why were targets introduced? The Government [UK, Labour] would have you believe it was to drive up standards; but in reality they were a means of showing that Labour "cared". They were a political device. Whenever ministers were challenged about high levels of offending or poor levels of literacy they could say: "But we have a target to reduce it/increase it/scrap it, so we must be good." Targets were ostensibly introduced to hold the Government to account, but were used as a means of deflecting criticism.

If such a situation could be shown to be the case (good ethnographic research would probably be needed) then this situation might be better understood as a product of Caulkin's observation that increasingly many government 'organisations are institutionally witless because they face the wrong way: towards ministers and target-setters, not customers or citizens'. Elsewhere I have observed that it may be useless railing against politicians for what they do or do not do because they themselves are trapped in a structure determined context in which it is really only possible to do what they do.[11]

Illustration 9.1

---

[11] A structure determined system is a delicate concept to get across. It should not be confused with causal determinism or pre-determinism. A system (or thing) can only do that which it has an appropriate structure for. I can't fly, no matter how you poke me. Thus something's structure determines what is possible for it.

Geoffrey Vickers [39, 40], using an analogy of the lobster pot, speaks of traps in our thinking and doing that are of our own making. The concept of a trap has been found useful by generations of OU systems students because it invites a practice of thinking about traps in our own being as well as in the social technologies we have invented. Western style democracy in its current bureaucratic and administrative form may be such a trap but my second concern is with something that is less obvious but none-the-less pervasive in its effects.

9.3 The Consequences of Living in a Projectified World

9.3.1 Projectification

As soon as you think about it, it becomes patently obvious that we live in a projectified world. I can hardly remember a time when a project was not part of what I did – whether at school or throughout my professional life. The word project has its origins in the Latin *projectum*, 'something thrown forth' from which the current meaning of a plan, draft or scheme arises. It would seem that the meaning, now common across the world, of a project as a special assignment carried out by a person, initially a student, but now almost anyone, is first recorded in 1916 [2]. From that beginning I am not really sure how we came to live with projects in the manner that led Simon Bell and Stephen Morse [5] to speak of a 'projectified-world order'. Perhaps mass education carried forth the project into all walks of life? Whatever this history, my experience suggests it is no longer tenable in a climate changing world to have almost all that we do 'framed' by our invention of 'the project'.[12]

Bell and Morse [6] describe a project as 'defined activities carried out by defined people with a defined end point in mind at a defined cost and over a defined period of time' (p. 97). They go on to outline how 'projects are popular with those responsible for spending money' and 'embrace a targeted set of activities with a clear aim (and hence cost), and hence accountability [that] can be maximized.' This allows, they argue, limited time-horizons for spending the budget and the achievement of targets allow a long-term commitment to be circumvented or even negated altogether. This 'fits neatly into the short-term time-frames that politicians inhabit' they claim (p. 98).

---

[12]Take this book for instance – for me it is really an exercise in reification of an ongoing inquiry into what it means to be an effective systems practitioner. However my framing does not hold for staff in my University or at the publisher who see it as a project – with all that that entails re deadlines etc.

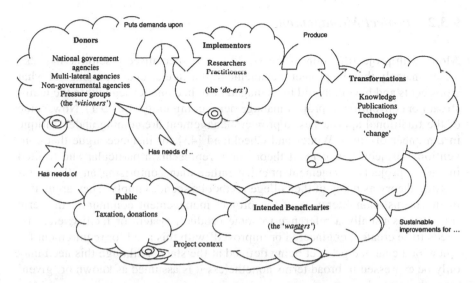

Fig. 9.1 The project process from the perspective of systems practitioners involved in doing sustainable development projects (Source: Bell and Morse [6] Figure 1, p. 98)

It is possible to capture the double sense of the meaning of *project* if one thinks of what we do when we *project our projects onto the world*. I have this image of shelves upon shelves, and now electronic files galore, of projects that have been labelled 'finished' and thus are hardly ever engaged with again. When I express my concerns in this way I imagine a grotesque expansion of the E-ball (discussed in Chapter 5) to the extent that our collective capacity to juggle the rest of the balls effectively and systemically is lost. In making this connection to the E-ball I am pointing out how the *project* has become one of the most pervasive of social technologies.

The contemporary project process has been characterised by Bell and Morse [6] in the form shown in Fig. 9.1. These authors describe Fig. 9.1 as the 'project ideal'. They note that 'while the diagram is circular in the sense that what the projects set out to do should have an impact in wider society, and society provides the funding, the circularity does not necessarily imply a continuation or longevity of the benefits that should accrue from the projects' existence' (p. 98). They express my own concerns and experience when they go on to say that their special concern is that there is rarely any lasting benefit to the situation from doing of the project. They observe that 'the projectified world order significantly fails to meet long-term needs and goals' (ibid).

Bell and Morse's experiences arise in the domain of sustainable development practice, primarily through 'development assistance' in poorer countries. What of other domains?

9.3.2 Project Management

Along with projectification of the world a new specialised discipline has also arisen, namely that of 'project management.' Particular understandings of what a project is and how it should be managed have become reified within the mainstream or conventional 'project management' community (Box 9.1). Now, some of the traditional approaches to project management are coming under critique in a number of areas. Winter and Checkland [43] for instance argue that conventional project management theory only represents a particular and limited image of project management practice rather than comprising an all-encompassing theory as many of the college textbooks seem to imply. They argue that in the mainstream literature of the project management community the term 'project' 'is usually a reference to some product, system or facility, etc. that needs to be created, engineered or improved' with this need 'or requirement for a new or a changed product being defined at the start.' Although this need may only be expressed in broad terms nonetheless it is assumed as known or 'given' from the outset.'

In their work, Winter and Checkland [43] seek to 'show that conventional project management theory embodies a particular way of seeing the practice, which is, simultaneously, a way of not seeing it.' For them this way of "seeing and not seeing is the paradigm of 'hard' systems thinking", which 'has been a prime influence on the development of project management ideas and practices over the last 40 years' (p. 188). The main characteristics of the 'hard systems' paradigm as reflected in the material in Box 9.1 are [43]:

Box 9.1 Examples of the 'Mainstream' Understanding of 'Projects' [43]

A project involves a group of people working to complete a particular end product, or to achieve a specific result, by a specified date, within a specified budget and to meet a specified standard of performance (quality) [23].

[A project is] an endeavour in which human, material and financial resources are organised in a novel way, to undertake a unique scope of work, of given specification, within constraints of cost and time, so as to achieve beneficial change defined by quantitative and qualitative objectives [38].

A project is a human activity that achieves a clear objective against a time scale [30].

A project is an endeavour to accomplish a specific objective through a unique set of interrelated tasks.... A project has a well-defined objective – an expected result or product. The objective of a project is usually defined in terms of scope, schedule, and cost [17].

- There is a clear objective or goal to be achieved, within some specified scope, schedule and cost
- Achieving the goal – the process dimension – is the primary task of project managemen
- The project is carried out through a sequence of stages as defined by the project life cycle, involving the application of various techniques such as critical path analysis, product-based planning and work breakdown structures[13]

The main thrust of this critique is that historically project management has been built on particular theoretical assumptions that have been found wanting or are no longer valid. In Winter and Checkland's [43] case they point an accusatory finger at hard systems thinking and approaches. In the language of earlier chapters the culprit is the overreliance on systematic, rather than systemic thinking. However, they are also at pains to say that both 'paradigms' are required. What is principally missing, they argue, is lack of awareness that the particular image of mainstream project management practice represents that conventional wisdom which the practice has itself generated. In turn the understandings upon which these practices have evolved have been reified into social technologies known as 'good project management'. In 'Images of Projects', Winter and Szczepanek [44] reject "outright the idea of a one 'best way' to view all projects and also the idea of following a prescriptive approach". They 'encourage a more pragmatic and reflective approach, based on deliberately seeing projects from multiple perspectives.'

I have my own experiences of project management practices that are more systematic than systemic and unsuited to the context in which they were employed. A particular example involved understandings that have been reified into project management procedures known as PRINCE2 as sponsored by the UK government. The acronym PRINCE stands for Projects IN Controlled Environments. The first PRINCE standard was published in 1990 and whilst it is subject to Crown Copyright it is available in the public domain. PRINCE2 was released in 1996 for use in more than just IT (information technology) projects [12]. My particular experience was based in the Environment Agency of England & Wales (EA), a large statutory body with about 12,000 employees, responsible for most aspects of environmental monitoring, regulation and compliance in England and Wales. At the time we were engaged in research related to the implementation of the European Water Framework Directive (WFD), an ambitious policy designed to improve the quality and ecological status of Europe's water in river basins over the period 2000 to 2027.

PRINCE2 was mandated for use within the UK civil service on projects above a certain size. PRINCE2, it was claimed, was needed in order to manage

---

[13] Winter and Checkland [43] also say that this core image of practice can also be seen in many of the college textbooks on project management and in many of the official sources of information about project management. It can also be seen operating in project management education and training programmes, and is generally the dominant image in much of the literature on project management, both academic and popular.

the complexity of the structure, including the complex inter-dependencies between individual projects and work packages that were employed in implementing the WFD. The establishment of these project management procedures was done with good intentions; without them WFD implementation would undoubtedly have come adrift very rapidly. In practice, however, this approach has not proved to be satisfactory. It was not satisfactory because, in implementation, PRINCE2 led to the systematic fragmentation of a very large and complex activity with the result that all involved lost sight of the whole. In addition no one was responsible for managing the connectivity between the different elements. On entering the organisation we encountered isolated individuals in seemingly discrete projects lacking in awareness of overall purpose and how what they were doing related to others.

The understandings on which PRINCE-type methods are built perpetuate and reproduce practices that privilege a 'technical rationality'. This rationality has been pervasive in organisations such as the EA, responsible for water policy and management. As I will explain below this rationality is not well suited to managing in situations of complexity and uncertainty [10, 20].

In other words we have arrived at a point where those who do project managing are not fully aware of what they do when they do what they do! Ironically this is largely due to the reification or projectification of project management itself. This has major implications for governance and, ultimately, how we respond in a climate change world.

9.3.3 Governance and the 'Project State'

In addition to the understandings of projects outlined in the previous two sections some practitioners conceive of projects as 'temporary organisations', though often embedded in or between permanent organisations. From the perspective of those who are trying to understand projects in a wider social setting, project proliferation is seen as a consequence of the shift from government to governance. It is argued [42] that 'the project has become a post-modern symbol of adaptability and contingency – it is thought of as a superior way of reacting to unforeseen and non-standard situations'. In part, it is argued, this has happened by opening up who participates in projects. Others argue it has arisen because of the moral weakness of the state.

Governance is a much broader idea than management, it encompasses the totality of mechanisms and instruments available for influencing social and organisational change, especially adaptation, in certain directions [13]. Sjöblom [35] claims that the shift from government, which many regard as associated with top down or command and control practices, to governance is 'one of the mega-trends in industrialised societies' (p. 9). Governance, as a concept, has of course the same origins as *cybernetics* (meaning steersman or helmsman). In practice it means adjusting to circumstances. Perhaps more significantly in the context of a climate-changing world the question becomes: how do we as a species chart a course within a rapidly changing co-evolutionary dynamic given we now live in a projectified world?[14]

Within these different discourses about the emergence of a 'projectified world' there are competing claims as to what a project enables or not. Thus some see them positively as a means to be more open to context – 'as mechanisms for joined-up governance with a horizontal approach to governing and organizing' [29, p. 67]. Yet others see them as part of a failed 'rationalistic dream' which creates a pervasive normative pressure on what it is we do under the rubric of 'a project' [35].[15] At its worst the project state has come to represent an 'unholy marriage between bureaucratic and managerialist rationalities, while pretending to privilege citizen engagement and direct participation in governance' [18].

What seems clear to me is that the pervasiveness of systematic thinking and practices associated with goals, targets and projects, what Winter and Checkland [43] call the 'hard systems paradigm', does not augur well for adapting in a climate change world. We need to invent something better.[16]

In the next section I offer an explanation as to why targets and projects have proliferated in society – that through the understandings and practices we have reified we collectively fail to realise that we have choices that can be made about the nature of situations.

9.4 Making Choices About Framing a Situation[17]

From the outset in this section I want to make it clear that in drawing attention to the choices we can make for framing situations I am not advocating engaging with situations with an *a priori* set of possible choices in mind. My primary concern is

---

[14] As noted by Bateson [3] systems and cybernetics 'can be a way of looking that cuts across fields, linking art and science and allowing us to move from a single organism to an ecosystem, from a forest to a university or a corporation, to recognise the essential recurrent patterns before taking action'.

[15] For further background see papers associated with the seminar 'Theory and Practice of Governance in the Project State', the Swedish School of Social Science at the University of Helsinki, October 2003.

[16] Winter and Checkland [43] propose the use of SSM as an alternative model for project conception and managing. They argue (p. 92) for: 'for a broader image of project management practice than that which has been dominant in the past.' They advocate a new perspective "with a focus on the process of 'managing', rather than the life-cycle process of 'project management', this new perspective seeks to enrich and enlarge the traditional life-cycle image of project management. It also offers to provide a new foundation for future research in the project management field."

[17] The work described in this section comes from a number of research situations, mainly in Europe in the period 2000–2009. A major component was work with the Environment Agency of England & Wales (EA) associated with implementation of the European Water Framework Directive. Over this period what I would once have referred to as 'research projects' or programmes were purposefully framed as systemic inquiries, although not all of them in contractual terms.

to privilege experience, understood as that which we distinguish in relation to ourselves, in a process in which one is as open to the circumstances as possible. This, as I have outlined in Chapter 5, involves attempting to be aware of the traditions of understanding out of which we think and act.

In the recent past my own concerns have been with situations associated with water, river and catchment managing. In recent history understanding and managing of rivers has been heavily influenced by hydrologists, engineers and physical geographers. In the past a river or a water catchment was rarely understood as if it were a human activity system. But having made this shift a river catchment or watershed exemplifies what some describe as a multi-stakeholder situation. But it is a multi-stakeholder situation of a particular type in that the connectivity, or lack of it, between humans and the biophysical dimensions are of critical importance. Thus some would choose to describe a catchment as a coupled socio-ecological system.[18] For those who are not aware there is a growing global water crisis that is manifest in similar yet specific ways in almost all countries. It is likely that in many areas climate change will make the current situation worse. Both globally and locally these situations have many or all the features of situations that others have described as wicked problems, messes or complex adaptive systems as described in Chapter 6.[19] Aware of this history and drawing on a literature associated with the 'framing' of such situations as 'resource dilemmas' [20, 34, 37] my colleagues and I now choose to characterise river catchments in terms depicted in Fig. 9.2 [36]. The figure draws attention to how we have selected a new lens with which to engage (understand) these situations. I elaborate on the terms that make up our new lens in Box 9.2.

In my experience there are many situations that could be usefully framed in the terms we now employ in relation to water catchments (Box 9.2). By useful I mean making them amenable to some form of action that leads to systemic improvement. My use of this framing is an example of how I juggle the E-ball. I do so with an appreciation of the history of the use of the terms 'interdependencies', 'complexity', 'uncertainty', 'controversy' and 'multiple stakeholding and/or perspectives' (Box 9.2) as well as that of 'messes', 'wicked problems' and the 'swamp of real life issues' as discussed in Chapter 6.

---

[18] In many ways a river catchment is no different to any business or other form of human activity in that they are, knowingly or not, coupled with a biophysical environment – it is just that in most circles this is not appreciated and all too often the environment is treated as an externality.

[19] Roux et al. [31] for example refer to social-ecological systems, as well as organisations, as complex systems. They go on to say that 'complex does not mean complicated. An engine is complicated. It is also predictable, at least by those who put it together. A complex system has particular properties that make it inherently unpredictable. Being able to recognise a system as complex allows one to better understand that system at least to the extent that one understands why, in a general sense it is the way it is. It is the unpredictability of such systems that has fundamental implications for their management.'

Fig. 9.2 Choices that can be made about the nature of a situation such as water governance and catchment management situations (Adapted from SLIM 2004 [36])

Box 9.2 Characterising Natural Resource Issues as Resource Dilemmas

1. Interdependencies [36]

The use of natural resources through one type of human activity affects ecological processes in ways that interact with other people's uses of natural resources, both across geographic and ecosystem boundaries and time scales. Integrated Catchment Managing and the sustainable use of water, for example, address interdependencies among:

Human activities, relative to:

- Their qualitative and quantitative effect on water
- Their water-related needs

Linked geographical areas:

- Such as upstream areas, lowland wetlands and estuaries
- Aquatic and terrestrial ecosystems

2. Complexity

Natural resources are under the influence of a complex mix of enmeshed natural, technical and social processes, including changes in public policy, organisations and a diversity of stakeholders, each with their own perceptions.

(continued)

Box 9.2 (continued)

When considering water as a resource for human uses as well as a part of nature, we are compelled to make the link between ecology and societal processes such as technological development, the market, public policies and interpersonal relations. Integrated Catchment Managing and the sustainable use of water operate within a set of interlinked and assorted elements that create a high level of complexity.

3. Uncertainty

The complexity of such circumstances makes them impossible to explain comprehensively and accurately, and the effects of proposed solutions cannot be forecast because of uncertainties. The realms of uncertainties are also diverse: Technical and ecological, regarding:

- The relationship between human activities and ecological processes
- Fragmented and sector-specific technical and scientific knowledge

Socio-economic, relative to:

- Market and consumer trends
- Changes in social demands
- The emergence of new sorts of crises
- The proliferation of institutional arrangements

Political, with respect to the increasing diversity and number of:

- Public policies generating contradictions
- Decision-making levels and organisations implementing these policies

4. Controversy

Uncertainty and interdependencies result in different perceptions and lasting disagreements on which issue is to be addressed. Controversies emerge from questioning the existence of problems, their origins, how cause-and-effect relations are understood, how they should be managed and by whom.

5. Multiple Stakeholders and/or Perspectives

In situations understood as 'resource dilemmas' there is likely to be a mix of people, each with multiple, partial views if a situation; some will have strong stakes (stakeholding) in what is at issue in the situation, others' stakes will be less well developed even though the implications or potential impacts may be equally great for both groups. Thus the nature of what is at issue and what constitutes an improvement is likely to be contested.

As I outlined earlier (Section 8.6), in my recent research with colleagues on systemic and adaptive governance of natural resource situations we have made a choice to understand sustainable and regenerated water catchments as the emergent property of social processes and not the intrinsic property of an eco-system [20]. That is, desirable water catchment properties arise out of interaction among multiple, inter-dependent, stakeholders in the water catchment as these stakeholders engage in issue formulation and monitoring, negotiation, conflict resolution, learning, agreement, creating and maintaining public goods, concertation of action. When it occurs in a complex natural resource arena we describe this overall set of interactions as *social learning* [37]. I will say more about this in Chapter 10. We have made a choice to perceive 'ecosystems' as bounded by the conceptualisations and judgments of humans as are agreements to what constitutes an improvement. This contrasts with the mainstream position wherein ecosystems are regarded as having an existence of their own [9].[20]

Making a choice about a situation that appropriately acknowledges complexity and uncertainty is a key starting point for managing (juggling the M-ball). Failure to account for complexity and uncertainty leads, all to often, to treating situations, whether consciously, or unconsciously, as difficulties, 'tame problems' or amenable, only, to scientific explanation.[21]

As I have outlined we have choices that can be made as to how to engage with situations. The particular framing we chose in our work on river catchment managing is one of many choices that could be made.

The use of particular tools and techniques are important in preventing premature boundary closure and thus in remaining emotionally and conceptually open to the circumstances. Such an approach also helps to avoid premature framing of situations. For example I have used metaphor analysis (as described

---

[20]Many scientists and non-scientists alike hold the view that "a natural system is a whole created by nature". These few simple words represent ideas that have been the subject of many books. In the way in which I experience use of these terms I understand the users to reify "nature" (as if nature existed) and system "as in the world".

[21]There is a generic problem of privileging science, in the sense that science normally treats what it studies as a discrete, separate-from-humans object i.e., as "objective". This led Maturana [25] to characterise his concerns on the privileging of particular aspects of science and technology in the following terms: "In our modern Western culture we speak of science and technology as sources of human well-being. However, usually it is not human well-being that moves us to value science and technology, but rather, the possibilities of domination, of control over nature, and of unlimited wealth that they seem to offer.... We speak of progress in science and technology in terms of domination and control, and not in terms of understanding and responsible coexistence.... What science and the training to be a scientist does not provide us with is wisdom.... Wisdom breeds in the respect for the others, in the recognition that power arises through submission and loss of dignity, in the recognition that love is the emotion that constitutes social coexistence, honesty and trustfulness and in the recognition that the world that we live is always, and unavoidably so, our doing."

Fig. 9.3 A rich picture generated as part of a two person dialogue in a situation where traditional project managing was not working

in Reading 5 in Chapter 7). This can be a particularly insightful approach when dealing with published reports and policy documents or professionals immersed in particular lineages with strong organising and foundational metaphors. Other techniques involve simple shifts of understanding and language – such as a move away from use of 'the problem', or 'problem situation' to 'the situation of concern'.

In SSM (soft systems methodology), Checkland and co-workers advocate the use of 'rich pictures' a particular form of diagramming as a means of engaging with situations of concern. As a form of diagram they have particular advantages (see Table 6.1). Generating rich pictures like Fig. 9.3 can be used to mediate conversations between pairs or small groups and in the process surface multiple perspectives, deeply held views and underlying conceptions and emotions. Surfacing this material – i.e. bringing it into conversation is a useful precursor to inviting a shift in reframing of a situation. Underlying emotions are often significant, but rarely admitted in conversation. I will discuss this in the next section.

9.5 Breaking Down an Apartheid of the Emotions

Being open to one's circumstances is a matter of emotion more than anything else. The underlying emotional dynamics in a situation can be understood as what makes possible, or not, the emergence of new distinctions and thus experiences.

This phenomenon, perhaps more than any other, opens up the spaces for change in what we as humans do.[22] I exemplified the type of emotional apartheid that concerns me in Section 8.1.1. It has its apotheosis of course in academic discourse and practice.[23]

As noted by Russell and Ison [32] the proposition that our experience and subsequent action is shaped by a particular emotion is not a new one in experimental psychology. Beginning in the late 1800s William James [21] suggested that "my experience is what I agree to attend to" (p. 402). Research by Arne Öhman and his colleagues (see Öhman [28], for a review of the relevant experimental studies) clearly shows how attention is controlled by the currently activated emotional system, that emotion appears to drive attention, and that emotions are assumed to be functionally shaped by evolution. Öhman presents evidence that emotions, particularly those of fear and anxiety, can be aroused by events that are "outside the spotlight of conscious attention" (p. 265). The finding that an emotional change can be elicited by a pre-attentive, automatic analysis of a stimulus, with an absence of any conscious recognition of that stimulus, is particularly relevant to any model of conversational behaviour.

Following Maturana et al. [26] emotioning is a process that takes place in a relational flow. This involves both behaviour and a body with a responsive physiology that enables changing behaviour. Thus, 'a change of emotion is a change of body, including the brain. Through different emotions human and non-human animals become different beings, beings that see differently, hear differently, move and act differently. In particular, we human beings become different rational beings, and we think, reason, and reflect differently as our emotions change'. Maturana (ibid) explains that humans move in the drift of our living following a path guided by our emotions. 'As we interact our emotions change; as we talk our emotions change; as we reflect our emotions change; as we act our emotions change; as we think our emotions change; as we emotion... our emotions change. Moreover, as our emotions constitute the grounding of all our doings they guide our living'.

I find examples of what Maturana means all around me. Let me give an example of systemic practice in which, as you will see, emotions play a part.

9.5.1 An Example of Emotionally Aware Systemic Practice

I, like many others around the world, have been impressed by the actions and words of US President Barack Obama. I have experienced his words as profound,

---

[22]I do not claim that some people are more open to circumstances than others but I do claim that our capability is a product of our history (structural coupling) and the relational milieu we find ourselves in at any moment, which of course also includes language or conversation. Thus different relational dynamics bring forth different emotions.

[23]I do not use apartheid to be deliberatively emotive but as a descriptor for my experience of professional academic practice in particular and organisational life in general. By apartheid I mean separateness as in the Afrikaans use of the term.

sincere and aspirational – a set of characteristics that is sometimes described in the learning literature as authentic. I know something about 'authenticity' because in the early 1980s I took part in developing a radical education programme that threw traditional approaches to curricula out of the door and rebuilt a whole degree programme based on experiential learning, systems agriculture and effectiveness in communication [33].[24] From that experience I came to understand and appreciate the power and utility of authentic communication. Later, amongst the millions of words that have been written about the President, I came across an article by Jonathon Freedland that gave me insights into Barrack Obama's practice (i.e. what he did when he did what he did). Not surprisingly, to me, many of these practices are similar to those I ascribe to a systems practitioner in my juggler isophor. These are some of the practices that Freedland describes that caught my attention [16]:

- '...he was a good listener, often spending hours with individuals at a time to hear the full story of their lives. It is as if he wanted to learn from them as much as to help them..' (p. 6)
- He was efficient. 'He once arranged for 600 residents to talk with officials about contaminated water. He stood at the back, clipboard in hand with a diagram setting out the names of all those who would speak and what points they would make.' (p. 6)
- 'He was learning the centrality of preparation – and organization – to making political change.' (p. 6)
- 'He won [election as the first African-American president of the Harvard Law Review] thanks, in part, to the votes of conservatives on the Review. They did not agree with him on issues, but they were impressed that he truly listened to them, that he seemed to take them seriously' (p. 6)
- 'On one occasion he made a speech defending affirmative action that effectively articulated the objections to it. Rightwingers believed Obama had shown them deep understanding and respect. It was a mode of discourse that Obama would employ again and again...' (p. 6)
- 'A former teacher at Harvard, Martha Minow, has said that 'he spoke with a kind of ability to rise above the conversation and summarise it and reframe it"..
- 'He always listens, and he might not agree with you, but you never felt he was brushing you off..' (p. 6)
- '...Obama learned a crucial lesson in Springfield – that progress wouldn't come through smart policy papers or stirring speeches. Relationships were the key....' (p. 7)

---

[24] A concern in the many versions of experiential learning is that of 'authenticity' – the relationship of learning to the world of practice. The concept, it is argued, lies at the heart of the attempts by educators since John Dewey to address the relationship between learning and life [24].

- '…at least one aspect of Obama's modus operandi should travel with him into the White House. By all accounts, it's the same working method he employed at the Harvard Law Review. He would ask his policy advisers to convene the top experts in a given field for a dinner. Obama would make introductory remarks, then sit back and listen – hard. Similarly when convening his own staff for a key decision he might stretch out on a couch on his office, his eyes closed, listening… he asked everyone in the room to take turns sharing their advice, insisting on the participation of even his most quiet, junior staffers. He particularly encouraged internal argument among his advisers thrashing out both sides of an argument' (p. 7).

To anyone familiar with practice in fields such as community or rural development, grassroots activism, social work and organisational change management, the list of practice characteristics attributed to President Obama in Freedland's article will not be that surprising. What is surprising of course is that these attributes are held by someone who is now US President. As Freedland observes, 'after eight years of a president who ostracised those advisers who dared tell him what he didn't want to hear, the Obama style will be quite a change' (p. 7).

If asked to explain, on the basis of this sample of attributes, what the key elements of Barack Obama's practice have been that contribute to his success, I would point to his:

- Encountering of the other as a legitimate other[25]
- Predisposition to learning (which in itself is a way of abandoning certainty)
- Capacity for listening – such that he creates for those in the conversation the experience of being actively listened to
- Capacity and technique of 'mirroring back' his understanding of the position of others[26]
- Understanding and valuing of multiple perspectives in respect to a situation or issue of concern
- Ability to move between different levels of abstraction and to synthesise different strands of an argument
- Awareness that change comes through relationships[27]
- Ability, knowingly or not, to be both systemic and systematic (the latter typified by his being organised)
- Use of diagrams as a 'mediating object' in his practice

---

[25] This is how Maturana explains the arising of love – thus when enacted it generates an underlying emotional dynamic that brings forth 'love'.

[26] We use 'mirroring back' as a form of practice in our research that acknowledges that what we say following, for example, a series of interviews, is our interpretation of what we heard, not a statement of 'how things are' (see [41]).

[27] Geoffrey Vickers referred to this as an appreciative system in which choices about relationship making and relationship breaking are made, through which one's standards of fact and value also change.

Underpinning several of these attributes is a systems practice skill that David Russell and I have described as the choreography of the emotions [33]. In this work we draw on Maturana's biology of cognition and claims that each conversation is shaped interactively by a particular flow of emotion. Our contention is that with practice we are capable of being aware of exactly which emotion is being enacted at any one moment and thus are free to maintain or change the nature of the conversation, and of the relationship in which the conversation is embedded, by modifying the emotion [33, p. 134].[28] An analogy is Donella Meadows dancing with systems as an exemplar of dancing with the emotioning.[29] How one dances becomes part of the flow – as we emotion our emotions change. Evidence of the underlying emotional flow of Donella's own practice can be gleaned from Reading 3 in Chapter 4.

Having accepted this understanding it became a guiding influence for our research and consulting activities for over 15 years. The notions of chorographer (one practised in the experiencing of territory, or situations) and choreographer (one practised in the design of a dance arrangement) become a way of describing our concerns. Mapping the initial relationships locates which emotions are getting which results and offers reflections on how a particular workplace, social, and/or personal culture (pattern of relationships embedded over time) has come about. Designing conversations and actions, itself an ongoing process, is thus an essential role for the systems practitioner. This role, as creative as it is responsible, is at its heart the strategic management of emotions where an emotion is defined as that flow of desire predisposing one towards a particular action. The emotion determines the nature of the action: it is emotions not resources that determine what we do!

9.5.2   Generating a Choreography of the Emotions

What is striking for me about President Obama is the seeming congruence between what he espouses and what he does. That said, there is no doubt that he has entered politics to effect change for the better, which of course raises the questions of: (i) better from whose perspective? and (ii) how is change effected?[30] Within systems

---

[28] We do not mean exact in the sense of a universal set of categories but exact in relation to the history of that person, their manner of living.

[29] See Meadows [27] in which she describes the following dance: (i) Get the beat, (ii) Listen to the wisdom of the system, (iii) Expose your mental models to the open air, (iv) Stay humble. Stay a learner, (v) Honour and protect information, (vi) Locate responsibility in the system, (vii) Make feedback policies for feedback systems, (viii) Pay attention to what is important, not just what is quantifiable, (ix) Go for the good of the whole, (x) Expand time horizons, (xi) Expand thought horizons, (xii) Expand the boundary of caring, (xiii) Celebrate complexity, (xiv) Hold fast to the goal of goodness.

[30] The situation in which the US President operates can also be understood as a structure determined situation so what President Obama can and cannot do is not merely reliant on a set of personal attributes, unfortunately!!

practice in general, and systemic inquiry in particular, the surfacing and valuing of multiple partial perspectives is an important means to address the question of what constitutes change for the better. There is never one single right answer or perspective in relation to complex and uncertain issues. Hence processes of decision making that employ and value different perspectives are likely to lead to decisions that are more robust and fit for purpose. They achieve this because in part they have a more effective grasp on what, in the circumstances, constitutes better. In my experience the act of acknowledging other perspectives also profoundly changes the underlying emotional dynamics.

Too often change is understood systematically rather than systemically. From a systemic perspective change takes place in a relational space, or dynamic, including the space of one's relationship with oneself (i.e. through personal reflection). Russell and Ison [32, 33] have argued that it is a shift in our conversation and the underlying emotional dynamics that more than anything else brings about change in human social systems. We devised the following procedures as part of our systemic practice as a means to engage with the desires, wishes, fears, interests (the full gamut of emotions) of participants in the situations of concern with the aim of achieving an experience of systematic reflection through which there is either: (i) a change in the emotion shaping a particular behaviour or set of behaviours, or (ii) a maintenance of that behaviour because the circumstances have not been conducive to a change in the underlying emotion(s). From this perspective having a choice is understood as creating the circumstances for choosing between alternative emotions. The procedure included:

1. Offer the invitation to tell of one's experience vis-à-vis a specific set of circumstances (What is happening to you? What is your interest in what's happening?)
2. From the above account, identify the dominant metaphors and image schemas
3. Ascribe determining emotions to the imaginative structures (metaphors; organising image). Assisting a participant to become aware of a determining emotion is clearly a crucial step
4. Reflect the emotions back to the participant embedded in the same or in amplified imaginative structures and couched as an invitation to further engage
5. The sequence begins over again and finishes when either party considers that there is something better to do elsewhere

You may recognise this as a refinement of the process that I described in Reading 5 (Chapter 7) in which active listening was a key element of my systemic inquiry practice which also involved mirroring back my understandings as well as those metaphors in use with their entailments.

Another 'choreographic opportunity' exists through inviting others to reflect on what they do when they do what they do! In this regard I have found it helpful to invite others to explore how they understand practice (e.g. Fig. 3.5) and to recognise that we have choices that can be made about the nature of situations.

Illustration 9.2

Choreography is concerned with dance which is a common practice across all human societies. Significantly dance is one of the most obvious of embodied practices and in the doing and observing (as part of an audience) its emotional flow is readily apparent. In reflecting on the sensibilities of her parents (Gregory Bateson and Margaret Mead) Mary Catherine Bateson observed that 'both Margaret and Gregory grew up to regard the arts as higher and more challenging than the sciences. This sense of humility in relation to the arts lasted right through their lives' [3].

Unfortunately the realisation of an holistic artistic practice, in the sense imagined by Mary Catherine Bateson, is significantly constrained by a misplaced targets culture, the uncritical acceptance of projectification, our collective failure to be open to circumstance and its contingent nature, as evidenced by inadequate awareness of the choice we make or do not make in framing situations, and the self-imposed apartheid we place on the role of emotions in our doings. These all combine to both create a need for systems practice but at the same time make the circumstances for its uptake and enactment less than conducive. We thus need to invent new social technologies better suited to our circumstances. In the next chapter I explore the opportunities that investment in 'systemic inquiry' might create.

References

1. Ackoff, R. (undated) Why few organizations adopt systems thinking. Keynote Address, UK Systems Society Annual Conference, Hull.
2. Barnhart, R. (2001) Chambers Dictionary of Etymology. Chambers: USA.
3. Bateson, M.C. (2001) The wisdom of recognition, Cybernetics & Human Knowing 8(4), 87–90.
4. Bawden, R.J. (2005) Systemic development at Hawkesbury: Some personal lessons from experience. Systems Research Behavioral Science 22, 151–164.
5. Bell, S. and Morse, S. (2005) Delivering sustainability therapy in sustainable development projects. Journal of Environmental & Management 75(1), 37–51.

6. Bell, S. and Morse, S. (2007) Story telling in sustainable development projects. Sustainable Development 15, 97–110.
7. Blackmore, C.P. (Ed.) (2010) Social Learning Systems and Communities of Practice. Springer: Dordrecht.
8. Caulkin, S. (2009) This isn't an abstract problem. Targets can kill. Buzz up! Digg it. The Observer Sunday 22 March 2009.
9. Collins, K.B. and Ison, R.L. (2009) Living with environmental change: adaptation as social learning. Editorial, Special Edition, Environmental Policy & Governance 19, 351–57.
10. Collins, K. B., Ison, R. L. and Blackmore, C.P. (2005) River basin planning project: social learning (Phase 1) Environment Agency, Bristol (see www.environment-agency.gov.uk).
11. Fairtlough, G. (2007) The Three Ways of Getting Things Done. Hierarchy, Heterarchy & Responsible Autonomy in Organizations. Triarchy Press: Axminster.
12. Field, M. (1999) Project Management. Unit 7. Standard Methods. Computing for Commerce and Industry Program, The Open University: Milton Keynes 72pp.
13. Fisher, D.E. (2006) Water resources governance and the law. Australasian Journal of Natural Resources Law & Policy 11(1), 1–41.
14. Fortune, J. and Peters, G. (1995) Learning from Failure: The systems approach. Wiley: Chichester.
15. Fortune, J. and Peters, G. (2005) Information Systems. Achieving Success by Avoiding Failure. Wiley: Chichester.
16. Freedland, J. (2008) The Obama Story. The Improbable Journey. The Guardian, Thursday, November 6 2008, pp. 1–8.
17. Gido, J. and Clements, J. (1999) Successful Project Management. International Thompson Publishing: Boston.
18. High, C., Ison, R., Blackmore, C., and Nemes, G. (2008) Starting off right: Reframing participation though stakeholder analysis and the politics of invitation. Proc. Working Group 13 'The OECD's New Rural Paradigm', XII World Congress of Rural Sociology, Seoul, Korea.
19. Ison, R.L. and Armson, R. (2006) Think, Act & Play im Leadership der Kybernetik zweiter Ordnung. Lernende Organisation. Zeitschrift fur systemishes Management und Organisation No 33, September/Oktober (www.lo.isct.net) ISSN 1609-1248 pp. 12–23. [Published by: Institut für Systemisches Coaching und Training, Zielorientierte Entwicklung von Menschen, Teams & Unternehmen GmbH, Lange Gasse 65 A-1080 Wien, Austria].
20. Ison, R.L., Röling, N. and Watson, D. (2007) Challenges to science and society in the sustainable management and use of water: investigating the role of social learning. Environmental Science & Policy 10 (6) 499–511.
21. James, W. (1890/1950) The Principles of Psychology Vol 1, Dover: New York.
22. Johnston, P. (2009) The ultimate turnaround from Labour, the dying Government. By abandoning targets, Labour is admitting the depth of its failure. Published: 6:41AM BST 29 Jun 2009 – see Telegraph.co.uk
23. Levene, R. (1997) Project management, in Concise Encyclopaedia of Business Management. International Thompson Business Press: Boston, pp. 578–597.
24. Maharg, P. (2002) Authenticity in learning: Transactional learning in virtual communities. (See: http://zeugma.typepad.com/Publications/Prof20Paul20Maharg20ISAGA20article1.doc, Accessed 1st October 2009).
25. Maturana, H.R. (1991) Response to Berman's critique of The Tree of Knowledge. Journal of Humanistic Psychology 31(2), 88–97.
26. Maturana, H.R., Verden-Zöller, G., and Bunnell, P. (Ed.) (2008) The Origin of Humanness in the Biology of Love. Imprint Academic: Exeter.
27. Meadows, D. (2001) Dancing with systems. Whole Earth, winter.
28. Öhman, A. (1997) On the edge of consciousness: Pre-attentive mechanisms in the generation of anxiety, In A Century of Psychology: Progress, paradigms and prospects for the new millennium. Edited by Ray Fuller, Patricia Noonan Walsh and Patrick McGinley, Routledge: London.
29. Pollitt, C. (2003) The Essential Public Manager. Open University Press: Glasgow.
30. Reiss, G. (1992) Project Management Demystified. E & FN Spon: London.

31. Roux, D. J., Murray K. and Hill, L. (2009) A learning strategy framework for natural resource management organizations. Water Research Commission: South Africa.
32. Russell, D. B. and Ison, R. L. (2004) Maturana's intellectual contribution as a choreography of conversation and action. Cybernetics & Human Knowing 11 (2) 36–48.
33. Russell, D. B. and Ison, R. L. (2005) The researcher of human systems is both choreographer and chorographer. Systems Research & Behavioural Science 22, 131–138.
34. Schön, D. A. and Rein, M. (1994) Frame Reflection: Toward the Resolution of Intractable Policy Controversies. Basic Books: New York.
35. Sjöblom, S. (2006) Introduction: Towards a projectified public sector – project proliferation as a phenomenon. In Sjöblom, S., Andersson, K., Eklund, E. and Godenhjelm, S. (eds) Project Proliferation and Governance – the Case of Finland. Helsinki: Helsinki University Press.
36. SLIM (2004) SLIM Framework: Social Learning as a Policy Approach for Sustainable Use of Water (available at http://slim.open.ac.uk) 41p.
37. Steyaert, P. and Jiggins, J. (2007) Governance of complex environmental situations through social learning: a synthesis of SLIM's lessons for research, policy and practice. Environmental Science & Policy 10 (6), 575–586.
38. Turner, R.J. (1999) Handbook of Project-Based Management, 2nd edn. McGraw-Hill: London.
39. Vickers, G. (1965) The Art of Judgment: a Study of Policy Making, Chapman and Hall: London.
40. Vickers, G. (1970) Value Systems and Social Process, Penguin Books: London. [First published by Tavistock Publications, London in 1968].
41. Webber, L. (2000) Co-researching: braiding theory and practice for research with people. In R. L. Ison and D. B. Russell (Eds.), Agricultural Extension and Rural Development: Breaking out of Traditions (pp. 161–188). Cambridge University Press: Cambridge, UK.
42. Wikstrom, K. and Rehn, A. (1999) As cited in Sjöblom, S (2006).
43. Winter, M. and Checkland, P.B. (2003) Soft systems: a fresh perspective for project management. Civil Engineering 156 (4), 187–192.
44. Winter, M.C. and Szczepanek, T. (2009) Images of Projects, Gower: London.

Chapter 10
Systemic Inquiry

10.1 Clarifying What Systemic Inquiry Could Be

The failings highlighted in the previous chapter associated with goal-focused thinking, a targets mentality and a project culture suggest two related needs. Firstly, inventing a different way of organising how things are done in certain situations, something that is more than a project or a collection of projects as a programme. Once this way of organising has been invented, the second need is to have different thinking and practice employed which, in its enactment, encompasses making choices about framing of situations and breaking out of the apartheid of the emotions. Users of such an approach must be capable of:

1. Understanding situations in context (both current and historical)
2. Appreciating multiple stakeholders and thus perspectives
3. Addressing and clarifying questions of purpose
4. Distinguishing what, how, and why, and clarifying when it is appropriate to address each
5. Facilitating action that is purposeful and which can be judged as systemically desirable and/or culturally feasible[1]
6. Developing a means to orchestrate understandings and practices across space and time in a manner that continues to address social concerns when it is unclear at the start as to what would constitute an improvement (i.e., to adaptively manage a co-evolutionary dynamic)
7. Institutionalising on-going use of the approach in a manner that does not trivialise and instrumentalise the premises on which it is built

---

[1] The idea that something is 'systemically desirable' is a key feature of SSM thinking as articulated by Checkland and colleagues [11, 13]. I use it to mean a possible action(s) that is understood to be systemically coherent as the result of some form of inquiry or investigation into a situation, including, but not only conceptual and/or quantitative modelling. What is systemically desirable however may not be culturally feasible to implement. However, if the right people learn through a process of systemic inquiry then what is culturally feasible, or not, can change.

R. Ison, *Systems Practice: How to Act in a Climate-Change World*,
DOI 10.1007/978-1-84996-125-7_10, © The Open University 2010.
Published in Association with Springer-Verlag London Limited

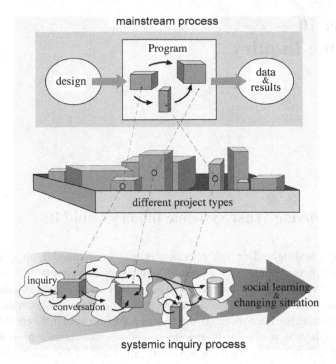

Fig. 10.1 A way of conceptualising systemic inquiry as a meta-form of purposeful action able to make better use of contextualised programmes and projects of different institutional form

In recent research we have experimented with Systemic Inquiry as an approach that meets these needs, recognising that, as outlined in Part II of this book, it is not the approach alone that is important but how it is enacted by a user in context specific ways [17, 32]. We understand systemic inquiry as a meta-platform or process for 'project or program managing' (Fig. 10.1) as well as a particular means of facilitating movement towards social learning (understood as concerted action by multiple stakeholders in situations of complexity and uncertainty).

Inquiry-based practice has been a concern within different systems practice lineages for many years [27]. Blackmore [4] traces some of the antecedents in the systems field to current practices associated with 'inquiries'. She identifies the following lineages:

1. What Schön [43] calls 'Deweyan Inquiry' [22] i.e. thought is intertwined with action and inquiry begins with 'problem situations'
2. Inquiry based on Vickers' idea of appreciative systems [10, 48, 49]
3. Churchman's [9] inquiring systems, particularly in the sense of recognising that there are many possible worldviews and perspectives

As was discussed in Chapter 6, in the 1950s and 1960s concern within some systems lineages developed around effectiveness for planning and decision-making in uncertain and complex situations. This led West Churchman [9], for example, to address what he called the 'design of inquiring systems'. He reflected that the tendency, then prevalent, was to bolster science and its research as the paradigmatic exemplar of an inquiring system. He rejected this and observed that 'in every age when men [sic] have struggled to learn more about themselves and the universe they inhabit, there have always been a few reflective thinkers who have tried to learn how men learn, and by what right they can claim that what they profess to learn is truly knowledge. This is reflective learning in the literal sense: it is the thinking about thinking, doubting about doubting, learning about learning, and (hopefully) knowing about knowing' (p. 17). He defined 'inquiry' as an activity which produces knowledge (p. 8); put another way inquiry facilitates a particular way of knowing which, when enacted, makes a difference. Churchman [9] recognised the central role of the practitioner in any process of inquiry as exemplified by his exploration of the metaphor of a 'library of science'. He argued that the common definition of science as a systematic collection of knowledge is 'almost entirely useless for the purposes of designing inquiring systems… in other words knowledge resides in the user not in the collection… it is how the user reacts to the collection... that matters' (p. 10). For Churchman an 'inquiring system' was a descriptor of a process of inquiry discernable in a given situation. As I will describe in more detail below an *inquiring system* or a *learning system* can both be seen as products of the enactment of a systemic inquiry.[2]

Figure 10.1 depicts how systemic inquiry could be conceptualised as a meta- form of purposeful action that, with appropriate praxis and institutional arrangements, could provide a more conducive, systemic setting for programmes, and projects with a diversity of forms of practice and institutional arrangements that are appropriate to the context (i.e. contextualised) e.g. scientific projects; action research projects, systemic action research projects, systemic interventions etc. Why is this form of innovation needed? Because:

1. Of the nature of situations we are having to engage with
2. We live in a 'projectified-world' and there is increasing evidence that 'projects' deal poorly with complex, long-term phenomena e.g. PRINCE2 project management package
3. There is considerable rhetoric about being more joined-up, holistic, 'integrated' …but theory-informed practice is often weak
4. An inquiry-based approach enables managing or researching for emergence – adaptive managing
5. In systemic inquiry ethics arise in context-related action (they are not reduced to a code – as discussed in Box 5.2 in Chapter 5)

The broader rationale for such an innovation is to better manage our ongoing structural coupling with the biosphere in a climate changing world in a manner that could be understood as a form of on-going systemic development [19, 36].[3]

---

[2] Systemic inquiry builds on and extends Churchman's epistemological assumptions; it is concerned with the design of inquiring (or learning) systems and is grounded in various traditions of systems scholarship including second-order cybernetics [39, 51] and applied systems studies [11, 12, 40]).

[3] I do not plan to expand upon systemic development here – see Bawden [2].

10.2   The Opportunity for Systemic Inquiry

Systemic inquiry is a device for enabling systems practice that acknowledges and addresses uncertainty. It is a conceptual and institutional framing to avoid the worst excesses of living in a projectified and programmatic world (Fig. 10.1). Systemic inquiry can thus be understood as a compound noun – a systemic inquiry can be invested in, set up, funded – as well as a framing for doing systems practice as systemic inquiry – i.e. a process. As yet there is no definitive advice to give on how to set up, or institutionalise a systemic inquiry, though I will provide a case study later in this chapter of one way of doing it.

Having created a systemic inquiry there is also no one way to enact it. But, as I see it, appreciating and embodying, over time, the different aspects of the juggling practices I described in earlier chapters is a key requirement. I have already exemplified some aspects of how a systemic inquiry might be enacted in Reading 5 in Chapter 7.

Systemic practice can be distinguished from other forms of inquiry (e.g. appreciative inquiry [20, 26], action inquiry [47], first, second and third person inquiry [41, 42], networked systemic inquiry [8]) in that those who pursue it purposefully connect with the different lineages of doing systems (Chapter 2).[4] Also, I am not aware that other inquiry practitioners are as concerned as I am about systemic inquiry as a social technology or institutional device (i.e., something that is meta to projects or programmes). In process terms all forms of inquiry have the potential to be mutually supportive as ways of understanding practice. All have as primary concerns reflexivity, the abandoning of certainty and being open to circumstances.

Systemic inquiry is above all else an approach to practice which is adaptive to changing circumstances and which draws explicitly on understandings of systems thinking, theories of learning, action research, cooperative inquiry and adaptive management. Like many inquiry processes systemic inquiry can be conceptualised as a cyclic process but it is much more than a linear sequence drawn as a circle!

Of course much more than systemic inquiry needs to be done to build systemic practice as an alternative form of institutionalised practice. In doing this, however, there are many potential pitfalls for the unwary, including the mistaken belief that my claim that systemic inquiry is conceptually meta to programmes or projects means that they are always bigger or longer lasting. They do not have to be.[5] Inquiry is as much about predisposition as it is about institutional form.

---

[4] The lineage of doing systemic inquiry that we enact is not the only lineage amongst systems scholars. Burns [8 p. 8] refers to networked systemic inquiry as a more organic form of action inquiry, sets of inquiry practices that underpin systemic action research. For him an 'inquiry stream is a series of linked meetings which explore issues and constructs action over a period of time.' On the other hand Klein [38] describes Systemic Inquiry as they enact it as a methodology for organisational development on the basis of applied narratives.

[5] For example thinking of a systemic inquiry as akin to a 'Royal Commission' or 'Board of Inquiry' would be a mistake. Both of these institutional arrangements have become reified in ways that mean they are either open to political manipulation (e.g. by specifying terms of reference in a Royal Commission) or are closed to changing context. Another way of thinking of a systemic inquiry is as a device to enact deliberative decision making, or as a cornerstone of more deliberative democracy.

So, for example, systemic inquiry could be part of everyday managing in circumstances where we choose to acknowledge uncertainty and complexity. Or it could be a useful replacement, both conceptually and practically, for 'problem structuring methods' as currently understood in the operations research community (as discussed in Chapter 8). Equally it could be built into government's practices as a way of avoiding high staff turnover, the loss of organisational memory and thus the propensity to 'reinvent the wheel' in the face of long-term issues that need managing beyond the current time frames of elections and other human constructed political and economic cycles. To achieve such a shift means that those inclined to innovate in this way will need to appreciate a little more what might be entailed in doing systemic inquiry.

10.3 The Basic Process of Systemic Inquiry

Peter Checkland [12] has described the enactment of SSM (soft systems methodology) as a form of systemic inquiry (Fig. 10.2). Checkland's articulation of the use of SSM as a systemic inquiry can be understood as an 'ideal type' (as discussed for my juggler isophor in Chapter 3). He conceptualises the enactment of SSM as an iterative process in which deciding when it is complete is a matter of contextual judgment. It is an action-oriented approach – that is to say, the intention of enactment is to produce a change. It is not about introducing or creating 'a system'. Change is mainly manifest as changes in understandings and practices (Fig. 1.3), changes in social relations amongst those involved, changes in process or changes in structure. The products of SSM enacted as a systemic inquiry are conceptual (systems) and sometimes quantitative models which are both prescriptive models and devices to make sense of the situation of concern. [6]

From Checkland's perspective the role of the systems practitioner sits somewhere between being a consultant, brought in from outside to analyse the situation and advise on change, and a facilitator, who helps the participants understand their own situation so as to make changes for the better. In his work with Winter, Checkland [14, 52] marshals many of the arguments for understanding systemic inquiry as particular form of practice that could be institutionalised within the 'project management' field.

Figure 10.2 depicts an activity model for conducting a systemic inquiry. The focus is identifying the linked activities. Above all else systemic inquiry is a practical activity that can enable situations to be improved. A number of systems levels are depicted in Fig. 10.2. The large system (the main shape) has two main activities:

---

[6]Figure 10.1 is itself a conceptual systems model – sometimes also referred to as an activity model. At the core of SSM-style systems practice is an appreciation that the 'modeling language' is all of the verbs in the [English] language – hence the focus within SSM on activity [13].

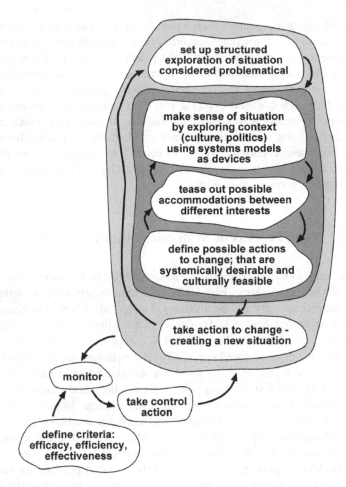

Fig. 10.2 An activity model of a system to conduct a systemic inquiry (Adapted from Checkland [12])

1. Set up structured exploration of a situation experienced as complex or uncertain
2. Take action to change in the situation

However to operate as a system this larger system depends on the activities of one sub-system (represented by an inner circle within the large shape). The sub-system has three activities depicted by the verbs (actions):

1. 'Make sense of'
2. 'Tease out... accommodations'
3. 'Define possible actions'

In Fig. 10.2 systemic inquiry begins with a process of sense-making of differing contexts, identifying areas where differences can be accommodated and moves on to defining possible actions. The overall inquiry (system) has to be monitored, measures of performance articulated against acceptable criteria (the three e's depicted here) and control action taken. It is important to appreciate that the measures of performance are always present but always contextualised, not imposed from the outside. In practice the steps in this process are never systematic. Iteration and concurrent action in different stages are common. When joint action to change is taken as a result of key stakeholders learning their way to an understanding of what needs to be done then social learning has occurred because both changes in understanding and practices result and the situation is changed or transformed (Fig. 1.3).[7] Box 10.1 gives a very brief summary of how we have come to understand social learning through our recent research.[8]

Box 10.1 Social Learning

Social learning can be interpreted as one or more of the following processes [15]:

- The convergence of goals (more usefully expressed as agreement about purpose or purposes), criteria and knowledge leading to awareness of mutual expectations and the building of relational capital (a dynamic form of capital that integrates the other forms, viz: artificial, natural, social and human – see SLIM [45])
- The process of co-creation of knowledge, which provides insight into the history of, and the means required to transform, a situation. Social learning is thus an integral part of the make-up of concerted action
- The change in behaviours that results from the understanding gained through doing ('knowing') that leads to concerted action
- Arising from these, social learning is thus an emergent property of the process of transforming a situation (Fig. 1.3); [16, 46]

The metaphor of an orchestra helps to reveal what we mean by both social learning and systemic inquiry: an orchestra is something that can be invested in; it is thus referred to and understood as an entity. At the same time what is being invested in is the on-going capacity to create, adapt and deliver performances by a group of people with different instruments, skills, perspectives, histories and so on, that satisfy some socially determined purpose – such as a performance that people pay to attend, or as an iconic investment that

(continued)

---

[7] This section draws extensively on Collins et al. [17].

[8] In presenting Checkland's model for enacting a systemic inquiry I am not presenting a blueprint or plan for doing systemic inquiry. As I have, or will outline, I am not committed to all the verbs that Checkland uses in his model nor am I committed to the problem metaphor. Of course consistent with SSM practice one should always feel free to act with awareness and change the verbs (activities) in the model recognising that in the process a new system of interest is created.

Box 10.1 (continued)

communicates a city's artistic and cultural commitments. In terms of progressing climate change adaptation, for example, social learning can also be understood as a governance mechanism or policy instrument and systemic inquiry as an incipient social technology.[9]

With awareness systemic inquiry invites a consistent way of being within an on-going inquiry process (e.g. living life as inquiry). Systemic inquiry has another primary concern – that of taking 'a design turn', the products of which can be understood as 'learning or inquiring systems'.

10.4 An Example of Setting Up a Systemic Inquiry

Beginning in 2003, with colleagues, I was engaged in a systemic inquiry within the Environment Agency of England and Wales (EA). The purpose of our engagement was to undertake research to support the implementation of the Water Framework Directive (WFD) through introduction and use of social learning approaches. Integration was and remains a key theme for the WFD. This includes not only integration across media – land, surface and ground waters – but also between different stakeholder interests: environmental, economic and social. Within the Agency, tackling this task of integration was initially supported through an ambitious WFD Program structure. This included a large number of projects and within some of these, such as River Basin Planning (RBP), a large number of work packages.

At the same time, a strategy for engagement with diverse stakeholder interests at multiple geographical levels had to be developed through a Public Participation work package. Subsequent to our initial scoping work there was growing recognition that project management protocols which set out in detail each stage of a project were hindering learning and effective achievement of that project's overall objectives. This was because the focus on 'doing' the activities identified in the project specification prevented attention being given to what was actually being learned during the project. We argued that apart from being an emerging issue in the RBP Project, in the longer term, this was an important lesson for the Agency's approach to the WFD Program as a whole and to its operational implementation.

We noted that a systemic, learning approach required a shift in thinking/practice and the development of new skills. This is because "social learning" for concerted action depends on the perceived interdependencies of stakeholders. This meant that continuing to operate as individuals (or individual functions) was unlikely to enable these interdependencies to be perceived and acted upon. In contrast a systemic learning approach can help stakeholders explore and make sense of their interdependencies and work out how their collective roles can be complimentary.

---

[9] Our theoretical approach to social learning is discussed in full elsewhere (see Blackmore et al. [6]; Collins et al. [18]; Collins and Ison [15]; Collins et al. [19]).

10.4.1 *Contracting a Systemic Inquiry*

The work we did with the EA between 2003 and 2008 was framed by us, and in an initial contract, as a *systemic inquiry*. Some extracts from the contract which framed our initial research work with the EA are given in (Box 10.2).[10]

Box 10.2 Details of the Systemic Inquiry Set Up Under Contract with the Environment Agency (England and Wales)

OBJECTIVES

The overall objective of this work is to inform the development of the River Basin Planning strategy and more broadly, to make a timely improvement to the effectiveness of the Agency's WFD (Water Framework Directive) Programme, drawing on social learning practices and concepts.

The specific objectives are:

1. To introduce the concept of "social learning" to the Environment Agency, applying it initially as a tool to facilitate the development of, and decision-making within, the River Basin Planning (RBP) and Programme of Measures (PoMs) projects. This will also inform the emerging Public Participation strategy and help ensure its ability to evolve over successive WFD cycles to meet external aspirations, as well as the requirement to achieve "active involvement" of stakeholders in the operational implementation of the WFD
2. To develop capacity building for social learning at Programme Board/ Functional Head level. The Agency has expressed a preference for "evolution rather than revolution" in WFD implementation, but in the medium term, difficult decisions will have to be made if implementation is to be fully effective in achieving an integrated approach to environmental management. Social learning could provide a critical aid to effective decision making in this context
3. To begin to build capacity for social learning at the local level through the Business Implementation project. *Making it Happen* commits the Agency to increasing partnership working with external bodies. The concept of social learning will ultimately be embodied in the WFD Public Participation strategy and social learning approaches will not only ensure that "active involvement" is demonstrable, but also help develop local skills that will be increasingly important outside our WFD activities

(continued)

---

[10] I have already provided detail on the ethics statement included in the contract in Chapter 5 – see Box 5.2).

Box 10.2 (continued)

CONTENT AND OUTPUTS

Building on SLIM-UK's original tender to the Agency for the Public
Participation (PP) strategy work and the subsequent Agency 'Form A' on
social learning derived from this, a meeting was held on 25 November 2003
between the Agency's RBP project and the Open University's SLIM-UK
team (hereafter referred to as the 'OU team').

At this meeting five content areas[11] (referred to hereafter as "work
streams") were agreed as a way of structuring the research. We also agreed
that for each of these work streams there would be activities and outputs we
could specify in advance; and others that by the very nature of this work,
would be emergent. These five work streams were further discussed at a
meeting on 1 December and agreed as follows:

Work stream 1: Intellectual framing of social learning

This first work stream will focus on literature (including 'grey literature')
reviews of social learning, its origins and its application, for example in the
context of water management and planning. The purpose being to support
intellectual framing and understanding of this area of practice. For example,
this could be focused on recovering Agency stories that are illustrative of
social learning and setting them in the context of a theoretical discussion.

Reports to be produced for key (internal and/or external) audiences, as and
when specified under work stream 4.

Focus of literature reviews also to be specified, as above.

Output: report.

Work stream 2: Participant conceptual advice and feedback

This second work stream is concerned with specific pieces of conceptual
advice and feedback where we consider that the advice of the OU team would
'add value' to work currently being undertaken/planned in different parts of
the public participation (PP) work package/RBP project/WFD programme.
The term 'participant advice' refers to the supporting and co-development
rather than leading nature of this advice.

An initial specification would need to be worked up around the targeting of:

* Key elements of the Public Participation work package, including those
  concerning social learning, but potentially also including others e.g. models
  of public participation
* Other elements within the RBP project

(continued)

---

[11] The fifth work stream was the production of a refereed scientific paper with our
co-researchers – thus I only deal with the first four here.

- Elements within the wider WFD programme e.g. skills and capacity building within the Business Implementation project

Outputs: papers and editorial work to be mutually agreed.

Work stream 3: Systemic inquiry

This refers to the process of embedding social learning within the RBP and PoM projects (objective 1) the Business Implementation project (objective 3) and the wider WFD programme (objective 2). 'Systemic inquiry' is conceived of as a cyclic process of learning by doing (experiencing), reflecting, conceptualising (and then planning a further cycle of learning). For each project we will work with a bounded 'community of inquiry' (within which there might be several cycles and orders of learning occurring). The purpose of these systemic inquiries would include:

- Learning about the benefits and risks of social learning, especially in supporting more effective River Basin Planning; Programme of Measures; and Business Implementation
- Learning how social learning can be extended to the engagement between Agency staff and stakeholders, and between Agency staff and contractors
- Evaluating how social learning approaches have benefited RBP tasks

We anticipate that the bulk of research will fall within this work stream.

Output: a brief (evaluative) report of the process (including recommendations for further inquiry cycle(s)) to be produced on completion of one or more bounded inquiry cycles [see 17].

Work stream 4: Project management

The first three work streams should be nested within a project management work stream which itself involves a cycle of reflective evaluation and management. The aim of the fourth work stream is to gain a better understanding of the constraints and opportunities for social learning within the Agency culture. One or more management cycles should be designed within the total budget.

The project will be managed by a small Steering Group. Steering Group meetings will be held as needed and should also include enthusiastic individuals who wish to contribute to the development of insights about social learning under this work stream. The OU team will be responsible for preparing minutes of these meetings that shall be a simple record of agreed actions, decisions and emerging learning points.

At the start of each management cycle there should be agreement as to what work is specified within that cycle and how much room there is for emergent work within that cycle.

Outputs: Minutes of steering group meetings. Each cycle should end with a brief evaluation report plus a more comprehensive report at the end of the final cycle.

10.4.2   Interpreting Our Contract

All contracts arise out of certain circumstances and ours was no different. The contract detail in Box 10.2 therefore requires some interpretation. If you read the material in Box 10.2 closely some lack of clarity between the terms 'inquiry' and 'project' will be apparent. I would argue that this was not lack of conceptual clarity on our part but indicative of the circumstances i.e., an organisation in which all external supplies of services were organised through contracts and where the dominant form of service was to 'undertake a project'. Thus for us the fourth inquiry 'project management' was actually 'systemic inquiry management' but as we were already challenging the boundaries of common practice we were obliged to stay with the known language. Work streams four and three can thus be understood, in our terms, as nested inquiries conceptualised in the manner depicted in Fig. 10.1.

Of course contracts do not just happen – there is always the work leading up to the issuing and framing of a contract. As with any form of purposeful activity the initial starting conditions usually determine what is, or is not feasible. It thus becomes interesting to ask: when did our systemic inquiry begin? My answer would be: 'well before the contract formulation and signing stage'. With this in mind I want to outline some of the main activities that comprised the enactment of our systemic inquiry.

10.4.3   Enacting Our Systemic Inquiry

In this section I want to highlight some of the main activities we undertook as part of our systemic inquiry, remembering that systems practice is the key praxis element in enacting a systemic inquiry. In reading this section it is important to appreciate that there was never at any stage a grand plan – a systematic procedure or recipe to follow because at the same time as enacting our systemic inquiry we were trying to work out what doing a systemic inquiry was, or could be! We were thus always trying to juggle our thinking and doing at two conceptual levels. Fortunately we were also guided by Checkland's articulation of a system to do a systemic inquiry (Fig. 10.2) though at no stage did this become a blueprint.

In the case of our EA systemic inquiry this is what we did:

1. Through a combination of circumstances we provided the head of Social Science research in the EA and some of his colleagues, through the conduct of semi-structured interviews, the experience of being actively listened to – in the sense discussed in Chapter 9
2. When invited to tender for a project to do with EA concerns about public participation and social learning we critically examined their expression of need and found it wanting (from our perspective). We wrote back but did not submit a tender bid against the specifications. Instead we wrote saying what we would do and why if given the chance. In this sense our practice resembled that of Patricia

Shaw outlined in Chapter 5 (Section 5.6). The outcome of this action was an invitation from the EA to submit a proposal based on what we proposed

3. Before finalising a contract we rapidly immersed ourselves in the context of the EA's RBP programme. This included meetings with key stakeholders within the EA and then conducting an initial workshop which was designed to 'make sense of the situation' by surfacing multiple partial perspectives and to build a conversational milieu between those who we found isolated from each other through the unintended consequences of the PRINCE2 project management procedures. This workshop included the use of group-based systems diagramming e.g. Fig. 9.3. This experience helped to shape the final form of our contract and convinced us that it made sense to frame the situation we were entering as one characterised by complexity, uncertainty, interdependencies, controversy, and multiple stakeholdings[12]

4. Over the period of a year we 'teased out accommodations' (e.g. by using an understanding of the politics of the situation to design workshops) and

5. 'Defined possible actions' (e.g. by orchestrating debate about the congruence, or lack of it, between systemic models and what was happening or not)

The overall inquiry (system) was monitored, measures of performance articulated against mutually acceptable criteria that worked specifically in this situation (the three e's of efficacy, efficiency and effectiveness) and control action taken (see Collins et al. [17] for more details).

The engagement with the EA around river basin planning (RBPlg) began in late 2004 and continued until early 2006. During this period we convened six workshops on various aspects or topics relating to RBPlg such as 'decision-making'; 'integration'; convergence [of organisational practices]; and 'stakeholding'.

The exact format, methods and tools utilised in the workshops varied according to the issues under discussion, the audience, the purpose of the event and its duration. As outlined in Ison et al. [37] 'in broad terms, the workshops were divided into four main parts. The first part of the workshops aimed to expose differences in understanding among the participants. This was done in activities using non-linear ways of presenting, using and analysing information (through, for example, developing rich pictures, metaphors, conversation maps). The second part of the workshops helped define the nature of the issue emerging from the earlier discussions. This was often done through plenary discussion and reflection on what had emerged from the first part, with some element of distillation of the core themes. The third part identified a series of activity models to enable participants to gain more systemic understanding of the issues and enable staff to progress the situation. The workshops ended with a plenary, in which proposals for next steps and review of learning and evaluation occurred.'

---

[12] Although this particular framing has its origins in natural resource dilemmas our experience is that it can be equally useful as a framing of situations in large and complex organisations such as the EA.

10.4.4 Changing Understandings Can Change Practices Can Change Understandings

The contract outlined in Box 10.2 was a first for the EA. In follow up evaluation at the end of the first phase (after 1 year) some of the EA managers with whom we worked, who in the main had a technical (e.g., engineering) or scientific background, were able to say:

- After months of work, it was such a relief finding out that the WFD was difficult and it was not just me being stupid
- We use the word integration all the time. But this is the first time we've ever had a conversation about its meaning!
- We need to learn our way into this because we've never done anything like this before; and it's obviously a new form of learning that involves doing it together. So it's social rather than technical
- We know through the work of the RBP project that the only way to do this thing called river basin planning is through a learning and adaptive management process. We will never be able to get sufficient agreement (either internally or from our stakeholders) to operate in a stable environment. There are consequences of this reality that we seem unable to face up to... there is a fundamental issue somewhere that we as managers have failed to manage. It's broken. We need to fix it

What happened that these managers felt able to say what they said? From my perspective they engaged in a systemic inquiry which created the circumstances in which they came to realise the need to abandon certainty and to value the benefits of a social learning approach. It is my contention that systemic inquiry and social learning are important elements of a much needed form of governance that is systemic and adaptive and fit for managing our co-evolutionary trajectory.[13] Whilst a governance framework moves beyond management, our experience is that it is important that the praxis (theory informed practical action) elements associated with enacting governance are not lost. For example, Collins and Ison [16] argue the case for building a praxis of 'integrated catchment managing' rather than the more widely accepted 'integrated catchment management' because they see the latter as mainly descriptive and missing an effective form of praxis.

There is still more to be understood about how best to set up and institutionalise a systemic inquiry. Our experience with the EA was a positive one for all concerned. At the same time I and other colleagues had initiated a significant systemic inquiry within the Open University. Our experience over 5.5 years with this inquiry provides other evidence to support a case for further investment in, and capability building for, establishing and enacting systemic inquiries.

---

[13] Of course I do not make the claim that these are the only innovations necessary.

10.4.5   Other Evidence

The PersSyst Project represented an investment of about £1 million over 5 years by The Open University (OU) to develop systems thinking and practice skills as a basis for 'distributed leadership' for managing complexity. An initial aspiration was to explore whether it was possible to provide evidence for claims that the University could be understood as a 'learning organisation'.[14] PersSyst, a title coined from the words Personnel and Systems, was an ambitious collaboration between staff in an academic unit (formerly Systems) and staff within the organisational development arm of the Personnel, now HR (Human Resources) Division. PersSyst was not set up as a traditional project (and thus managed in a typically first-order project managing manner) but rather as a systemic inquiry encompassing systemic action research elements. Consistent with second-order cybernetic understandings the academics involved recognised that as researchers, facilitators, etc. they were part of the situation – there was no external 'objective' position.

Illustration 10.1

PersSyst was set up as a systemic inquiry as we did not know how to introduce systems thinking and practice capability into a large and complex organisation like the OU. We realised that we had to learn our way towards effective action. Space does not permit me to describe the many features of this endeavour – see Armson et al. [1] and Ison and Armson [33] – but I would like to outline some of our initial design principles which remain relevant:

---

[14] I have sadly come to the conclusion that as currently constituted within a globalising Higher Education sector it is virtually impossible for any University to become a learning organisation.

- Someone able to enact systems practice – a 'good juggler' is a central need[15]
- It is possible to work with others in a large organisation to develop capability (and responsibility) for managing complex, messy situations and in the process improving the situation
- Well crafted workshops can foster enthusiasm and build capability – thus creating a new cadre of systems practitioners (but this cannot be done quickly)

Over the period that PersSyst ran, the issues for which PersSyst interventions were requested from OU staff included:

- Lack of team working
- Lack of a clear sense of purpose
- Issues around leadership
- Tensions between individuals
- Lack of synergy between different groups or projects and
- The need to deal with substantial changes

In an independent evaluation all PersSyst activities were shown to have had a number of positive outcomes. These varied across interventions and included:

- More effective team working and performance
- Highlighting of important issues which could subsequently be addressed
- Improved collaboration or synergy between different teams, units, or departments
- Changes in work approaches
- Improved sense of clear purpose
- Improved leadership and
- Improved strategic thinking

In HR terms PersSyst was not a large investment yet it brought significant returns to the University including a cadre of systems practitioners distributed throughout the organisation responding to their circumstances more effectively. In a celebration of PersSyst in September 2009 the former Director of HR was able to reflect back on his experience of nearly 20 years of HR innovation and claim that PersSyst was in the top two. Unfortunately in praxis terms it proved exceptionally difficult to hold on to the systemic inquiry framing for our activities beyond the first 3 years. 'Mainstreaming' all but spelt the death of the systemic inquiry bringing as it did all the organisations 'sunk' resources in 'projectification'. Ironically the act of mainstreaming commissioned PersSyst to the margins!

---

[15] We need to 'apprentice' systems practitioners as jugglers. They can't be "trained" except for some basic concepts and techniques. My experience is that they need to learn on the ground with an experienced 'juggler' understood perhaps as the progression from apprentice, to journeyman juggler and then master juggler!

10.5 'Institutionalising' Systemic Inquiry

I want to provide two recent examples of institutional innovation that, from my perspective, might more usefully have been set up as systemic inquiries. I offer these examples in the spirit of Anthony Giddens [24, p. 8] who argues that 'to develop a politics of climate change, new concepts are needed'. One might equally add that to develop a praxis of climate change adaptation new, conducive, institutional arrangements are also needed.

My first example is that of the Australian Murray-Darling Basin Authority (MDBA) established at the end of 2008 as the organisation responsible for overseeing the Water Act 2007 which transfers many powers previously held by State Governments to the Australian Federal Government. The Murray-Darling, Australia's largest 'river system', is severely degraded and suffering the consequences of climate change (i.e., increased temperatures and reduced rainfall which seems beyond the normal drought cycle that is common in Australia). As noted by Foerster et al. [23]:

> A key element is a new level of planning and regulation. Federal water legislation sets tight parameters for a new round of water allocation planning at the scale of the MDB with plans for individual catchments to be accredited within this framework, with plans to set environmentally sustainable limits on diversions. The Basin plan will also include rules for the operation of water markets basin-wide, and for the delivery of environmental water to environmental priorities across the MDB. The primary regulatory planning responsibility is given to a new Commonwealth body, the Murray Darling Authority, which will replace and absorb the MDB Commission. The Basin plan is intended to be prepared in consultation with Basin States and communities, and the Act does make some provision for consultation during the planning process, including the establishment of a Basin Community Committee to assist in a process of community engagement. The depth of the proposed consultation, particularly the extent to which it will move beyond consultation with high level stakeholder groups, is not yet clear. There is much scepticism among commentators over whether the new Commonwealth package will indeed achieve its noble aims of reforming governance arrangements in the MDB to enable sustainable water management in the national interest.

The new basin Plan is expected to be available for preliminary consideration in 2010 but for various reasons it seems unlikely that any on-the-ground enactment of the new plan will begin in earnest before 2017–2019. In the face of a crisis of nationwide significance this seems unacceptable. In using this example I wish to make the point that in the face of the uncertainties, complexities, interdependencies and multiple stakeholdings in the MDB an approach to its managing is needed that is adaptive and contingent. In such situations a national systemic inquiry could have been chosen as an alternative governance mechanism instead of a traditional regulatory, legislative and planning approach. An effectively constituted systemic inquiry – after all the issues are unlikely to go away in the short to medium term, if ever – could become a vehicle for the deployment of social learning approaches and the adoption of systems practices.

Current forms of government in most western democracies despite their many strengths are not well suited for managing long-term complex issues. My second example is that of the recently formed 'Climate Change Committee' set up in Britain to oversee recently introduced legislation (the Climate Change Bill 2008). It has the task of advising government on the level of carbon budgets and thus the optimal path towards emission reduction targets [24, p. 81]. The terms of reference for this committee seem anything but systemic; seemingly it is also not fit for purpose because its role is only advisory to government. Giddens [24, p. 116] argues that it should have 'capacity to intervene in legislation by, for example, having clearly specified rights to take government to the courts if it has gone back on its obligations'. He also notes that its composition is crucial and appointments should not be made by ministers of the day. I would go further and claim that this is exactly the situation for a national on-going Systemic Inquiry that institutionalises and facilitates the key learning and strategic intelligence gathering and responding necessary to enact climate change adaptation as a systemic co-evolutionary practice.[16]

Capability building in systems practice throughout the public sector – the necessary wherewithal to set up and enact a systemic inquiry – is an urgent need. Seddon [44, p. 196] argues, based on UK experience, that to change the current regimes of government requires a new philosophy: 'Instead of compliance we need innovation, and to foster innovation we need freedom. People need to feel free to act in the best interest of their stakeholders or, in Moore's terms, to do what is best in terms of their particular circumstances. To achieve that, we have to make public-sector managers responsible. They have to be able to choose what to do, free from the obligation of compliance. The way to foster innovation is by changing the locus of control from the regime, which compels compliance, to the public sector manager, who is the person who actually needs to change.'[17] One way to achieve the transformation Seddon refers to would be to equip civil servants and citizens as practitioners to take the 'design turn'.

10.6   Systemic Inquiry and the 'Design Turn'[18]

There is no prescription for a systemic inquiry, they have to be designed. It can be argued that in the context of systems practice, *design* is an involvement in an activity that has many players and that translates human culture, technology and aspiration into form [21, 29, 37]. My concern for design grew initially from a recognition that the future form of Australia's semi-arid rangelands was more a question of design than the application of rationalistic planning or science. My design focus

---

[16] In using these two examples I am trying to make the point that systemic inquiry is something that can be run at the level of the nation or as an international 'platform'. Examples of systemic inquiry that can be set up and institutionalised as part of personal practice or in organisational settings have already been given. Nor am I claiming that there are no examples already, perhaps of international commissions, or the Club of Rome, that could be understood in terms of a 'systemic inquiry' but the point is they have not been institutionalised as such and it is not an option that is currently considered by decision makers.

[17] Mark Moore at the Harvard Kennedy School of Government coined the term 'Public Value' in the mid 1990s. Seddon (p.162–171) outlines how the concept of 'public value' has been taken up and fostered in the UK [44].

[18] This section draws heavily on Ison et al. 2007b.

was in part a response to Hooker's [28] observation that: "The direct consequence of the profound changes in the character and role of organised knowledge is that the future must now be regarded as increasingly a human artefact – an art-in-fact. The future can no longer be regarded as a natural object, a fact already there or objectively determined by present trends. Rather it must be chosen."

What followed from these understandings was concerns, in our teaching and research, for participation in designing (e.g. of research questions; research projects; management plans, environmental decision making etc.) and more recently processes of deliberative systemic and adaptive governance, including social learning. Our current understanding of design is similar to that of Glanville [25] who argues that 'all research and all knowing/knowledge is a matter of design' (p. 120). In recent years the primary vehicle for enacting our understandings of design has been through designing and developing 'learning systems' [3, 16, 30, 31, 34], including curricula, and constituent courses, as well as 'research-based inquiry', situation-improving action and for the 'education' of the 'systems practitioner' [5, 31].

What do I mean by a 'learning system'? For Blackmore [3] a learning system comprises interconnected subsystems, made up of elements and processes that combine for the purpose of learning. The placement of a boundary around this system depends on both perspective and detailed purpose. From a first-order perspective the design of a learning system might seemingly involve combining elements and processes in some interconnected way as well as specifying some boundary conditions – what is in, what is out – for the purposes of learning. The specification of learning outcomes (often expressed as aims and/or objectives) in the absence of any real contextual understanding about learners or stakeholders predisposes, or restricts, most first-order learning system designs to this approach (e.g. most Open University (UK) distance-learning courses). However, in our own learning system design practice we have made the shift described by Bopry [7] as moving from prescription of instructional methods and means to the development of cognitive tools to provide support for the activity of the learner. With this shift we see a 'learning system' as moving from having a clear ontological status (e.g. a course or a policy to reduce carbon emissions) to becoming an epistemic device, a way of knowing and doing (*sensu* Maturana – see [39]).

The shift in perspective by moving to a second-order understanding is consistent with Blackmore's [3] claim that appreciative systems (*sensu* Vickers e.g., [50]) are learning systems. This in turn suggests a systemic design practice perspective that is more organic and observer dependent viz: let us consider this situation as if it were a *learning system*.[19]

Reflecting this turn, Ison and Russell [35] suggest it is a first-order logic that makes it possible to speak about, and act purposefully to design or model a 'learning system'. A second-order logic appreciates the limitations of the first-order position and leads to the claim that a 'learning system' exists when it has been experienced

---

[19] Thus Vickers' retirement work could be framed as 'I have found it useful to think of my life's work in terms of appreciative systems'.

through participation in the activities in which the thinking and techniques of the design or model are enacted and embodied. An implication of this logic is that a 'learning system' can only ever be said to exist after its enactment – that is on reflection. The second-order perspective is not a negation of the first – they can be understood as a duality. The same logic applies to the design of a systemic inquiry.

First-order designing is synonymous with first-order cybernetic understandings, in which goal seeking behaviour is the norm, control is considered possible and designs have a blueprint quality. This parallels systematic, or goal seeking, 'hard systems' approaches [11], rather than systemic practice. Second-order designing arises when the designer acts with awareness that they and their history are part of the design setting. First-order design delivers an output, second-order design delivers a performance [19].

These distinctions can be considered through the metaphor of a symphony orchestra that I introduced earlier: one reading of the metaphor is that the music is an output of the design of a set of instruments, another is that a musical performance is an emergent property of a set of interacting factors – musicians each with a history, orchestral practice, a score, and audience etc. In the second-order case it is understood that each practice setting has (i) a context in which a performance is enacted; (ii) a person or persons – the practitioner(s) and (iii) tools, techniques, methods, methodologies etc. As outlined in Chapters 5–8 there is also a fourth aspect which is not always so apparent – each element in the performance has a history which can be explored and understood. There is always a history of the context, the practitioners (each is a unique individual and thinks and acts differently even though they may come from similar cultures) and the tools, techniques etc. (Fig. 3.5). There is also a history of performing in a particular way – what is recognised as good practice in one setting may not be the same in another setting. The systemic connections between these elements are important if a performance that is effective is to ultimately emerge.

These design considerations apply in any arena of practice, including that of policy makers. Just as an OU academic is concerned with a student at a distance, a civil servant policy maker is concerned with citizens at a distance (those who will take up or be affected by a policy). In both cases the pedagogy and other elements of the practice setting can trigger a first-order response (utilitarian or instrumental learning or a knowledge transfer strategy), or a second-order response – creating the circumstances whereby the learner/stakeholder in context is able to take responsibility and orchestrate their own evolving praxis.

In the next chapter I turn to systemic action research which is an additional way to frame systems practice and is complementary to systemic inquiry.

References

1. Armson, R., Holwell, S.E., Ison, R.L., Lobley, K., Marsh, P. and Widdows D.C. (2010) Managing organizational complexity: project development for building a university-based leadership community for systems thinking and systems practice. Human Relations (in preparation)
2. Bawden, R.J. (2005) Systemic development at Hawkesbury: Some personal lessons from experience. Systems Research & Behavioral Science 22: 151–164.

3. Blackmore, C. (2005) Learning to appreciate learning systems for environmental decision making – a "work-in-progress" perspective. Systems Research & Behavioural Science 22: 329–341.

4. Blackmore, C.P. (2009) Learning systems and communities of practice for environmental decision making. Unpublished Ph.D Thesis, Communication & Systems Department, The Open University, UK.

5. Blackmore, C.P. and Morris, R.M. (2001) Systems and environmental decision making—Postgraduate open learning with the Open University. Systemic Practice & Action Research 14(6): 681–695.

6. Blackmore, C.P., Ison, R.L. and Jiggins, J. (2007) Social learning: an alternative policy instrument for managing in the context of Europe's water. Environmental Science & Policy 10, 493–498.

7. Bopry, J. (2001) Convergence toward enaction within educational technology: Design for learners and learning. Cybernetics & Human Knowing 8, 47–63.

8. Burns, D. (2007) Systemic Action Research. A Strategy for Whole System Change. Policy Press, Bristol.

9. Churchman, C.W. (1971) The Design of Inquiring Systems: basic concepts of systems and organizations, Basic Books: New York.

10. Checkland, P. (1994) Systems theory and management thinking, In Blunden, M. and Dando, M. (eds) special issue of American Behavioural Scientist 38 (1), 75–91.

11. Checkland, P.B. (1999) Soft Systems Methodology: A 30 Year Retrospective. Wiley: Chichester.

12. Checkland, P. (2002) 'The role of the practitioner in a soft systems study', notes of a talk given to OuSyS and UKSS, Saturday 8th December 2001, in Quarterly Newsletter of the Open University Systems Society (OUSyS), Open University: Milton Keynes, No. 27, March 2002, pp. S5–S11.

13. Checkland, P.B., Poulter, J. (2006) Learning for Action. A short definitive account of soft systems methodology and its use for practitioners, teachers and students. Wiley: Chichester.

14. Checkland, P.B. and Winter, M. (2006) Process and content: two ways of using SSM. Journal of the Operational Research Society 57, 1435–1441.

15. Collins, K.B. and Ison, R.L. (2009a) Jumping off Arnstein's ladder: social learning as a new policy paradigm for climate change adaptation. Environmental Policy & Governance 19, 358–373.

16. Collins, K.B. and Ison, R.L. (2009b) Trusting emergence: Some experiences of learning about integrated catchment science with the Environment Agency of England and Wales. Water Resources Management (Published online: 27 May 2009) see http://www.springerlink.com/content/f262271h16517072/

17. Collins, K.B., Ison, R.L. and Blackmore, C.P. (2005) River basin planning project: social learning (Phase 1). Environment Agency, Bristol, UK (http://publications.environment-agency.gov.uk/epages/eapublications.storefront/461d27cd009bdf4e273fc0a80296067e/Product/View/SCHO0805BJHK&2DE&2DE#).

18. Collins, K., Blackmore, C., Morris, D. and Watson, D. (2007) A systemic approach to managing multiple perspectives and stakeholding in water catchments: Some findings from three UK case studies. Environmental Science & Policy 10(6): 564–574.

19. Collins, K.B., Colvin, J. and Ison, R.L. (2009) Building 'learning catchments' for integrated catchment managing: designing learning systems and networks based on experiences in the UK, South Africa and Australia. Water Science & Technology 59(4) 687–693.

20. Cooperider, D.L. and Srivastva, S. (1987) Appreciative inquiry in organizational life, In: R. Woodman and W. Passmore (Eds) Research in Organizational Change and Development, Vol. 1: JAI Press: Greenwich, CT.

21. Coyne, R. and Snodgrass, A. (1991) Is designing mysterious? Challenging the dual knowledge thesis. Design Studies 12 124–131.

22. Dewey, J. (1933) How We Think. A Restatement of the Relation of Reflective Thinking to the Educative Process (revised edn), Boston, D.C. Heath.

23. Foerster, A., Ison, R.L. and Godden, L. (2008) Systemic and adaptive water governance: Reconfiguring institutions for social learning and more effective water managing? Concept

Paper for a Program of Research prepared for a seminar to explore the role of social learning in water policy and law Friday 5 December 2008 (unpublished).

24. Giddens, A. (2009) The Politics of Climate Change. Polity: Cambridge.

25. Glanville, R. (2002) A (cybernetic) musing: some examples of cybernetically informed educational practice. Cybernetics & Human Knowing 9 117–126.

26. Gotches, G. and Ludema, J. (1995) An interview with David Cooperider on appreciative inquiry and the future of OD, Organization Development Journal 13, 5–13.

27. Helme, Marion (2002) Appreciating metaphor for participatory practice: constructivist inquiries in a children and young people's justice organisation. Unpublished PhD Thesis. The Open University: Milton Keynes.

28. Hooker, C.A. (1992) Value and system: notes toward the definition of agriculture. Proc. Centenary Conference, Agriculture and Human Values. UWS-Hawkesbury.

29. Ison, R.L. (1993) Changing community attitudes. The Rangeland Journal 15, 154-66.

30. Ison, R.L. (1994) Designing learning systems: How can systems approaches be applied in the training of research workers and development actors? Proceedings International Symposium on Systems-oriented Research in Agriculture and Rural Development Vol. 2. Lectures and Debates. pp. 369–394. CIRAD-SAR, Montpellier.

31. Ison, R.L. (2001) Systems practice at the United Kingdom's Open University. In J. Wilby and G. Ragsdell (Eds.). Understanding Complexity, (pp. 45–54). Kluwer Academic/Plenum Publishers: New York.

32. Ison, R.L. (2005) Traditions of understanding: language, dialogue and experience. In Meg Keen, Valerie A. Brown and Rob Dyball (eds), Social Learning in Environmental Management. Towards a Sustainable Future, pp. 22–40. Earthscan: London.

33. Ison, R.L. and Armson, R. (2006) Think, Act & Play im Leadership der Kybernetik zweiter Ordnung. Lernende Organisation. Zeitschrift fur systemishes Management und Organisation No 33, September/Oktober (www.lo.isct.net) ISSN 1609-1248 pp. 12–23. [Published by: Institut für Systemisches Coaching und Training, Zielorientierte Entwicklung von Menschen, Teams & Unternehmen GmbH, Lange Gasse 65 A-1080 Wien, Austria].

34. Ison, R.L. and Russell, D.B. (2000a) Exploring some distinctions for the design of learning systems. Cybernetics & Human Knowing 7 (4) 43–56.

35. Ison, R.L. and Russell, D.B. (Eds.) (2000b) Agricultural Extension and Rural Development: Breaking Out of Traditions (pp. 239). Cambridge University Press: Cambridge, UK.

36. Ison, R.L., Bawden, R.D., Mackenzie, B., Packham, R.G., Sriskandarajah, N. and Armson, R. (2007a) From sustainable to systemic development: an inquiry into transformations in discourse and praxis, Invited Keynote Paper, Australia New Zealand Systems Conference 2007, "Systemic development: local solutions in a global environment" 2–5 December 2007, Auckland, New Zealand.

37. Ison, R.L., Blackmore, C.P., Collins, K.B. and Furniss, P. (2007b) Systemic environmental decision making: designing learning systems. Kybernetes 36, (9/10) 1340–1361.

38. Klein, L. (2005) Systemic inquiry – exploring organizations. Kybernetes 34 (3/4), 439–447.

39. Maturana, H. and Poerkson, B. (2004) From Being to Doing: the Origins of the Biology of Cognition. Carl-Auer: Heidlleberg.

40. Open University (1998) Environmental Decision Making. A systems Approach (T860) Open University: Milton Keynes.

41. Reason, P. and Rowan, J. eds (1981) Human Inquiry: a sourcebook of new paradigm research. Wiley: Chichester.

42. Rowan, J. (2001) The humanistic approach to action research. In Handbook of Action Research. P. Reason and H. Bradbury, eds., Sage: London.

43. Schön, D.A. (1995) The new scholarship requires a new epistemology. Change November/ December, 27–34.

44. Seddon, J. (2008) Systems Thinking in the Public Sector: the failure of the reform regime... and a manifesto for a better way. Triarchy Press: Axminster.

45. SLIM (2004a) Developing Conducive and Enabling Policies for Concerted Action. SLIM Policy Briefing No. 5. SLIM (available at http://slim.open.ac.uk/objects/public/slimpb7final.pdf). 4p.
46. SLIM (2004b) SLIM Framework: Social Learning as a Policy Approach for Sustainable Use of Water (available at http://slim.open.ac.uk). 41p.
47. Torbert, W.R. and Cook-Greuter, S.R. (2004) Action Inquiry: The Secret of Timely and Transforming Leadership. Berrett – Koehler Publishers: San Francisco, CA.
48. Vickers, G. (1965) The Art of Judgement: a study of policy making. Chapman and Hall: London.
49. Vickers, G. (1970) Freedom in a Rocking Boat: changing values in an unstable society. Penguin: Harmondsworth.
50. Vickers, G. (1983) Human Systems are Different. Harper and Row: London
51. von Foerster, H. and Poerkson, B. (2002) Understanding Systems. Conversations on Epistemology and Ethics. IFSR International Series on Systems Science and Engineering, Vol 17. Kluwer/Plenum: New York.
52. Winter, M. and Checkland, P.B. (2003) Soft systems: a fresh perspective for project management, Civil Engineering 156 (4), 187–192.

Chapter 11
Systemic Action Research

11.1 Changing Your Situation for the Better

In Chapter 6 I offered some reflections by the late Donald Schön in which he talked about the 'dilemma of rigor or relevance' that all professional practitioners face at some time or another. Some, he said, 'stay on the high ground of technical rationality' whereas others descend into 'the swamp of real-life issues' [42]. When I first read Schön's words I was surprised by his perspicacity. In the 1980s I began the process of making such a choice; from 1984 I began to run my research on parallel tracks – traditional plant biological/ecological research on the one hand and an action research project called the Australian Seed Industry Study on the other [23, 31]. Now, having considerable experience of both types of research I am in a good position to appreciate the strengths and limitations of both.

Much has been written about action research. It is not my aim to add to this literature but to make the case as to why systemic action research (a significant elaboration of action research as I outline below) constitutes a useful and under-exploited vehicle for enacting systems practice in situations of complexity and uncertainty. I have been engaged in action research projects since the mid-1980s and have seen the acceptance of action research in most, but not all, circles grow during that period. Along with the now vast literature is a plethora of approaches to action research praxis [2, 8, 16, 19, 33]. On the other hand less has been written about systemic action research (see [4, 9, 20, 35]).

In this chapter I want to begin by reflecting on research practice because that is the form of practice that most people associate with the production of new knowledge. As I have outlined in several chapters this common view is not one that I hold personally (I consider knowledge arises through engagement and reflection as an ongoing process, whether or not it is explicitly sought as new knowledge [21]). On the other hand I do value the forms of practice that set about purposefully to generate new knowledge, recognising as I do, that all knowing is doing (see Chapter 5).

As I do not want to trigger a defensive response on the part of the reader let me state from the outset that I value highly the contributions that scientific and other forms of research can make to our society. On the other hand, although I see the doing of science and the conduct of evidence-based research as necessary for dealing

R. Ison, *Systems Practice: How to Act in a Climate-Change World*,
DOI 10.1007/978-1-84996-125-7_11, © The Open University 2010.
Published in Association with Springer-Verlag London Limited

with our contemporary circumstances, I don't think it is sufficient.[1] From my perspective systemic action research, my focus in this chapter, can be understood as an antidote to non-reflexive research practice.

I start from a premise that much, if not most, that is done under the rubric of research practice has a concern on the part of the practitioner with changing situations for the better. From this normative position any practice that only concerns itself with the so-called 'discovery of new knowledge' falls short of responsible practice.[2] It does so, whether implicitly or explicitly, through the failure to recognise that any research practice is first and foremost a socially embedded practice. As Law and Urry [26] observe the 'social sciences work upon, and within, the social world, helping in turn to make and to remake it.' The same applies, I would argue, to the so-called 'natural sciences'. In this chapter I will briefly look at some of the ways that social reality can be understood before characterising mainstream research practice as a first-order practice. Drawing on Russell and Ison [36] and Ison and Russell [22], I will then outline the fundamentals of a second-order research practice exemplified by systemic action research. I will pose the question: what makes "systemic action research" systemic? In the final part of the chapter I will exemplify from my own practice what is entailed in doing systemic action research.

11.1.1 The Nature of the Social World

The question of what constitutes the 'social' is, in my experience, rarely engaged with – it is a sort of taken for granted assumption that we all know! Having concerned myself with this question the answer I find most satisfying is, following Maturana, those instances when another arises as a legitimate other. From this perspective I could claim that when President Obama does what he does (if Freedland's descriptions are accurate) then in his interactions with others the social emerges.

Within the logic of bringing forth our worlds, as discussed in Chapter 5, it follows that those who describe themselves as social and natural scientists could all be understood to bring forth the social when they do what they do i.e., in bringing forth understandings and descriptions of other humans, other species or the biophysical world. From this perspective the common understandings of social and natural create an unhelpful dualism. Such a dualism is not conducive to enacting and enhancing the quality of a co-evolutionary dynamic between humans and the biophysical world as described in Chapter 1.

---

[1]I could at this point have provided a critique of evidence-based decision making from a systemic perspective but space does not allow – for some of the elements of such a critique see Mitchell [30] who argues for nursing 'that the notion of evidence based practice is not only a barren possibility but also that evidence-based practice obstructs nursing process, human care, and professional accountability.'

[2]In establishing the Open Systems Research Group at the Open University we explicitly recognised that research was pursued for a social purpose, in our case the pursuit of social justice.

John Law and John Urry [26] claim to be concerned with the power of social science and its methods, arguing that social inquiry and its methods are productive because 'they help to make social realities and social worlds – that is they do not simply describe the world as it is, but also enact it'. They speak of future social investigation getting involved in 'ontological politics' but now being stuck in a nineteenth century nation-state based politics that produces nineteenth century realities.' They go on to claim that the 'social sciences' need to re-imagine themselves, their methods and their 'worlds' for a twenty-first century where social relations appear increasingly complex, elusive, ephemeral and unpredictable. Also that there is a need for 'messy methods' that deal more effectively with 'the fleeting, distributed, multiple, non-causal, chaotic, complex, sensory, emotional and kinasthetic'.[3] From the perspective of their arguments *systems practice* could be considered to exemplify a 'messy method'.

Law and Urry [26] also explore some implications for social science arising from complexity theory particularly as a source of 'productive metaphors and theories for 21st century realities'. In their concerns for a new form of social science research practice they note:

- 'There is no innocence. But to the extent social science conceals its performativity from itself it is pretending to an innocence that it cannot have'
- 'If methods are not innocent then they are also political. They help to make realities. But the question is which realities?'

Law and Urry implicitly invite a (re)consideration of what we have to experience to claim that something is 'social' as well as what is involved in generating a scientific explanation and thus doing 'social science'? I will return to these points below, but first I want to characterise mainstream research as a form of practice – what I term first-order R&D.

11.1.2 The First-Order Research Tradition[4]

As Russell and Ison [36] note: 'Knowledge' and 'applying knowledge' are the very language of R&D (research and development),[5] a language that does not acknowledge its dependency on interpretation. The notion of 'information' as it is commonly used implies that an 'external world' is knowable in a way that is independent

---

[3] Kinesthetic learning is when someone learns things from doing or being part of them. It is claimed that learners have different learning styles which include visual learners, kinesthetic learners, and auditory learners (see http://en.wikipedia.org/wiki/Kinesthetic_learning Accessed 30th September 2010).

[4] The material in this section and the following is edited and extracted from Russell and Ison [36] and Ison and Russell [22].

[5] We use R&D as a noun to break out of the trap of the linear conception of research...and then development.

of the user of the language. In the mainstream rationalistic tradition,[6] the information and the knowledge are 'out there' and one can collect more and more information about the external world and the greater the 'knowledge base', the greater the chances of useful technology and better interventions. The current trend to make technology 'user-friendly' is indicative of the questioning of the naive equation that more information equates with better results. While this questioning might not lead to a questioning of the theoretical paradigm itself, it will lead to the increased development of a technology designed "to facilitate a dialogue of evolving understanding among a knowledgeable community" [49, p. 76].

In a review of rural extension [40, 41] the existing model of agricultural or rural extension was shown not to work well at all. It constituted neither good practice nor good theory. Promotion of innovative technology to the rural community has been based predominantly on the linear extension 'equation':

$$\text{research} \rightarrow \text{knowledge} \rightarrow \text{transfer} \rightarrow \text{adoption} \rightarrow \text{diffusion}$$

A study of the effectiveness of this model showed that research results were adopted by only a specific minority of farmers and that for the majority it was not a viable strategy for agricultural improvement. Experience of the deficiencies of this model in actual practice has led to the emergence of a very different conceptual system based on the idealised "farmer-led" model [10]. Despite the very real differences, both models incorporate current ways of thinking about and doing 'extension'. We concluded, on the basis of experiences contributing to our book in 2000 [21], that it was time to abandon the term 'extension' altogether because of what it has come to mean in practice along with a network of faulty assumptions embedded in its core.

Illustration 11.1

[6] See *Understanding computers and cognition: a new foundation for design* by Terry Winograd, Fernando Flores [49].

The term 'extension' arose from a particular tradition – from the North American land grant university model meaning 'to extend knowledge from a centre of learning to those in need of this knowledge.' Extension in its practice has remained captive of this initial western conception despite differences culturally apparent in say the German "beratung" (to counsel or deliberate) and the French move from 'vulgarisation' (to render popular) to 'development agricole' (involving the whole farming community). More recently in Victoria, Australia the term *extension* has been dropped in favour of 'practice change' but it is far from clear that the conceptual underpinnings, and thus practice, have changed significantly.

What I find particularly disturbing is that the experiences in agriculture and rural development are not well appreciated in other domains of practice such as medicine, but especially in those practices now emerging under the banner of climate change adaptation. It is possible to spot praxis traps whenever one encounters the language of information, knowledge or technology transfer.

The belief that knowledge could be 'transferable' has derived from the associated belief that 'communication' was the process of transmitting information. The media is convinced that we are now in the "Information Age" so it is not surprising that the most widely used metaphor for the practice of extension is that of 'information transfer'. So embedded is this notion, so pervasive has been the obviousness of electronic communication, that challenging the appropriateness of continuing to use this metaphor, is to risk being considered absurd. Risky or not, it must be done! The effectiveness of current practice continues to be judged and to be judged negatively [40, 43]. Not only has the simple notion that knowledge can be transferred from one person to another, as if it were a case of one computer 'talking' to another, been shown to not work in practice, but biologically (as was discussed in Chapter 5), it is clearly not possible.

Shannon and Weaver [44] were the first to use the model of electronic information transfer to refer to human communication. Simply put, they proposed that ideas were coded into signals, the messages (by the sender), and then transmitted to another person (the receiver) who then decoded the message back into the original ideas. The root metaphor has had numerous elaborations in its application such as those variously described as evidencing the conduit metaphor [34] or the hypodermic metaphor. In the first instance, ideas were seen as being packaged into words so as to gain access to the original ideas. The hypodermic understanding was obvious when there was an intention to persuade the other to follow a certain course of action. The effective communicator could 'get under the skin' of the other if he or she could present the information 'persuasively'. David Sless [45] analysed a number of communication models showing how the basic 'information transfer' metaphor still dominated the thinking of many communication theorists.

The prevalence of this established way of understanding communication, despite all the evidence to the contrary [24, 25, 45] shows how difficult it is to unearth a deeply embedded metaphor when it has taken root in the society's unconscious. The process of constructing more fitting metaphors will initially be awkward and cumbersome because we will inevitably have a foot in the old camp of fixed reality, a condition of the knowledge transfer idea, and a foot in the camp of multiple realities, the prerequisite for any new constructions.

The stages of the mainstream, first-order tradition can be characterised as:

- The 'problem' is seen as a mismatch between what is scientifically known and technically feasible, and what is current practice. The new technology is designed by research scientists and is then transferred to the end-users who put it into action to address the problem
- Built into the belief of a technological solution is a conception of the benefits that could be derived from better 'production or innovation systems' or, in the case of the environment, a return to the 'natural ecosystem' state, without consideration of who participates in defining "better" nor how what is perceived as "natural", by some, has come to be constructed
- Social and political insights are specifically added to the R&D equation – usually *ex poste* and in ways that desire social scientists to enhance the uptake or adoption of research results

First-order R&D in its enactment is heavily influenced by the four factors I outlined in Chapter 9 as inimical to the flourishing of systems practice. First-order R&D usually embraces the language of goals and targets. The project is usually the main vehicle for practice. Awareness that there are choices to be made as to how to frame situations rarely enters practice contexts and it is an arena in which the apartheid of the emotions is firmly entrenched. Within research practice second-order R&D and systemic action research offer an alternative and complementary framing for practice. With awareness, first and second-order R&D can be understood as a duality – as potentially gaining the best of the systemic with the systematic (Fig. 2.1).

11.1.3 Creating a Second-Order Research Tradition

Second-order R&D is built on our scientific understanding that human beings determine the world that they experience (see Chapter 5). The application of science demands that we reflect upon how we operate as perceiving and knowing 'observers' who bring forth their experiential worlds through the actual functioning of their nervous systems and the cognitive operation of making distinctions: You have to look in order to see!

The characteristics of second-order R&D can be summarised as:

- The doing (the praxis) is grounded in the extending of an invitation to, and the willing acceptance by, another to join in making a space for mutually satisfying action
- The reality that is brought forth includes the researcher. The relational process of 'bringing forth' constitutes a duality. Thus it is not subjectivity – subjectivity belongs to objectivity
- All participants share the responsibility associated with every outcome
- It involves the study of relationships, particularly their nature and quality rather than entities or objects (in its doing, relational capital is built)

- As science, it is grounded in the explanation of what is experienced (observed) and, unlike philosophy, is not concerned with adherence to, or the explication of, principles. It has no imperative character

The need for explicit contextual grounding is at the heart of this conceptual development. This contextual grounding has to do with an increasing understanding of the social construction of the very concepts of the 'research-development relationship', or any other form of practice and any situation to be understood or improved.

As outlined by Ison and Russell [22] 'each manner of engagement can be considered as if a conversation – from the Latin, con versare, meaning to turn together. From this perspective a conversation is a form of action because one of the implications of being human, and living in language, is that conversation is the primary means by which we coordinate our behaviours. In a coupled first/second order R&D system each manner of conversation can be usefully expressed as a metaphor, and underpinning each metaphor is an epistemology (or theory of knowledge) that has implications for understanding and action. For example, a first-order engagement is often based on a conversation between the stakeholders which is shaped by an emotion of satisfying a known need (information transfer to solve a first order problem). In biological terms, this is the 'seeking' and 'reward' sub-system. The root metaphor in this case is hunger/feeding relationship – the need to satisfy some espoused need'.

The second-order features arise in a conversation between stakeholders that is shaped by a very different emotion, namely, the desire to honour the other's world-of-experience as 'other'. The overarching metaphor in this case is communication understood in terms of the dance-ritual metaphor which is characterised by:

1. Continuity and repetitiveness
2. Its cooperative nature and
3. Its after effect ... it is individually satisfying to all participants [39]

In epistemological terms, the second-order system builds relational capital and follows a second-order logic. The practical implication is that the aware professional, working in the field, needs to be fluent (able to operate) in the two 'language' systems and the two emotional-flow systems in order to be able to deal with both first and second order issues.

11.2 What Makes Action Research Systemic?

My understanding of and approach to systemic action research has been heavily influenced by collaborative work with David Russell and within the Hawkesbury milieu [3, 39]. Working with stakeholders in the semi-arid pastoral zone of New South Wales, Australia [21] we used our understanding of systems thinking and systemic action research to develop an approach to doing R&D (research and development) relevant to the context of the lives of pastoralists in semi-arid Australia. Our experience had been that many action researchers, whilst espousing a systemic epistemology, often privileged in practice a systematic epistemology without awareness that that was what they were doing i.e. in practice they wished to conserve the notion of a fixed reality and the possibility of being objective. My distinction between action research and systemic action research is grounded in the ideas contained in Table 8.1. Expressed simply, the difference is that within systemic action research the 'researcher' understands and acts with awareness that they are part of the researching system of interest under co-construction, rather than external to it.

The motivation for distinguishing systemic action research from action research was to draw attention to the need, we believed, for the researcher to take responsibility for their epistemological commitments. Following Bateson [5] we understood epistemology as the exploration of the underlying premises of our knowing, deciding and action. From this understanding it follows that any practice that claims to be a form of research practice must account for shifts in the bases of knowing, deciding and effective action. Action research is transformed into systemic action research whenever those involved act, or strive to act, with epistemological awareness. It is this which ensures that what is done under the rubric of action research is more than mere activity.

Burns [9] argues also, "that systemic action research offers a 'learning architecture' for change processes that draw on in-depth inquiry, multi-stakeholder analysis, experimental action and experiential learning" (p. 1).

11.3 Doing Systemic Action Research

An outcome of our project was the design of a process to enable pastoralists to pursue their own R&D activities – as opposed to having someone else's R&D outcomes imposed on them. Our design was built around the notion that given the right experiences peoples' enthusiasms for action could be triggered in such a way that those with similar enthusiasms might work together. We understood enthusiasm as [20, 38]:

- A biological driving force (enthusiasm comes from the Greek meaning 'the god within'. Our use of 'god' in this context has no connection with organised religion – our position was to question the commonly held notion that 'information' comes from outside ourselves rather than from within in response to non-specific triggers from the environment)
- An emotion, which when present led to purposeful action
- A theoretical notion – in the sense of explanatory principles
- A methodology – a way to orchestrate purposeful action

We spent a lot of time designing a process that we thought had a chance to trigger peoples' enthusiasms. Our process did in fact enable peoples' enthusiasms to be surfaced and led to several years of R&D activity on the part of some pastoralists supported by ourselves but never determined by us – see Dignam and Major [15], for an account by the pastoralists of what they did. The process we designed did not lead to R&D actions (purposeful activity) in any cause and effect way, rather the purposeful activity taken was an emergent property of peoples' participation in the systemic, experiential learning process that we had designed. Our work has led to a four stage model for doing systemic action research grounded in second-order cybernetic understandings (Fig. 2.3). In summary these were:

1. Stage 1: Bringing the system of interest into existence (i.e., naming the system of interest)
2. Stage 2: Evaluating the effectiveness of the system of interest as a vehicle to elicit useful understanding (and acceptance) of the social and cultural context[7]
3. Stage 3: Generation of a joint decision-making process (a 'problem-determined system of interest') involving all key stakeholders
4. Stage 4: Evaluating the effectiveness of the decisions made (i.e., how has the action taken been judged by stakeholders?)

The way we went about designing the process (i.e., of doing each stage) is described in detail in Russell and Ison [37]. The enactment of the four stages requires awareness of the systemic/systematic distinctions in action i.e. as practice unfolds – they are not just abstracted descriptions of traditions. Our experience is that this is not easy as our early patterning predisposes us to take responsibility for someone else (tell them what to do), and to resort to an assumption about a fixed reality and to forget that my world is always different from your world. That is, we never have a common experience because even though we may have the same processes of perceiving and conceptualising it is biologically impossible to have a shared experience – all we have in common is language (in its broadest sense) with which to communicate about our experience.

In Box 11.1 I describe the process for triggering enthusiasm which was generated from our research [20, 36–38]. I might add that in the doing of this systemic action research project we also provided a scientific explanation of how enthusiasm operates. We did so by following the steps identified by Maturana and Varela [28] and Maturana [27]. Based on their neurobiological research and their concern for the question "how is it that we can know" they proposed that doing science, or providing a scientific explanation for a phenomenon, can be best described as:

1. Describing a phenomenon that has been experienced and doing this in a way that allows others to agree or disagree as to its existence
2. Proposing an explanation for the existence of this phenomenon. This explanation functions as a "generative mechanism" in the sense that when the mechanism operates the phenomenon appears
3. Deducing from the first experience other experiences that are coherent with the first, and which would result from the operation of this mechanism that has been proposed as an explanation and
4. Experiencing the other phenomena that were deduced in step 3[8]

---

[7]In expressing it in this way it is not my intention to exclude the 'ecological' but to recognise that what we regard as ecological is always brought forth in specific instances and contexts.

[8]Maturana notes that while quantification is not essential to this process it may be useful in the deductive phase.

Box 11.1 Triggering of Enthusiasm for Action as a Four Stage Process

(i)   When potential stakeholders are invited to talk (to tell of their experience past, present and anticipated future), the emotional connection occurs through active listening – "I really want to hear what you have got to say". Genuine concern for the other is manifest in the conversation (language in all its manifestations). Practices to create the environment for enthusiasm to emerge through narrative were also developed [38].

(ii) In the course of being listened to (showing respect, not prejudging, nor pushing one's own agenda) an invitation is made consciously or unconsciously. This provides space for options which lead to mutually satisfying action to be generated. In this process one is not naming these options for the other. In this form of interaction there is the possibility of self awareness and triggering of latent ideas or concepts.

(iii) Action is taken in the domain of interactions – to maintain the conversation. Processes (i) and (ii) will not lead to all individuals joining the conversation. Action does and continues to occur in other domains. For some however, new ways of being are triggered, the possibilities for which existed all the time. Thus there is no transfer of anything nor working towards predetermined plans or goals. What is triggered is what Maturana [27, p. 42] describes as emotions and moods or body dispositions for actions and where he distinguishes 'moods as emotions in which the observer does not distinguish directionality or possibility of an end for the type of actions that he or she expects the other to perform'.

(iv) External resources (e.g., money, technology) become amplifiers or suppressors of enthusiasms for action. In our research we have encountered both. A further two stages are necessary for the development of enthusiasm as R&D methodology:

(v)   Careful attention is necessary in the design of processes to bring people together who share common enthusiasms for action. We came to understand that consensus suppresses enthusiasm for action and that at many levels of activity the search for consensus is an inappropriate objective. We experience this in everyday life when in our relationships we are often forced to compromise and lose our energy and vitality for action. The alternative to consensus is to value diversity or difference [7].

(vi) Prior to about 1829 the word "enthusiasm" was associated with the emergence of the new radical religions and was seen as an emotional driving power devoid of reason and rationality. To avoid the prospect of enthusiasm giving rise to 'disordered intellect and action', we propose the need for cycles of critical reflection to be an essential part of the use of enthusiasm as methodology. There are no doubt many ways which this could be achieved (see Russell and Ison [38]).

I include a description of this process for generating a scientific explanation to make the point that one does not have to appeal to an independent reality to pursue the generation of scientific explanations. In my experience systemic action research is also capable of generating rigorous, publically verifiable 'knowledge'.

11.4 Enhancing Action Research with Systems Thinking and Practice[9]

Many action researchers, including Kurt Lewin, often regarded as the originator of Action Research (AR), have been influenced by systems thinking but what is not always clear is the extent to which this is done purposefully – with awareness of the different theoretical and practical lineages depicted in Fig. 2.3. Engaging with systems offers a set of conceptual tools which can be used to good effect in AR (e.g. Table 2.1). There are other potential advantages for AR practitioners. Firstly, systemic understandings enable reflections on the nature of research practice, including AR practice itself. This, I suggest, can be understood by exploring purpose. Secondly, there is a rich literature of how different systems approaches or methodologies, including systems tools and techniques, have been employed within AR projects to bring about practical benefits for those involved (e.g. [13] – see also Chapter 12). I explore some of these potential benefits in this final section.

The distinctions between what constitutes research (within the phrase systemic action research or action research) and how it might be differentiated from 'inquiry' or 'managing' is, I suggest, contested. AR has been a concern within the 'applied systems' lineage (Fig. 2.3) for over 30 years [11]; within this lineage Holwell [18] proposes three concepts that constitute action research as legitimate research: recoverability, iteration, and the purposeful articulation of research themes. She exemplifies her claims with a description of 'a program of action research with the prime research objective of understanding the ... nature of the contracting relationship [within the UK National Health Service] with a view to defining how it could be improved' (p. 5). The project was 'complex in execution, including several projects overlapping in time' covering work from different bodies of knowledge, and was undertaken by a seven-member multidisciplinary team with different intellectual traditions. The issues explored crossed many organisational boundaries; the work was done over a 4 year period and followed a three-part purposeful, but emergent design [12].

Within the Checkland and Holwell lineage they emphasise that the research process must:

1. Be recoverable by interested outsiders – 'the set of ideas and the process in which they are used methodologically must be stated, because these are the means by which researchers and others make sense of the research'

---

[9] This section is an edited version of parts of Ison [19].

2. Involve the researcher's interests embodied in themes which are not necessarily derived from a specific context. 'Rather, they are the longer term, broader set of questions, puzzles, and topics that motivate the researcher [and] such research interests are rarely confined to one-off situations' (I assume here they might also claim that themes can arise through a process of co-research or 'researching with' [29] and thus can be emergent as well)
3. Involve iteration, which is a key feature of rigour, something more complex than repetitions of a cycle through stages 'if thought of in relation to a set of themes explored over time through several different organizational contexts' [18] and
4. Involve the 'articulation of an epistemology in terms of which what will count as knowledge from the research will be expressed' [12]. As with Russell [35] they further claim that the 'literature has so far shown an inadequate appreciation of the need for a declared epistemology and hence a recoverable research process' (p. 20)

What is at issue here, as described in Chapter 8 (Section 8.2.1) are the differences between what I have called big 'R' (a particular form of purposeful human activity) and little 'r' research (something that is part of daily life, as is learning or adopting a 'researching or inquiring' attitude) although the boundaries are not always clear. Take recoverability. How in practice is this achieved? The most common form is to write an account of what has happened ensuring that certain elements of practice and outcome, including evidence are described (e.g. FMS in Fig. 3.5). But writing is itself a form of purposeful practice (done well or not well, as the case may be) that is always abstracted from the situation – it is always a reflection on action and is never the same as the actual doing. Of course recoverability could be achieved by other means – by participation (i.e. apprenticeship and the evolution of 'craft' knowledge) or through narrative, which may or may not be writing. It seems to me the key aspiration of recoverability is to create the circumstances where an explanation is accepted (by yourself or someone else – see Fig. 1.2) and as such to provide evidence of taking responsibility for the explanations we offer. It has a 'could I follow a similar path when I encounter a similar situation' quality about it. The alternative, as von Foerster [47] puts it is to avoid responsibility and claim correspondence with some external or transcendental reality.

As I outlined at the beginning of this chapter, in my own case I came to action research through my awareness that my traditional discipline-based research was not addressing what I perceived to be the 'real issues.' I had a crisis of relevance and rejected the high ground of technical rationality for the swamp of real-life issues. Warmington [48] was a major initial influence but my purpose was to do more relevant big 'R' research for which I sought and successfully gained funding [31]. It was during subsequent work on the CARR (Community Approaches to Rangelands Research) project, as reported in Ison and Russell [21], that my own epistemological awareness shifted – something that I claim is necessary for the shift from action to systemic action research. My experience is that such a shift has an emotional basis; mine stemmed from an emotionally experienced crisis of identity in which I had to confront the question of whether I cared more for what worked or for the belief system I had implicitly accepted.

I have now come to an appreciation that the core concerns for systemic AR practice are (i) awareness; (ii) emotioning; and (iii) purposefulness. As I outlined in Chapter 9 the epistemologically aware researcher can be seen as both chorographer (one versed in the systemic description of situations) and choreographer (one practised in the design of dance arrangements) of the emotions [35]. As also acknowledged in the distinctions between participatory action research and action science [1, 14] and first, second and third person inquiry [32] there is a need to be clear as to who takes responsibility for bringing forth a researching system. Any account of big 'R' research needs to ask the question 'who is the researcher at this moment in this context? Is it me, us or them?' Answers to this question determine what is ethical practice, bounding for example, what is mine from what is ours and what is yours [6, 17, 46].

References

1. Argyris, C. and Schön, D. (1991) Participatory action research and action science compared. A commentary, pp. 85–96. In William Foote Whyte ed. Participatory Action Research, Sage: Newbury Park.
2. Barton, J., Stephens, J. and Haslett, T. (2009) Action research: Its foundations in open systems thinking and relationship to the scientific method. Systemic Practice & Action Research 22, 475–478.
3. Bawden, R.J. (2005) Systemic development at Hawkesbury: Some personal lessons from experience. Systems Research & Behavioral Science 22: 151–164.
4. Bawden, R.J. Packham, R.G. (1991) Improving agriculture through systemic action research, in V. Squires and P. Tow (eds.) Dryland Farming Systems: A Systems Approach, Sydney University Press: Sydney, Chapter 20, pp. 262–71.
5. Bateson, Gregory (1999/1972) Steps to an Ecology of Mind. University Press: Chicago.
6. Bell, S. (1998) Self-reflection and vulnerability in action research: bringing forth new worlds in our learning. Systemic Practice & Action Research 11, 179–191.
7. Blackmore, C.P., Ison, R.L. and Jiggins, J. (2007) Social learning: an alternative policy instrument for managing in the context of Europe's water. Environmental Science & Policy 10, 493–498.
8. Bradbury, H., and Reason, P. (eds.) (2001) Handbook of Action Research. Participative Inquiry and Practice. Sage: London.
9. Burns, D. (2007) Systemic Action Research. A Strategy for Whole System Change. Policy Press: Bristol.
10. Chambers, R., A. Pacey and L. Thrupp (eds.) (1989) Farmer First. Intermediate Technology Publications: London.
11. Checkland, P.B. and Holwell, S. (1998a) Action research: Its nature and validity. Systemic Practice & Action Research, 11, 9–21.
12. Checkland, P.B. and Holwell, S. (1998b) Information, Systems and Information Systems, Wiley: Chichester.
13. Checkland, P.B. and Poulter, J. (2006) Learning for Action. A short definitive account of soft systems methodology and its use for practitioners, teachers and students. Wiley: Chichester.
14. Dash, D.P. (1997) Problems of action research – as I see it. Working Paper No. 14, Lincoln School of Management, University of Lincolnshire and Humberside, p. 9.
15. Dignam, D., and Major, P. (2000) The graziers' story. In Ison, R.L. and Russell, D.B. (eds.) Agricultural Extension and Rural Development: Breaking out of Traditions. Cambridge University Press: Cambridge, UK. pp. 189–204.

16. Flood, R.L. (2000) The relationship of systems thinking to action research. In Hilary Bradbury and Peter Reason (eds.) Handbook of Action Research, Sage: London, pp. 133–144.
17. Helme, M. (2002) Appreciating metaphor for participatory practice: constructivist inquiries in a children and young people's justice organisation. PhD Thesis, Systems Department, The Open University.
18. Holwell, S. E. (2004) Themes, iteration and recoverability in action research. Information Systems Research: Relevant Theory and Informed Practice (IFIP working group 8.2 conference), Manchester, 15th–17th July 2004, Kluwer.
19. Ison, R.L. (2008) Systems thinking and practice for action research. In Reason, P., and Bradbury, H. (eds.) The Sage Handbook of Action Research Participative Inquiry and Practice (2nd edn). Sage: London, pp. 139–158.
20. Ison, R.L. and Russell, D.B. (2000a) Exploring some distinctions for the design of learning systems. Cybernetics & Human Knowing 7 (4) 43–56.
21. Ison, R.L. and Russell, D.B. (eds.) (2000b) Agricultural Extension and Rural Development: Breaking Out of Traditions. Cambridge University Press: Cambridge, UK. p. 239.
22. Ison, R.L. and Russell, D.B. (2010) The worlds we create: designing learning systems for the underworld of extension practice. In Jennings, J., Packham, R.P. and Woodside, D. (eds.). Australasian Extension Publication (AEP), ASEN (in press).
23. Ison, R.L. Potts, W.H.C. and Beale, G. (1989) Improving herbage seed industry productivity and stability through action research. Proc. XVI International Grassland Congress, Nice, pp. 685–86.
24. Krippendorff, K. (1993) Major metaphors of communication and some constructivist reflections on their use. Cybernetics & Human Knowing 2, 3–25.
25. Krippendorff, K. (2009) Conversation. Possibilities of its repair and descent into discourse and computation. Constructivist Foundations 4 (3), 138–150.
26. Law, J. and Urry, J. (2004) Enacting the social. Economy and Society 33 (3) 390–410.
27. Maturana, H.R. (1988) Reality: the search for objectivity or the quest for a compelling argument. Irish Journal of Psychology 9: 25–82.
28. Maturana, H.R. and Varela, F.J. (1988) The Tree of Knowledge: The Biological Roots of Human Understanding, Shambhala: Boston, MA.
29. McClintock, D., Ison, R.L. and Armson, R. (2003) Metaphors of research and researching with people. Journal of Environmental Planning & Management, 46 (5) 715–731.
30. Mitchell, Gail J. (1999) Evidence-based practice: Critique and alternative view. Nursing Science Quarterly 12 (1) 30–35.
31. Potts, W.H.C. and Ison, R.L. (1987) Australian Seed Industry Study. Occasional Publication No. 1, Grains Council of Australia, Canberra. Vol. 1. 239 pp. Vol. 2 (Appendices). 316 pp.
32. Reason, P. (2001) Learning and change through action research. In Creative Management J. Henry, ed. Sage: London.
33. Reason, P., and Bradbury, H. (eds.) (2008) The Sage Handbook of Action Research Participative Inquiry and Practice (2nd edn). Sage Publications: London,
34. Reddy, Michael J. (1979) The conduit metaphor – A case of frame conflict in our language about language. In Metaphor and Thought, Andrew Ortony (ed.), Cambridge University Press: Cambridge.
35. Russell, D.B. (1986) How we see the world determines what we do in the world: Preparing the ground for action research. Mimeo, University of Western Sydney: Richmond.
36. Russell, D.B. and Ison, R.L. (2000a) The research-development relationship in rural communities: an opportunity for contextual science. In Ison, R.L. and Russell, D.B. (eds.) Agricultural Extension and Rural Development: Breaking out of Traditions. Cambridge University Press: Cambridge, UK. pp. 10–31.
37. Russell, D.B. and Ison, R.L. (2000b) Designing R&D systems for mutual benefit. In Ison, R.L. and Russell, D.B. (eds.) Agricultural Extension and Rural Development: Breaking out of Traditions. Cambridge University Press: Cambridge, UK. pp. 208–218.

38. Russell, D.B. and Ison, R.L. (2000c) Enthusiasm: developing critical action for second-order R&D. In Ison, R.L. and Russell, D.B. (eds.) Agricultural Extension and Rural Development: Breaking out of Traditions. Cambridge University Press: Cambridge, UK. pp. 136–160.
39. Russell, D.B. and Ison, R.L. (2005) The researcher of human systems is both choreographer and chorographer. Systems Research & Behavioural Science 22, 131–138.
40. Russell, D.B., Ison, R.L., Gamble, D.R. and Williams, R.K. (1989) A Critical Review of Rural Extension Theory and Practice. Australian Wool Corporation/University of Western Sydney (Hawkesbury) 67 pp.
41. Russell, D.B., Ison, R.A., Gamble, D.R. and Williams, R.K. (1991) Analyse Critique de la Theorie et de la Pratique de Vulgarisation Rurale en Australie. INRA, France. 79 pp.
42. Schön, D.A. (1995) The new scholarship requires a new epistemology. Change November/December, 27–34.
43. Scoones, Ian and Thompson, John (1994) Beyond Farmer First: Rural People's Knowledge, Agricultural Research and Extension Practice. Intermediate Technology Publications: London.
44. Shannon, C. and Weaver, W. (1949) The Mathematical Theory of Communication, University of Illinois Press: Urbana, IL.
45. Sless, David (1986) In Search of Semiotics, Croom Helm: London.
46. SLIM (2004) SLIM Framework: Social Learning as a Policy Approach for Sustainable Use of Water (available at http://slim.open.ac.uk) 41 p.
47. von Foerster, H. (1992) Ethics and second-order cybernetics, Cybernetics & Human Knowing 1, 9–19.
48. Warmington, A. (1980) Action research: its methods and its implications. Journal of Applied Systems Analysis 7, 23–39.
49. Winograd, T. and Flores, F. (1987) Understanding Computers and Cognition: A New Foundation for Design, Addison Wesley: New York.

Chapter 12
Systemic Intervention

12.1 Systems Practice in the National Health Service (UK)

Many thousands of mature age students have studied systems courses at The Open University (UK) but apart from examination results we know surprisingly little about them – who they are, whether they use Systems and if not, why not! I did not find this very satisfactory so I set out to find out something about some of those students who had studied Systems [2, 4]. The focus of the inquiry that was initiated was *how the study of Open University Systems courses made a difference to the professional and personal lives of individuals who are working or have worked in the National Health Service (NHS)*.[1] The main conclusions from the interviews in relation to the focus of the inquiry were:

1. People from a very wide range of posts and responsibilities in primary and secondary NHS Trusts usefully apply Open University Systems course learning at work, and they would all recommend Open University Systems courses for people in the NHS, including for those doing the same job as themselves
2. Studying Systems makes a difference by opening up a different way of thinking about situations to the way that some people learn in other educational experiences, for example, professional clinical training. For some people Systems thinking was a revelation; for others Systems thinking 'fitted' and supported how they thought already, but previously experienced as 'different' from other people
3. Systems concepts, methodologies and techniques are particularly useful for interdisciplinary work and projects. Those specifically mentioned include patient pathway and patient access improvement, patient/service user community health care involving social workers and health professionals, the management of challenging behaviour in hospital, and risk assessment
4. Systems thinking is also particularly useful for preparing clinicians for management roles, especially for those working in hospitals (secondary care), and Open University

---

[1]The report that was produced was based on 20 individual semi-structured interviews in August and September 2002. All of the people interviewed were working or had worked for the NHS in England, Wales or Scotland. All had experience of OU Systems courses. Most (16) held the OU Diploma in Systems Practice and studied OU Systems courses at level 2, level 3 and Summer School.

R. Ison, *Systems Practice: How to Act in a Climate-Change World*, DOI 10.1007/978-1-84996-125-7_12, © The Open University 2010. Published in Association with Springer-Verlag London Limited

Systems teaching could be a bridge between current NHS short introductory sessions for prospective managers, and management training for those in post

5. This usefulness for practice and management in NHS work derives from learning from Open University Systems courses concerning:

    *   A way of thinking that is based on a holistic understanding and appreciation of interconnectivities in complex situations, such as those experienced in the NHS, but also draws on systematic approaches
    *   Recognising and attending to different perspectives, specifically those of different professions in interdisciplinary work and management roles, and the perspectives of patients and carers
    *   A tool set of methodologies and techniques – those identified included diagramming and modelling, Soft Systems Methodology, Systems Failures and material concerning change management (hard and soft systems approaches)

6. Practice-based exercises and assignments enabled people to relate the material to their own current experiences in and out of work. For example, several people associated their Systems courses project work with improving practices and services at work as well as significant personal learning

Other insights emerged through this inquiry. For some, Systems has a latency in terms of learning and change. This was illustrated by examples of people dusting off their Systems course notes when they had a tricky problem at work, and then going on to adapt the techniques for other work situations. People continued to develop and adapt their Systems practice long after their study had concluded. There were some glimpses of contextual issues that prevented systems thinking making a difference in people's practice, for example uninterested managers, limited power (actual and perceived).

As I have explained earlier, Open University students have been taught historically to engage with situations of concern through the use of some form of systems diagramming. This may be part of a particular method or methodology (e.g., rich picturing as part of SSM practice) but more often than not it is as a precursor to choice about method or tool. In this way the Open University pedagogic approach has attempted to facilitate those learning Systems to be as open to their circumstances as possible. Systems diagramming is an approach worthy of inclusion in systemic inquiry, systemic action research or systemic intervention.

Under climate-change scenarios human health issues will emerge to add to the already complex issues associated with technological innovation, cost, over servicing, inequalities etc. Health is an arena or domain in need of systems practice skills and understandings and, as evidenced by my small survey, health professionals find systems thinking and practice useful for what they do. The next Reading adds to the evidence base of how Systems can contribute to situation improvement in the health domain through a third 'framing' for systems practice – *systemic intervention* as pioneered by Gerald Midgley.

12.2 Systemic Intervention

In this Reading Gerald offers a response to the many calls that have been made for a systems approach to public health. His response is to offer a methodological approach for systemic intervention that (i) emphasises the need to explore stakeholder values and boundaries for analysis, (ii) challenges marginalisation, and (iii) draws upon a wide range of methods (from the Systems literature and beyond) to create a flexible and responsive systems practice. He presents and discusses several well-tested methods with a view to identifying their potential for supporting systemic intervention for public health. As with earlier readings I invite you to read with a critical awareness. The question I would pose at the outset is: what is it that is done when doing systemic intervention?

Reading 8 Opportunities and demands in public health systems: Systemic Intervention for public health[2,3]

Gerald Midgley

Introduction

Because of the Complexity of many public health issues, where numerous interacting variables need to be accounted for and multiple agencies and groups bring different values and concerns to bear, it is not uncommon for people to call for a systems approach [1–5]. This should not be surprising, as the whole concept of public health is founded on the insight that health and illness have causes or conditions that go beyond the biology and behavior of the individual human being. If I can give an overly simplistic definition of systems thinking as "looking at things in terms of the bigger picture" (not a definition I would want to defend in a rigorous academic fashion, but adequate for my purposes), then it should be immediately apparent that public health is already founded on a systemic insight.

Because many public health professionals are calling for a systems approach, I offer a set of methodological concepts that I have found useful in my own practice to frame systemic inquiry. Of course, many different systems methodologies have been developed over the years. There are far too many to list, let alone review (see Midgley [6] for a wider set of readings). However, the methodology I want to introduce here, which I have called systemic intervention (more detailed information can be found elsewhere [7]), has the advantage of taking a pluralistic approach to the design of methods. It provides a rationale for creatively mixing methods from a variety of sources, yielding a more flexible and responsive approach than might be possible with a more limited set of tools.

(continued)

[2]Midgley, G. (2006) 'Opportunities and Demands in Public Health Systems: Systemic intervention for public health', *American Journal of Public Health*, Vol 96, No. 3.

[3]Gerald Midgley is with the Institute of Environmental Science and Research, Christchurch, New Zealand; and the School of Management, Victoria University of Wellington, New Zealand; and the Centre for Systems Studies, Business School, University of Hull, England.

I will outline this methodology before reviewing a selection of other systems approaches that have been designed for different purposes. We can borrow some useful methods from these approaches, which can then be woven into systemic intervention practice (and more traditional scientific methods plus methods from other sources can be drawn upon in the same way). Two brief practical examples of systemic intervention illustrate my argument.

Systemic Intervention

I define intervention as "purposeful action by an agent to create change." [7 p. 8] (I accept that this definition raises questions about purpose and agency, but these are addressed in other writings [7, 8]). Note that this emphasis on intervention contrasts with the usual scientific focus on observation – although, unlike some authors who champion intervention [9], I do not regard it as incompatible with scientific observation. Methods for observation can be harnessed into the service of intervention [8].

Building on the above definition, I characterize systemic intervention as "purposeful action by an agent to create change *in relation to reflection upon boundaries* [italics in original]" [7 p. 8]. One common assumption made by many systems thinkers is that everything in the universe is either directly or indirectly connected with everything else [10–18]. However, human beings cannot have a "God's-eye view" of this interconnectedness [19]. What we know about any situation has limits, and it is these limits that we call boundaries [19, 20]. Comprehensive analysis is therefore impossible [19–21]. Nevertheless, by acknowledging that this is the case, and by explicitly exploring different possible boundaries for analysis, we can, paradoxically, achieve greater comprehensiveness than if we take any single boundary for granted [7, 20–22]. I call this process of exploration "boundary critique," which, for me, is the crux of what it means to be systemic [7].

Boundary Critique

As far as I am aware, the term boundary critique was first coined by Ulrich [23] to refer to his own methodological practice, but here I am using it more broadly as a label for the concern with boundaries that is present in the writings of several authors, starting with Churchman [19].

Churchman's [19] basic insight is that boundary judgments and value judgments are intimately linked. Values direct the drawing of the boundaries that determine who and what is going to be included in an intervention, so the most ethical systems practice is one that involves pushing out the boundaries as far as possible so that a wide set of stakeholder values and concerns can be accounted for (but without compromising comprehension through over inclusion).

(continued)

Reading 8 (continued)

However, Ulrich [20] argues that in practice it is often difficult to push out the boundaries in this way: time, resource, and other constraints can intrude. Ulrich therefore stresses that boundary critique should involve the justification of choices among boundaries, and should be a rational process (the widest boundary not necessarily being the most rational, given practical considerations). For Ulrich [20] (following Habermas [24]), all rational arguments are expressed in language, and language is primarily a tool for dialogue, so a boundary judgment is only truly rational if it has been agreed upon in dialogue with all those involved in and affected by an intervention. Stakeholder participation (i.e., all those involved or affected) in decision making is therefore crucial to boundary critique.

In my own research on stakeholder participation, I am interested in what happens when 2 or more groups of people make different value/boundary judgments and the situation becomes entrenched. As an aid to understanding such situations, I offer a generic model of marginalization processes that explains the persistence of conflict between stakeholders [7]. Stakeholders and issues can both be marginalized, and this marginalization can become institutionalized. The generic model and some detailed examples of marginalization have been published elsewhere [7, 25–27].

I suggest that the focus of boundary critique on stakeholder participation and marginalization makes it strongly relevant to the "new public health," [28] which is particularly concerned with addressing disadvantage and social exclusion. For details of how boundary critique can be operationalized in health-related and other interventions, see some of the practical examples in the literature [7, 29–32].

As a brief illustration, in the late 1990s I worked with 2 colleagues (Alan Boyd and Mandy Brown) on a project to facilitate the design of new services for young people (aged less than 16 years) living on the streets [32]. We recognized (and all the relevant stakeholders concurred) that it was crucial for young people to be core participants in the research. This was a boundary judgment about participation that would have important consequences for the issues to be considered in the design process. The young people had quite specific concerns that they wanted to be addressed, and some of these would almost certainly have been omitted if participation had been limited to professionals alone.

However, when involving young people, we had to be aware that there was a double danger of marginalization: in general, young people under 16 years of age are viewed as less "rational" than adults. Also, these particular young people could easily have been stereotyped as troubled and untrustworthy teenagers (because, in order to survive on the streets, many of them had to resort to begging, petty crime, or prostitution). Therefore, in setting up design workshops, we gave the young people space out of the hearing of professionals to develop their

(continued)

Reading 8 (continued)

ideas (an empowerment technique) and we used exactly the same planning methods as we used with the adult participants to generate proposals for change. This allowed a direct comparison to be made between the ideas from the young people and adults, and prevented the kind of marginalization that might have occurred if we had used a more "playful" approach with the young people and a more traditional "rational planning" method with the professionals. It would have been easy, if we had done the latter, for the professionals to have viewed only their own output as the "proper" plan. This was just 1 of many issues that we explored and addressed through our boundary critique.

Methodological Pluralism

In addition to boundary critique, I also advocate 2 forms of methodological pluralism. The first is learning from other methodologies to inform one's own. This way, each agent has a continually developing systems methodology. We no longer have to accept a situation where people build a methodology like a castle and then defend it against others who want to breach the castle walls. Rather, if people begin to see methodology as dynamic and evolving, they can learn from others on an ongoing basis [7].

The second form of methodological pluralism is about drawing upon and mixing methods from other methodologies. The wider the range of methods available, the more flexible and responsive our systems practice can be [7, 33–43]. No methodology or method (whether it comes from the systems tradition or elsewhere) can do absolutely everything people might want. Therefore, being able to draw upon multiple methods from different paradigmatic sources can enhance the systems thinking resource we have available for intervention. See Luckett and Grossenbacher [44] and Boyd et al. [32] for some practical examples of methodological pluralism in systemic public health planning.

As a brief illustration, the aforementioned project to facilitate the design of new services for young people living on the streets used a number of different inter-linked methods and techniques:

- Individual interviews with young people, foster caretakers, and retailers
- The use of photographs and cards with evocative pictures to stimulate ideas
- A focus group with staff working in a children's home
- Rich pictures (visual depictions of the problem situation using drawings and arrows showing the links between key issues – see the "Soft Systems Methodology" section of this article for the origins of this technique)
- A synergy of 2 systemic planning methods (see the "Interactive Planning" and "Critical Systems Heuristics" sections of this article for details) implemented in separate stakeholder and multiagency workshops

(continued)

Reading 8 (continued)

- Values mapping (a method we developed to visualize people's values and the logical connections between them)
- Small group, multiagency action planning
- The production of reports, magazines, and posters for multi-audience dissemination; and
- Formative evaluation (feedback questionnaires filled in by participants)
- In the view of the research team, [32] no existing methodology was able to provide all the methods needed for this project. Methodological pluralism was absolutely necessary

Added Value

Arguably, the main added value of systemic intervention compared with earlier systems approaches is its synergy of boundary critique and methodological pluralism [32]. If boundary critique is practiced on its own, it is possible to generate some interesting sociological analyses, but there is a danger that these will not effect change unless other more action-oriented methods are used too [37]. Also, embracing methodological pluralism without up-front boundary critique can give rise to superficial diagnoses of problematic situations. If a complex issue is defined from only 1 limited perspective without reflecting on values and boundaries, and issues of marginalization are neglected, then the outcome could be the use of a systems approach that misses or even exacerbates significant social problems [7, 45] The synergy of boundary critique and methodological pluralism ensures that each aspect of systemic intervention corrects the potential weaknesses of the other [7, 32].

Other Resources for Systems Thinking

Arguably, one of the great strengths of the systems movement is the variety of methods that have been developed to serve different purposes over the years [6]. If we can begin to harness this variety into a form of systems practice that still keeps the idea of reflecting on value and boundary judgments at its core, I believe we will have a great deal to offer public health in the coming years. Below I provide some examples of other systems approaches, which have methods that can be incorporated into systemic intervention. These have been widely applied in practice and offer tools that I have found useful in my own public health research.

System Dynamics

System dynamics [46–51] offers methods for modeling complex feedback processes and considering possible impacts of changes to the system of concern. By experimenting with a model, decision makers are able to anticipate possible emerging scenarios that could follow from a new policy initiative or intervention.

(continued)

Reading 8 (continued)

System dynamics has been used to address a number of significant public health issues [52, 53]. It gives public health professionals some useful tools to model feedback processes in a manner that can not only help to make transparent why certain health effects might occur at the population level, but can also help policymakers anticipate counterintuitive effects of public health initiatives. As Forrester [54] has demonstrated, some policies, introduced with the best of intentions, have the opposite effects of those that are desired. By modeling the feedback loops that stabilize and/or destabilize the system of concern, the approach can highlight surprising side effects of policy options that might not otherwise have been visible in advance of implementation.

The Viable System Model[4]

The second methodology of interest is the viable system model, [55–60] which proposes that for an organization to become and remain viable in a complex and rapidly changing environment, it must carry out each of the following 5 functions.

- Operations: the provision of products or services that address particular needs in the organization's environment
- Coordination: ensuring that the operational units work together and communicate effectively
- Support and control: especially with regard to distributing resources, providing training, gathering and distributing information about quality, etc.
- Intelligence: the forecasting of future needs, opportunities, and threats. This involves a comparison between the external requirements placed upon the organization and its internal capacity; and
- Policymaking: setting long-term goals and objectives

According to the viable system model, the key to effective organization is not only to make sure that all 5 functions exist, but also to ensure that communications among the functions are appropriate and effective. Together, these functions manage the information and decision flows necessary for effective organization, and consequently each function is of equal importance. The model can be used to diagnose current organizational failings or to design entirely new organizations.

Given the complexities of public health policymaking and service delivery, organizational viability is an important factor. For professionals to be able to respond adequately to the issues they face, they need to have an effective organizational infrastructure behind them. The viable system model could make a useful contribution to organizational development.

(continued)

---

[4]See Fig. 8.5 – the viable system model.

Interactive Planning

Although system dynamics and the viable system model involve modeling ecological, social, and/or organizational systems, other methodologists have moved away from modeling to focus on the facilitation of dialogue among stakeholders who bring different insights to bear on complex issues. An example is Ackoff [61–63], whose methodology of interactive planning seeks to liberate and harness the knowledge and creative abilities of everybody in (and often including stakeholders beyond) an organization to produce a plan of the ideal future that the organization can work toward. The plan may take some time to implement, perhaps many years, but it offers a feasible set of targets for the longer term. A key idea is that the plan should be wide enough and creative enough to "dissolve" any disagreements among participants. The transformation it proposes should result in the commitment of all concerned.

The approach can be represented in the form of 3 stages: (1) establishing planning boards (every role in the organization should be represented in planning, with participation as widespread as possible); (2) generating desired properties of the organization's products and/or activities (this is "ends planning," conducted under conditions of minimum constraint with only technological feasibility, viability, and adaptability limiting proposals); and (3) producing the plan itself ("means planning," where all sections of the organization agree on how to move forward).

I have used aspects of Ackoff's work in my own public health research, for example, to look at how the mental health and criminal justice systems would have to be changed to prevent people with mental health problems from inappropriately ending up in prison [7, 64]. If organizations are willing to commit the resources to participative planning, I believe this is a useful approach that can help people move beyond everyday fire fighting toward the formulation of inspiring (but still feasible) long-term visions of how public health can be improved. My only caveat is that in the area of public health it will usually be important to extend participation beyond the boundaries of a single organization to take in other agency representatives and community groups. I have always used interactive planning in this wider participative manner, and it puts some responsibility on the systems practitioner to ensure that marginalized groups are properly included [7].

Soft Systems Methodology

Another approach that can be used to facilitate dialogue among stakeholders is soft systems methodology [65, 66]. This encourages participants in intervention to generate issues through ongoing explorations of their perceptions, allowing people to model desirable future human activities. These models of

(continued)

Reading 8 (continued)

future human activities can then be used as a basis for guiding actual human activities in the world. However, to ensure that the models will indeed be useful, it is necessary for participants to relate them back to their perceptions of their current situation. In this way, possibilities for change can be tested for feasibility.

The methods of soft systems methodology, which are often operational-ized in a workshop format, can be summarized as follows: (1) Consider the problem situation in an unstructured form; (2) Produce a "rich picture" (a visual representation – with pictures and arrows to represent links between issues – of the current situation); (3) Identify possible "relevant systems" that might be designed to improve the situation, and harmonize understandings of these by exploring who should be the beneficiaries of a proposed system change, who should carry it out, what the transformation should be, what worldview is being assumed, who could prevent the change from happening, and what environmental constraints need to be accepted; (4) Produce a "con-ceptual model" for each relevant system (a map of the interconnected human activities that need to be undertaken if the system is to become operational); (5) Refer back to the rich picture to check the feasibility of the ideas; (6) Produce an action plan; and (7) Proceed to implementation. Of course, par-ticipants need to move backward and forward among these activities, harmo-nizing the outputs from each one with the others – the activities should not be implemented mechanistically in a linear sequence.

Soft systems methodology has been used in several public health and health management interventions [67, 68]. It provides a useful language to ensure that ongoing planning retains a systemic focus, and can support people in making accommodations to find acceptable ways forward when they have different perspectives on an issue. I have found it particularly useful for mul-tiagency planning – for example, when facilitating a debate among 19 agency representatives who wanted to cooperate on the design of a counseling ser-vice that could be activated in the event of a major disaster, but their different perspectives were obstructing progress. Over 6 days, the agencies came to an agreement that resulted in the design, funding, and implementation of the counseling service [7, 69].

Critical Systems Heuristics

The final methodology I want to review is Ulrich's critical systems heuristics [20, 70]. As mentioned previously (in the section on "Boundary Critique"), Ulrich asks, when people make decisions on who to consult and what issues to include in planning, how can people rationally justify the boundaries they use? [20] An important aspect of Ulrich's thinking about boundaries is that boundary and value judgments are intimately linked [20]: the values adopted will direct

(continued)

Reading 8 (continued)

the drawing of boundaries that define the knowledge accepted as pertinent. Similarly, the inevitable process of drawing boundaries constrains the values that can be pursued. Being concerned with values, boundary critique is an ethical process. Because of the focus on dialogue among stakeholders in dealing with ethical issues, a priority for Ulrich is to evolve practical guidelines that planners and ordinary citizens can both use equally proficiently to conduct boundary critique [20]. For this purpose, he offers a list of 12 questions that can be employed by those involved in and affected by planning to interrogate what the system currently is, and what it ought to be. These 12 questions cover 4 key areas of concern: motivation, control, expertise, and legitimacy.[5]

In my view, there is significant potential for using Ulrich's 12 questions in public health planning, not least because they cut to the heart of many issues that are of fundamental concern to people in communities who find themselves on the receiving end of policies and initiatives that they either do not agree with or find irrelevant. In my own research, I have used these questions with people with mental health problems recently released from prison, [7, 64] older people in sheltered housing [7, 29, 71], young people who have run away from children's homes [7, 32, 72], and others. Ulrich claims that his questions can be answered equally proficiently by "ordinary" people with no experience of planning as they can by professionals [20], and my experience tells me that he is right – with the caveat that the questions should be made specific to the plans being discussed, and also need to be expressed in plain English. If the questions about what ought to be done are asked early on in planning a new public health initiative, I have found that "ordinary" people are usually able to think just as systemically as professionals (indeed, sometimes more so) [7].

A Practical Example

To further ground this presentation of methodology, I briefly outline another systemic intervention that I undertook, again with 2 colleagues (Isaac Munlo and Mandy Brown). Only a sketch of this intervention is provided, and therefore many of the social dynamics that were important to it have been omitted. However, more detailed expositions can be found elsewhere [7, 29, 71, 73].

The initial remit of the project, funded by the Joseph Rowntree Foundation, was to work with local governments in the United Kingdom to find out how information from assessments of older people applying for health, housing, and welfare services could be aggregated to inform the development of housing policy. However, some initial interviews with stakeholders quickly revealed that there were 2 major problems with the boundaries of our study.

(continued)

---

[5]These 'questions' are discussed in Chapter 7 (see 7.3.2).

Reading 8 (continued)

First, it became apparent that if the housing "needs" expressed by older people fell outside local government spending priorities, they were not recorded. This meant that aggregating information from assessments would paint an artificially rosy picture, making it seem as if all needs were being met. Second, many urgent problems with service provision, assessment, and multiagency planning were being raised by stakeholders (including older people themselves). We felt that ignoring these would be unethical – especially as we had already come to the conclusion that the initial remit of the intervention was flawed. As a consequence, we worked with the funder to expand the remit of our study to look at the wider system of assessment, information provision, and multiagency planning for older people's housing, and what could be done to improve it.

Semi-structured interviews with 131 stakeholders from a wide variety of organizations (including older people themselves) yielded data that we used to create a "problem map." This is similar to a system dynamics model, except that problem mapping is purely qualitative. The purpose is to demonstrate to stakeholders that their problems are strongly interdependent and, therefore, to be resolved, they require changes to the wider system.

Having demonstrated the systemic nature of the issues, the next stage was to ask what kind of system change was needed. To answer this, we held a series of interactive planning workshops, asking what ideal (but still technologically feasible, viable, and adaptable) housing services would look like. We integrated the critical systems heuristics questions so we could explore issues of motivation (or purpose), control (including governance), expertise, and legitimacy. To prevent the marginalization of older people, we worked with them separately from professionals, allowing them more time and space to develop their views. The interactive planning/critical systems heuristics workshops demonstrated a widespread agreement among stakeholders on housing policy, with only a few relatively minor disagreements needing resolution.

We then brought together senior managers from health, housing, and welfare organizations to look at what kind of organizational system could deliver the housing services that the stakeholders had asked for. We introduced the viable system model as a template for the organizational design, and systematically evaluated this design using criteria derived from the earlier work with older people and frontline professionals (thereby ensuring that these perspectives were not marginalized now that participation had been narrowed to managers). In this way, we could be reasonably confident that the managers' proposals would either meet the stakeholders' requirements directly or would provide the organizational means to address them in future years.

This example of systemic intervention demonstrates the benefits of boundary critique: The initial problematic remit of the project was usefully expanded,

(continued)

Reading 8 (continued)

and the potential for marginalizing older people was identified and addressed. It also demonstrates the value of methodological pluralism. In my view, no single set of methods yet developed could have addressed all the issues in this intervention. It took a combination of semistructured interviewing, problem mapping, interactive planning, critical systems heuristics, and viable system modeling to support stakeholders in both defining the issue and responding to it systemically.

Conclusion

I have presented a methodology for systemic intervention (incorporating boundary critique and methodological pluralism), and have discussed several systems approaches from which we can borrow useful methods. I have also provided 2 practical examples of systemic intervention.

I suggest that this kind of approach is not only able to address issues of values, boundaries, and marginalization in defining complex problems (making it particularly relevant to the "new public health" [28]), but it also has the potential to deliver all the utility of other systems approaches because it explicitly advocates learning about and drawing methods from those approaches to deliver maximum flexibility and responsiveness in systemic interventions.

In my view, systems thinking has the potential to make a significant difference to public health, so (if you have not already done so) I invite you to try out some of the ideas and methods touched upon in this article, and share your experiences with others so that the whole public health research community can be enriched in the process.

References

1. Laporte RE, Barinas E, Chang Y-F, Libman I. Global epidemiology and public health in the 21st Century: Applications of new technology. Ann Epidemiol. 1996; 6:162–167
2. Holder HD. Prevention of alcohol problems in the 21st century: challenges and opportunities. Am J Addict. 2001; 10;1–15.
3. Best A, Moor G, Holmes B, et al. Health promotion dissemination and systems thinking: Towards an integrative model. Am J Health Behav. 2003; 27 (suppl 3):S206–S216.
4. Best A, Stokols D, Green LW, Leischow S, Holmes B, Buchholz K. An integrative framework for community partnering to translate theory into effective health promotion strategy. Am J Health Promot. 2003; 18: 168–176.
5. World Health Organization and World Bank. World Report on Road Traffic Injury Prevention. World Health Organization: Geneva, Switzerland, 2004.
6. Midgley G. Systems Thinking. Volumes I–IV. Sage Publications: London, England, 2003.

(continued)

Reading 8 (continued)

7. Midgley G. Systemic Intervention: Philosophy, Methodology, and Practice. Kluwer Academic/Plenum Publishers: New York, NY, 2000.
8. Midgley G. Science as systemic intervention: Some implications of systems thinking and complexity for the philosophy of science. Syst Pract Act Res. 2003; 16:77–97.
9. Seidman E. Back to the future, community psychology: Unfolding a theory of social intervention. Am J Community Psychol. 1988; 16:3–24.
10. Bogdanov AA. Bogdanov's Tektology. Dudley P, ed. Centre for Systems Studies Press: Hull, England, 1996 [first published in Russian 1913–1917].
11. Koehler W. The Place of Values in the World of Fact. Liveright: New York, NY. 1938.
12. Boulding KE. General systems theory – the skeleton of science. Manage Sci. 1956; 2:197–208.
13. Kremyanskiy VI. Certain peculiarities of organisms as a "system" from the point of view of physics, cybernetics and biology. General Systems 1958; 5:221–230.
14. Von Bertalanffy L. General Systems Theory. London, England: Penguin; 1968.
15. Bateson G. Steps to an Ecology of Mind. Jason Aronson Publishers: Northvale, NJ, 1972.
16. Miller JG. Living Systems. McGraw-Hill: New York, NY, 1978.
17. Prigogine I, Stengers I. Order out of Chaos: Man's New Dialogue with Nature. Fontana: London, England, 1984.
18. Laszlo E. The Interconnected Universe: Conceptual Foundations of Trans-disciplinary Unified Theory. World Scientific Publishing: London, England, 1995.
19. Churchman CW. Operations research as a profession. Manage Sci. 1970; 17:B37–B53.
20. Ulrich W. Critical Heuristics of Social Planning: A New Approach to Practical Philosophy. Verlag Paul Haupt: Bern, Switzerland, 1983.
21. Cilliers P. Complexity and Postmodernism: Understanding Complex Systems. Routledge: London, England. 1998.
22. Midgley G, Ochoa-Arias AE. An introduction to community operational research. In: Midgley G, Ochoa-Arias AE, eds. Community Operational Research: OR and Systems Thinking for Community Development. Kluwer Academic/Plenum Publishers: New York, NY, 2004.
23. Ulrich W. Critical Systems Thinking for Citizens: A Research Proposal. Centre for Systems Studies Research Memorandum #10. Centre for Systems Studies, University of Hull: Hull, England, 1996.
24. Habermas J. Communication and the Evolution of Society. English ed., 1979. Heinemann: London, England, 1976.
25. Midgley G. The sacred and profane in critical systems thinking. Syst Pract. 1992; 5:5–16.
26. Midgley G. Ecology and the poverty of humanism: A critical systems perspective. Syst Res. 1994; 11:67–76.
27. Yolles M. Viable boundary critique. J Operat Res Soc. 2001; 52: 35–37.
28. Baum F. The New Public Health: An Australian Perspective. Oxford University Press: Melbourne, Australia: 1998.
29. Midgley G, Munlo I, Brown M. The theory and practice of boundary critique: Developing housing services for older people. J Operat Res Soc. 1998; 49:467–478.
30. Córdoba J, Midgley G. Addressing organisational and societal concerns: An application of critical systems thinking to information systems planning in Colombia. In: Cano J, ed. Critical Reflections on Information Systems: A Systemic Approach. Idea Group Publishing: Hershey, Pa: 2003.
31. Baker V, Foote J, Gregor J, Houston D, Midgley G. Boundary critique and community involvement in watershed management. In: Dew K, Fitzgerald R, eds. Challenging

(continued)

Reading 8 (continued)

Science: Issues for New Zealand Society in the 21st Century. Palmerston North, New Zealand: Dunmore Press; 2004.

32. Boyd A, Brown M, Midgley G. Systemic intervention for community OR: Developing services with young people (under 16) living on the streets. In: Midgley G, Ochoa-Arias AE, eds. Community Operational Research: OR and Systems Thinking for Community Development. New York, NY: Kluwer Academic/Plenum Publishers; 2004.

33. Jackson MC, Keys P. Towards a system of systems methodologies. J Operat Res Soc. 1984; 35:473–486.

34. Flood RL. Liberating Systems Theory. New York, NY: Plenum Press; 1990.

35. Jackson MC. Systems Methodology for the Management Sciences. New York, NY: Plenum Press; 1991.

36. Flood RL, Jackson MC, eds. Critical Systems Thinking: Directed Readings. Chichester, England: John Wiley and Sons; 1991.

37. Flood RL, Jackson MC. Creative Problem Solving: Total Systems Intervention. Chichester, England: John Wiley and Sons; 1991.

38. Midgley G. Pluralism and the legitimation of systems science. Syst Pract. 1992; 5:147–172.

39. Flood RL, Romm NRA, eds. Critical Systems Thinking: Current Research and Practice. New York, NY: Plenum Press; 1996.

40. Gregory WJ. Discordant pluralism: A new strategy for critical systems thinking? Syst Pract. 1996; 9:605–625.

41. Mingers J, Gill A, eds. Multimethodology: The Theory and Practice of Combining Management Science Methodologies. Chichester, England: John Wiley and Sons; 1997.

42. Jackson MC. Systems Approaches to Management. New York, NY: Kluwer Academic/Plenum Publishers; 2000.

43. Jackson MC. Systems Thinking: Creative Holism for Managers. Chichester, England: John Wiley and Sons; 2003.

44. Luckett S, Grossenbacher K. A critical systems intervention to improve the implementation of a district health system in KwaZulu Natal. Syst Res Behav Sci. 2003; 20:147–162.

45. Ulrich W. Some difficulties of ecological thinking, considered from a critical systems perspective: A plea for critical holism. Syst Pract. 1993; 6: 583–611.

46. Forrester JW. Industrial Dynamics. Cambridge Mass: MIT Press; 1961.

47. Sterman JD. Learning in and about complex systems. Syst Dyn Rev. 1994; 10:291–330.

48. Morecroft JDW, Sterman JD, eds. Modeling for Learning Organizations. Portland, Ore: Productivity Press; 1994.

49. Vennix JAM. Group Model Building: Facilitating Team Learning Using System Dynamics. Chichester, England: John Wiley and Sons; 1996.

50. Coyle RG. System Dynamics Modelling: A Practical Approach. London, England: Chapman and Hall; 1996.

51. Maani KE, Cavana RY. Systems Thinking and Modelling: Understanding Change and Complexity. Auckland, New Zealand: Prentice Hall; 2000.

52. Sudhir V, Srinivasan G, Muraleedharan VR. Planning for sustainable solid waste management in urban India. Syst Dyn Rev. 1997; 13: 223–246.

53. Townshend JRP, Turner HS. Analysing the effectiveness of Chlamydia screening. J Operat Res Soc. 2000; 51:812–824.

54. Forrester JW. Counterintuitive behavior of social systems. Theory and Decision. 1971; 2:109–140.

(continued)

Reading 8 (continued)

55. Beer S. Cybernetics and Management. Oxford, England: English Universities Press; 1959.
56. Beer S. Decision and Control. Chichester, England: John Wiley and Sons; 1966.
57. Beer S. Brain of the Firm. 2nd ed. Chichester, England: John Wiley and Sons; 1981.
58. Beer S. The viable system model: Its provenance, development, methodology and pathology. J Operat Res Soc. 1984; 35:7–25.
59. Beer S. Diagnosing the System for Organisations. Chichester, England: John Wiley and Sons; 1985.
60. Espejo R, Harnden RJ, eds. The Viable System Model: Interpretations and Applications of Stafford Beer's VSM. Chichester, England: John Wiley and Sons; 1989.
61. Ackoff RL. Redesigning the Future: A Systems Approach to Societal Problems. Chichester, England: John Wiley and Sons; 1974.
62. Ackoff RL. Resurrecting the future of operational research. J Operat Res Soc. 1979; 30:189–199.
63. Ackoff RL. Creating the Corporate Future. New York, NY: John Wiley and Sons; 1981.
64. Cohen C, Midgley G. The North Humberside Diversion From Custody Project for Mentally Disordered Offenders: Research Report. Hull, England: Centre for Systems Studies; 1994.
65. Checkland P. Systems Thinking, Systems Practice. Chichester, England: John Wiley and Sons; 1981.
66. Checkland P, Scholes J. Soft Systems Methodology in Action. Chichester, England: John Wiley and Sons; 1990.
67. Checkland P. Rhetoric and reality in contracting: Research in and on the NHS. In: Flynn R, Williams G, eds. Contracting for Health. Oxford, England: Oxford University Press; 1997.
68. Fahey DK, Carson ER, Cramp DG, Muir Gray JA. Applying systems modelling to public health. Syst Res Behav Sci. 2004; 21:635–649.
69. Gregory WJ, Midgley G. Planning for disaster: Developing a multiagency counselling service. J Operat Res Soc. 2000; 51:278–290.
70. Ulrich W. Critical heuristics of social systems design. Eur J Operat Res. 1987; 31:276–283.
71. Midgley G, Munlo I, Brown M. Sharing Power: Integrating User Involvement and Multi-Agency Working to Improve Housing for Older People. Bristol, England: Policy Press; 1997.
72. Boyd A, Brown M, Midgley G. Home and Away: Developing Services with Young People Missing from Home or Care. Hull, England: Centre for Systems Studies; 1999.
73. Midgley G. Ethical dilemmas: A reply to Richard Ormerod. J Operat Res Soc. 1999; 50:549–553.

12.3   Other Possibilities for Contextualising Systems Practice

Changing circumstances creates new opportunities for different forms of systems practice. For example, the seeming failure of many public sector reforms in Britain over the period of the New Labour government has created opportunities for the proponents of an approach to practice called the 'lean systems approach'. The lean systems (LS) approach is said [5] to provide: 'a method for... achieving the ideals many managers aspire to: a learning, improving, innovative, adaptive and energised organisation.' According to Jackson et al. [3]:

> it provides the means to develop a customer-driven adaptive organisation; an organisation that behaves and learns according to what matters to customers'. It was developed by John Seddon, originally an occupational psychologist, who had become interested in change programs and why they often failed.[4] LS incorporates aspects of intervention theory and systems thinking (the work of Deming [1] and Senge [6] being particularly influential), together with lessons from Toyota's 'lean manufacturing' adapted for service organisations [p. 187].

From my own perspective the LS approach and systemic intervention, when in the hands of aware practitioners can exemplify taking the 'design turn' that I wrote about in Chapter 10. The latter particularly invites design considerations through the device of making and re-making boundary judgments to a given system of interest. LS practitioners claim that when appropriately enacted this method is based upon the systems principle that 'operations or organizations should be viewed as wholes serving a purpose' [3]:

> In Vanguard's case the purpose of a system is always seen in terms of its customer – 'what matters is what matters to the customer'. Once the customer's purpose has been established, attention can be given to how the parts or tasks must be fitted together in order to best achieve that purpose. Here, in line with systems thinking, it is the interactions between the parts that are viewed as being critical. LS insists that a customer focus remains central throughout an intervention. The design of support systems, such as IT systems, should follow design of the primary customer serving system. Evaluation must be in terms of overall system performance in pursuit of customer purposes. Inappropriate targets can distort the behaviour of the system in ways that are not beneficial to its customer's purposes [p. 187].

The emergence of new framings and labels for systems practice is to be expected even though it causes confusion for those outside the field. In part this is natural as any new field is being developed; and in part the proliferation arises as a consequence of the social context in which systems practitioners find themselves operating. Branding is a device to gain market entry and share in competitive fields, as is much of the consultancy world. Unfortunately the same is true of the academic world where academic practitioners, or their universities, pursue recognition and citation in an increasingly perverse world of performance measures. In this book I have striven to surpass these concerns by developing a generic 'ideal type' model for systems practice. Through the isophor of the juggler it should be possible to inquire of all forms and accounts of practice where claims for systems practice are made: what is it that they did when they did what they did? In pursuing this inquiry the question of value will arise. In Chapter 13 (Part IV), my concluding chapter, I turn my attention to the question of how we might value systems practice.

References

1. Deming W.E. (1982) Out of Crisis. Cambridge University Press: Cambridge.
2. Helme, M. (2002) Research Project: Systems Thinking, Systems Practice and the NHS. Summary and outcomes. Unpublished Research Report, The Open University: Milton Keynes.

3. Jackson, M.C., Johnston, N. and Seddon J. (2008) Evaluating systems thinking in housing. Journal of the Operational Research Society 59, 186–197.
4. Open University (2003) Waving not Drowning. How Systems Thinking and Practice benefit NHS practitioners. Open University: Milton Keynes.
5. Seddon J. (2003) Freedom from Command and Control: A Better Way to Make the Work Work. Vanguard Education Ltd: Buckingham.
6. Senge P. (1990) The Fifth Discipline: The Art and Practice of the Learning Organization. Random House: New York.

Part IV
Valuing Systems Practice
in a Climate-change World

Chapter 13
Valuing Systems Practice

13.1 The Emergence of Value

On what basis do we value what we do when we do what we do? Why is it that we value some practices above others? In this final chapter I want to consider the act of valuing based on an understanding of valuing as a social process from which value emerges.[1] Of course my primary interest is to ask and imagine what conditions and processes might create the circumstances for the emergence of value in relation to systems practice. The origins of the word value are related to that of 'valiant' in the sense of being strong or being well; in its earliest uses it was a verb, not a noun.

When writing about his experiences with Shell and his own biotechnology company Gerard Fairtlough made it clear that he valued a shift away from hierarchy. In the latter part of his life he became convinced of the need for contemporary organisations to be capable of radical transformation. He even went as far as setting up his own publishing company to help effect transformations in the thinking, and thus the practices of those who work in organisations. In doing what he did he created possibilities for a particular way of valuing organisational life. He had also made a choice about how to frame the situations – to understand the world as complex and to recognise that simplicity, though appealing, can be misleading and sometimes dangerous. He concluded that [12]:

> Hierarchy will not easily withdraw. Understanding inventiveness, balance and bravery will be needed to shift it. But there is good reason to hope that it can be shifted. Vast energy presently goes into propping up hierarchy. Releasing this energy for constructive use will bring great and clearly recognizable benefits. It will allow organizations to emerge that are more effective for getting things done and much better places in which to work [p. 101].

My introductory anecdote about Gerard Fairtlough exemplifies what I mean by actions that operationalise a process of valuing. In this concluding chapter I am not

---

[1]Another way of exploring this is to ask: "what is it we see when we say we value what is happening? What happens if we reify values?" Framing this as "you can create a map from the territory, but you can't create the territory from my map" – which is a subtly different concept than "the map is not the territory" – has helped to communicate my intended meaning (i.e., you can create the value from the happening, but you can't create the happening from the value).

R. Ison, *Systems Practice: How to Act in a Climate-Change World*,
DOI 10.1007/978-1-84996-125-7_13, © The Open University 2010.
Published in Association with Springer-Verlag London Limited

concerned with the traditional way of understanding values.[2] Consistent with my overall approach in this book I am concerned with the praxis of valuing – those practices through which what we can distinguish as "value" emerges. From this perspective value does not exist in and of itself, it always arises in praxis and is firmly grounded in the emotions that give rise to our doings. In Chapter 11 I described how the emotion of enthusiasm could be triggered. In its enactment – in the pursuit of something for which we have enthusiasm – value emerges because it satisfies some characteristic of what it is to be human. What these characteristics are is never fully generalizable because, as I have outlined in Chapter 5, we all think and act out of unique traditions of understanding (see Fig. 5.3).

In Chapter 1 I invited consideration of the question: what is it that you would have to experience that created the circumstances where you could experiment with thinking and acting systemically? Here, at the end of this book, it is appropriate to revisit this question. In essence this is a question about what you have come to value. In posing the question I said that I could not answer it for others but encouraged (i) the abandonment of certainty (or the acknowledgment of the certainty of uncertainty) and (ii) openness to circumstances.

I have purposefully taken as a backdrop to this book the emergence of human induced climate change and all the uncertainty this contested situation creates.[3] A key concern I have is whether as individuals, groups or nation states and beyond

Illustration 13.1

---

we will become open to these circumstances in time to develop practices that are more effective. A major claim of this book is that investment in systems practice, as both a generic and specific practice, offers significant opportunities, especially if embedded in more conducive systemic and adaptive governance arrangements (Chapter 9). In Chapter 4 I said I would develop arguments that support four claims:

- Systems practice has particular characteristics that make it qualitatively different to other forms of practice
- An effective and epistemologically aware systems practitioner can call on a greater variety of options for doing something about complex uncertain situations than other non-aware practitioners
- Being able to deploy more choices when acting so as to enhance systemically desirable and culturally feasible change has important ethical dimensions; and
- Our individual and collective capabilities to think and act systemically are under-developed and this situation is a strategic vulnerability for us, as a species, at a time when concerns are growing for our continued existence in a co-evolutionary, climate change world

You may or may not accept that these arguments have been made. If you have then the dynamic depicted in Fig. 1.2 is at play – you will have concluded that as explanations they are acceptable. But as I have said, the acceptance or not of my explanations will probably depend on your own circumstances and the relational dynamics from which what you have come to value has emerged.

In this chapter I first want to explore some different perspectives on valuing and situate these in the history of valuing and the related practice of evaluating. In doing

Illustration 13.2

this my concern is to illuminate how effectiveness comes to be judged in relation
to systems practice. Through the lens of a final reading I then want to explore the
transition from *being systemic* to *doing Systems*.[4] I do so because how these two
modes are understood is, it seems to me, contested within the Systems community.
At the core of this issue is how those in Systems understand their own intellectual
and praxis domain and thus how they value what they and others do. Within the
mode of 'being systemic' and 'doing Systems', there is a need to understand how
capability can be built. How capability-building processes are designed and enacted
is a reflection of valuing and evaluating. Three frameworks to consider these issues
are provided. I conclude by considering how the valuing of systems practice can be
understood in a context of *hope*.

13.2   Perspectives on Valuing

Valuing, and thus what emerges as value, happens at the personal, interpersonal,
organisational and societal level in an unfolding dynamic. Frequently it seems that
what each generation comes to value is undervalued by the preceding generation.
But it is dangerous to attribute values to groups or sectors rather than to be concerned with the question: how is the act of valuing conserved through our manners
of living? An answer to this question requires an appreciation that manners of living
are conserved as part of a lineage. This is the same process that operates in the
conservation of different manners of doing systems practice.

In Chapter 1 I claimed that failings in our governance arrangements called out
for transformation based on fundamental shifts in thinking and practice as well as
what we choose to value. But what value do we place on getting our thinking sorted
out – of doing different things rather than the current things more efficiently? Take
the field of human genetics and the revolution in biology. As Steve Talbott [42] has
observed, a few years ago biologists thought they had the basis of life all sorted out.
But they were wrong. Talbott explores how, in doing what they did, biologists put
at stake the 'nature of biological explanation – our understanding of understanding
itself.' Particularly at issue, he concluded 'are the distortions introduced by a one-
sidedly logical-causal habit of thinking – distortions worsened by the continuing
failure to enter into the more organic sort of understanding that so many have hoped
for over the years and even centuries.' His excellent essay reveals how misunderstood this whole field of endeavour is, not least by scientists themselves, for:

> nothing less than the dynamics of cell, whole organism, and environment can make sense
> of any particular tract of DNA—can interpret it and turn it into a fitting expression of its
> larger context. The genome, perhaps we could say, is not so much an instruction manual as

---

[4]In Chapter 5 I spoke of being and doing as recursively related. So, with awareness, to do one is
also to do the other. I have chosen to capitalise the word Systems here to denote the process of
making a connection with the intellectual and social history of doing Systems (as per Fig. 2.3) and
of acting with awareness of both systemic and systematic practice.

a dictionary of words and phrases together with a set of grammatical constraints. And then, from conception through maturity, the developing organism continually plays over this dictionary epigenetically, constructing the story of its destiny from the available textual (genetic) resources.[5]

What Talbott holds up for critical consideration is the explanations we offer and accept about the nature of life, and, as he says, the very understanding of understanding.[6] How we engage with such issues, the practices that we pursue, and the explanations that are offered frame what is or is not valued. One might reflect that our prospects for managing an on-going co-evolutionary dynamic between humans and the biosphere is not good if the fundamentals of how we understand life and understanding are built on shaky foundations. However, I would like to think that in Talbott's analysis lie seeds of transformational change. He connects with the systems biologists who, in the early part of the twentieth century, became concerned that other biologists, in doing what they did, had lost sight of the properties of whole organisms [9]. He also adds weight to the rationale for the re-emergence of systems biology as a legitimate area for research and understanding [34].[7]

Elsewhere there have been calls for critically rethinking approaches to interdisciplinarity in research practice in ways that value 'epistemological pluralism'. This is understood as an approach to practice that recognises that there are different ways of knowing that warrant valuing and accommodation [30]. Such calls are driven by an emerging framing which breaks away from the historical dualistic focus on 'ecosystems' and 'social systems' to seeing these in more relational terms as 'social-ecological systems'.[8]

When preparing to write this chapter I asked Peter Checkland what he had come to value from his 40 odd years of doing systems practice. His answer was: 'its the penny dropping in the people in organizations ... that something makes them realize that they are now thinking in a completely different way i.e., they are thinking about their own thinking and this is a huge step for most people..... This shift in thinking, when the penny drops, is when the paradigm shifts.'[9] Checkland's reflections

---

[5]Epigenesis is a concept worth understanding in the systems domain. In biology it literally means development on top of (epi) development. Thus it refers to the current development based sequence of further development in an organism. It was originally coined to refer to embryonic development of a plant or animal from an egg or spore through a sequence of steps in which each step creates the conditions for the next step so that cells differentiate, organs form, and the shape of the living system results.

[6]For a critique that reaches similar conclusions, though from a different worldview, see Hodges [16].

[7]A speculative thought is that perhaps his understanding of the operation of DNA could also be seen as an isophor for the systems practitioner, an embodied juggling act!

[8]Donald Schön of course made similar arguments quite early on in his long and productive career.

[9]This is a particularly British expression usually meaning to convey the moment when a previously confused or mistaken person finally understands something, generally something important and maybe something painfully obvious to others' (see http://horizon.bloghouse.net/archives/000204. html); it refers to the same phenomenon as my 'aha moment' referred to in Chapter 2.

parallel my own experience and that of many of my colleagues who are systems educators. But why is it that it is this particular facet of our doing that we have come to value so highly?

Before I respond to my own question I want to explore some aspects of the history of valuing so as to create a context for what I want to say.

13.2.1 Appreciating Some of the History of Valuing

In the late twentieth and early twenty-first centuries the dominant mode of practice in relation to valuing has been appropriated by the discipline, politics and ideology of economics. How this has come to pass is a topic worthy of a book in its own right, but the main points of contention can be summarised as [41]:

> The classical conceptualization of value [in economics] is related to supply; it is also objectivist, locating the origin of value in the things from which objects are made, such as land or labour..... In contrast, the modern neoclassical conceptualisation of value incorporates a convergence of supply and demand that produces an equilibrium or market-clearing price. Value is considered to originate in the minds of individuals, as revealed through their subjective preferences. In short, value is determined by the market-place. Ecological resources complicate the modern neoclassical approach to determining value due to their complex nature, considerable non-market values and the difficulty in assigning property rights. Application of the market model through economic valuation only provides analytical solutions based on virtual markets, and neither the demand nor supply-side techniques of valuation can adequately consider the complex set of bio-physical and ecological relations that lead to the provision of ecosystem goods and services [p. 402].[10]

It is not only ecological resources that complicate the neo-classical paradigm (some might claim ideology). As I outlined in Chapter 8 (Section 8.5), the understanding of human behaviour and rationality that is at the core of the neo-classical paradigm has been shown to be both limiting and flawed. In addition, situations such as climate change are best understood as non-equilibrial yet for decision making we rely on a paradigm that is built on concepts of equilibrium. This is significant as the treasuries of the world are, in the main, places where this type of thinking is institutionalised.

However, the failings of economics and the mainstream ways of valuing are not my concern here. I want to introduce some other perspectives, but before I do I need to turn to the process of evaluating.

---

[10]And these goods and services are distinguished specifically in relation to what we take/use (and hence value) from "the ecosystem" – i.e., they are not about ecology!

13.2.2 Evaluating

At its simplest the process of evaluating can be understood as a device to bring questions of value into the conversation. Expressed crudely it can be understood as a means to abstract value from a process. As in economics and project management a mainstream view has come to exist to which there are alternative conceptions and discourses [45]. Patton [35, p. 11] observed that:

> Human beings are engaged in all kinds of efforts to make the world a better place. These efforts include assessing needs, formulating policies, passing laws, delivering programs, managing people and resources, providing therapy, developing communities, changing organizational culture, educating students, intervening in conflicts and solving problems. In these and other efforts to make the world a better place, the question of whether the people involved are accomplishing what they want to accomplish arises. When one examines and judges accomplishments and effectiveness, one is engaged in evaluation. When this examination of effectiveness is conducted systematically and empirically through careful data collection and thoughtful analysis, one is engaged in evaluation research.

Wadsworth [43, p. 1] has a perspective on evaluation that exemplifies the process of valuing that is of concern to me. She argues that:

> We evaluate all the time. From the minute we meet someone new, or sift through the day's mail, or walk into a shop or office, or decide on the week's activities, we are evaluating. We decide whether things are valuable or unimportant, worthwhile, or not 'worth it', whether things are good or bad, right or wrong, are going OK or 'off the rails', are attractive, difficult, exciting, offputting, useful, undesirable, functional, effective, boring, expensive, too much, too little, just right, interesting, too simple, much too complex, or a disaster! Every time we choose, decide, accept, or reject – we have made an evaluation.

But it is only when we distinguish in language that we have done so, that we have "evaluated". Otherwise what happens just happens in the flow of living without awareness [11]. Wadsworth goes on to say [43, p. 1] that evaluation begins when we notice a discrepancy between what we expected (or did not expect) or wanted (or did not want) and what actually has occurred:

> A difference between an 'is' and an 'ought' (or an 'ought not'). Or more accurately, the difference between a valued (or it might be an unvalued) 'is' and a valued (or unvalued) 'ought' or expectation.

Wadsworth's and my own understanding differ from the mainstream understanding. For example, a common, mainstream, definition of evaluation used in environment and development contexts, particularly in relation to projects, is:

> An examination, as systematic and objective as possible, of an ongoing or completed project or program, its design, implementation, and results, with the aim of determining its efficiency, effectiveness, impact, sustainability, and the relevance of its objectives. The purpose of an evaluation is to guide decision makers [32].

Wadsworth's perspective challenges this narrow focus on evaluation. My own perspective is that evaluation needs to be practised as an ordinary, everyday part of what we do. If it is thought of otherwise it always ends up looming as the thing that has to be done at the end of a project, and which in practice there is never enough time left for, so it is often done only to satisfy a directive. If evaluation is left to the end, how do we know that we are still on track as we proceed? My approach also suggests that evaluation is not something that needs to be left to an outside expert to do. Obviously, there are cases where this might be desirable or the only practical way to evaluate but I am suggesting the need to break out of a trap in our thinking which sees evaluation as "difficult, uncomfortable, time consuming and done by someone else, preferably an 'expert'" [43, p. 7].

Wadsworth [43, p. 8] claims that evaluation is:

> meant to be self-consciously value-driven (although some evaluation pretends to proceed as if it isn't). This makes it both easier and also more difficult to sort out whose values are predominating. On the one hand it is easier because it is clear – even just from the term 'evaluation' – that values are being used to judge practices. On the other hand, it can be more difficult because evaluation can try to appear 'objective' (for purposes of legitimacy, certainty, agreement, etc.) but without real agreement around values. This may make it more difficult to realise that value is not inherent in what is being evaluated, but is ascribed by those observing it.

The underlining in the quote above is my added emphasis. McCallister [26] goes further, arguing (p. 280) that 'evaluation can rightly be considered a branch of ethics' (but a form of ethics that arises in the doing as I outlined in Chapter 5). How ethics arises can be understood through the difference between 'open inquiry evaluation' and 'audit review evaluation'. The former can be seen as systemic, illuminative and process- and learning-based. In contrast the latter is undertaken as systematic, objective and standards-based (see [43, p. 34]).

What emerges for me from the perspectives of Patton and Wadsworth is an appreciation and preference to understand evaluation, and thus effectiveness, in relation to systems practice as:

- Always context specific (relating to both performance and situation)
- Enacted through praxis
- Beginning when any purposeful behaviour begins (i.e., at the start)
- More robust when informed by multiple perspectives
- Valuing difference (or requisite variety) – the differences that make a difference
- Based on systemic, circular rather than linear models of causation
- Always socially constructed and thus relational
- Related to articulated measures of performance (of a system of interest), particularly effectiveness (why), or efficiency (how) or efficacy (what)
- Enabling the emergence of new and different narratives (accounts) which result in different emotional dynamics

Systemic evaluation is a form of systems practice that could well have been a topic of focus in Part III. Bob Williams explains it in the following terms [44]:

> The systems field comprises methodologies, methods and tools that are deeply evaluative. But it is not just about method. As a friend said recently, the biggest benefit of systems ideas in her work was that it enabled her to ask more powerful questions. And questions, especially powerful questions, are the lifeblood of good evaluation.

For me, systems concepts provide me with very powerful ways of exploring inter-relationships, perspectives and boundaries. These are important issues within evaluation:

- Inter-relationships are the key to understanding how programs behave
- Perspectives provide insight into motivations and thus how people behave
- Boundaries determine who wins and who loses from an intervention, what is "in" and what is "out" of an assessment of that intervention. In other words indicate value or worth

More than evaluation, the systems field has thought deeply about these three concepts and come up with approaches that can transform the way in which evaluation does its job.

However, because systemic evaluation has been the focus of a recent anthology edited by Williams and Imam [45], I have chosen not to replicate their efforts. I do however want to return to the dynamics of living in language, particularly conversations, and notions of authenticity and accountability. I do so because I want to widen the usual horizon of thinking about these matters, to create opportunities, hopefully, for making more choices and thus for the emergence of different ways of valuing.

13.2.3 Authenticity and Accountability in Conversations

Krippendorff [22] proposes an explanation of authenticity as:

the pleasure of participating in togetherness in which one is free to speak for oneself, not in the name of absent others, not under pressure to say things one does not believe in, and not having to hide something for fear of being reprimanded or excluded from further conversation [p. 141].

He points out that authentic conversation is not easily, if at all, identifiable from the outside and suggests that participants in authentic conversations, whether as speakers or listeners, may experience conversations as:

- Occurring in the presence of addressable and responsive individuals
- Maintaining mutual understanding – evident only in performance (since cognition cannot be observed) and sometimes indicated by statements such as "I understand", "tell me more", "I agree"[11]
- Self-organising and constituted in the contributions their participants make to each other (i.e., conversations are not abstract but embodied in real participants)

---

[11]Krippendorff [22] makes the important point that 'acknowledging understanding does not mean similarity or sharing of conceptions, its affirmation constitutes an invitation to go on, including to other subjects' [p. 142].

- Intuitive, not rule governed (e.g. children born into a community need to learn how to join in conversations not a set of rules about a conversation)
- Dialogically equal (i.e., everyone in the conversation has an equal opportunity to contribute)
- Creating possibilities of participation (e.g. opening up possibilities to increase the number of possibilities for subsequent action)
- Irreversible, progressive and unique (each turn is experienced as unique)
- Coordinating constitutions of reality (i.e., what becomes institutionalised as a consequence of being in a conversation)
- Continuable in principle

Ironically authentic conversation seems more obvious when it isn't present – when faced with breakdown and dysfunctionality. In a sense the same applies to systems practice – for those who understand what systems practice is, the lack is what becomes most obvious.

Why is this relevant to valuing and evaluating? From a personal perspective I gain a great deal of satisfaction from being in an authentic learning relationship, whether with students, family, colleagues or clients. I suggest something similar leads Checkland to value what he does when doing his systems practice. Krippendorff [22] notes that 'authentic conversation is typical among trusting friends but also among strangers who, having nothing to lose, feel alive in each other's presence' [p. 143]. In my experience the doing of systems practice creates the possibilities for the emergence of authenticity understood in the terms described above. Furthermore, I would claim that the seeds of a "good performance" amongst stakeholders in an issue are to be found in the experiencing of authenticity. Unfolding authenticity creates the circumstances for building and re-building relational capital, that form of capital that synthesises the other forms of capital (natural, manufactured, social, cultural and institutional) (see [40]) – and that is so easy to destroy in modern organisational life.

Krippendorff [22] illuminates another aspect of living in language relevant to my concerns here. He contends that: 'everything said is said not only in the expectation of being understood, but also in the expectation of being held accountable for what was said or done' [p. 143].[12] The most typical accounts are (i) explanations; (ii) justifications; (iii) excuses and (iv) apologies. Explanations are the least disruptive of conversations and best coordinate understandings and thus behaviour. Explanations are expansive. Justifications and apologies admit the speaker's agency, unlike excuses. Of these only apologies and explanations admit responsibility.

---

[12] See also Shotter [38, 39] who claims that speakers tend to articulate their contributions to a conversation not merely in response to other speakers but also with possible *accounts* in mind in case their contribution is challenged.

13.3 Valuing Being Systemic

In my explication of the B-ball (Chapter 5) I presented a range of factors that are worthy of valuing as part of one's systems practice and that, for me, seem central to being systemic. I think there is a case for fostering more systemic ways of being. In this final chapter I have chosen to synthesise what I mean by introducing a final reading (Reading 9). Having been fortunate to have met Mary Catherine Bateson, as well as having read her work, I experience her as exemplifying a systemic being. This can be understood most readily by reading her book 'Willing to Learn: Passages of Personal Discovery' [8].[13] Reflecting on the influences of her parents (Gregory Bateson and Margaret Mead) Mary Catherine Bateson said that she sometimes 'thought that she learned social engagement from her mother and abstraction from my father'. She also observed that their different styles eventually drove them apart. That a couple could be driven apart by their different manners of *being* is testimony to the power of ways of thinking and acting. This essay [7], based on a presentation to the American Society of Cybernetics, but also reproduced in her reflective essays [8] is chosen because of the authenticity I experience when reading it.[14]

Reading 9
_____

The Wisdom of Recognition

Mary Catherine Bateson

In spite of the tighter and tighter economic and electronic interlinking of the world we live in, our vision is ever more fragmented by specialization. Cybernetics can be a way of looking that cuts across fields, linking art and science and allowing us to move from a single organism to an ecosystem, from a forest to a university or a corporation, to recognize the essential recurrent patterns before taking action. I want you to have in the back of your minds today how this spectrum is built into the tradition of the American Society for Cybernetics, and how it has been obscured in activities that bear the label "cybernetics."

It has been my task to deal with the dual intellectual heritage of my parents, Margaret Mead and Gregory Bateson, which is also part of the heritage of the

(continued)

---

[13]As observed on her website Bateson's book is constructed around the proposal that lives should be looked at as compositions, each one an artistic creation expressing individual responses to the unexpected – see http://www.marycatherinebateson.com/reviews.html (Accessed 7th October 2009).

[14]I cannot claim that in your reading you will experience the authenticity that I do.

cybernetics movement—both belonged to the original Macy conferences. Their approaches to issues of aesthetics and action are part of our common history, and relevant today. Both Margaret and Gregory grew up to regard the arts as higher and more challenging than the sciences. This sense of humility in relation to the arts lasted right through their lives and seems to connect with their concern for whole systems. Their attitudes toward action in the world, however, differed sharply, and I believe the difference came out of World War II. Margaret's war work used her professional training to increase understanding between allies, while Gregory's role was sowing confusion among the enemy. This year is the Mead centenary, using her quote, "Never doubt that a small group of thoughtful, committed citizens can change the world." She believed that you could use social science to improve society. Gregory did not, and even in psychiatric contexts, he resisted the transformation of his ideas into specific strategies of intervention and looked at Margaret's activism very much askance. Both would have seen the observer as part of the system, but emotionally Gregory was an outsider, alienated from the society in which he lived, while Margaret was always very much engaged and felt that her commentary was made as a participant.

Margaret and Gregory shared an orientation toward whole systems. In those days, anthropologists spoke of "salvage anthropology." The concept was that every human group has developed an integrated and unique adaptive pattern passed on from generation to generation—myths, subsistence technology, family structure, language. The urgency, analogous to the danger today of species loss, was the loss of alternative ways of being human. If you went into a preliterate community that had never been studied, you knew you couldn't get the whole thing, but you got as much as you possibly could, because you might be the last person with the opportunity to write it down. Today we divide the world between groups of specialists, but anthropology in those days was a training ground for thinking about whole systems.

After the war, Gregory drifted away from ethnography, but the way he thought was grounded in biology and natural history, and he continued to think in terms of systems and analogies. He went from studying Bali to studying alcoholics and the families of schizophrenics, and dolphins, and octopuses, and philosophical issues around the biosphere. He seemed to be flitting from one interest to another and of course it was held against him. I think many of us have had the same problem. It was not until the late sixties, I believe, that it became fully clear to Gregory that he had been struggling to develop a way of thinking that would be transferable, systematically, from one subject matter to another, so that what looked like a group of random essays was recognizable as steps—to what he called an ecology of mind [3].

(continued)

Reading 9 (continued)

Margaret got into a different kind of trouble. She wrote for the general public, drawing the parallels from every society she studied to issues in western society. She was criticized for talking to ordinary people and for talking about a very wide range of topics. She seemed to have opinions about everything under the sun—but that's what anthropologists did, they looked at everything under the sun of whatever community they were in. (She did write specialized monographs, designed for expert colleagues, but most of her colleagues don't read them, they read the popular books.).

Mead wrote a column for Redbook Magazine, a women's magazine. The majority of readers in those days were housewives with a high school education, but Mead had been studying ordinary people, after all, in small societies where being an adult meant that you understood your culture in a way that none of us today can. Mead argued that "The way in which people behave is all of a piece" [28]. The lullaby that a mother sings, the classes in school, and the way marriages are negotiated... are interlocked and reinforced. There were really two approaches to the unity of cultural systems. One was "functionalism," which emphasized the causal interconnections between different aspects of culture, associated with Malinowski; and the second, associated with Ruth Benedict, emphasized the stylistic congruence of different aspects of culture, so they added up to a single ethos, and this was an aesthetic concept. Both of these approaches fit with the argument that you can neither understand a culture nor act on it effectively in terms of little pieces.

In 1968 Gregory convened the Conference on Conscious Purpose and Human Adaptation in Gloggnitz Austria, which focused on the damage done in pursuit of conscious purposes without a sense of the whole [6]. By that time Gregory was deeply worried about environmental degradation, but whenever a member of the group would propose action, Gregory would dig in his heels and say, "We don't understand well enough yet to feel that we can act without making things worse." At the end he called for a theory of action that would be moral in the sense of not disrupting the larger systems in which it occurs, and he suspected that this would have to do with aesthetic judgment that might transcend the need for a complete description of the state of a system and the implications of a particular action. Some people, he argued, have a "green thumb." They know how to look after a plant, just as some doctors simply have a sense of how to care for a patient, just as a painter or a composer or a poet produces a work of art without a complete and detailed theory of what she is doing, but recognizes the aesthetic value [4].

"Is it the first virtue of art," he asked, "...to force the player and the listener, the painter and the viewer, and so on, to surrender to that necessity which marks the boundary between conscious self-correction and unconscious obedience to inner calibration?" [5] That inner calibration corresponds to one

(continued)

Reading 9 (continued)

of the many meanings of Gregory's most famous phrase, "the pattern which connects." Not only do all living systems have certain characteristics in common, but this similarity can be a basis for relationship, so some individuals, not knowing fully in consciousness what it is to be alive, might perceive living systems on the basis of an internal template or calibration. This recognition, like the ancient mariner's recognition of the beauty of the tropical water snakes around his becalmed ship, makes it possible to act with wisdom (Bateson and Bateson 1987:73). Interestingly enough, Kant's third critique, the Critique of Judgment, makes the same kind of connection between the purposive and the aesthetic [31]. There too there is something like a template within the self, that makes possible the recognition of aesthetic order in the other. We reveal something about ourselves in judging something beautiful.

Groups, like organisms, are systemic, perhaps in ways more accessible to awareness than our individual systemic characteristics. Interestingly, both Margaret and Gregory saw the potential of groups for wiser decision making. At the Gloggnitz conference we finally felt that the best wisdom we had came out of the diversity of the members of the group and the mingling of thought and emotion in their interactions, so that together they constituted a whole that could model a kind of wisdom [6]. Mead had been creating groups to do just that, starting during World War II with carefully designed working groups of various sorts. From her point of view the cybernetics group was a model of precisely the kind of thinking organism composed of a diverse group of people that can take on a life of its own and spin that vitality into sensitivity [29].

We repeatedly slip back into the same kinds of fragmented visions, saying we are interested in facts instead of responding to patterns. Since my parents' time, anthropology, which was one of the only disciplines that could try to understand being human holistically, has deconstructed itself. In fact, the same thing has happened with cybernetics, so that increasingly we think not of cybernetics as a unifying way of seeing the world, but as a tool to be applied here and there, for purposes of manipulation. Society is becoming newly infatuated with simple causal models. Just within the last half decade, I see a worrying new wave of simplistic biological determinism coming along not only as a form of explanation, but also as a model for intervention.

All of us here have clues to addressing this problem, part of a common heritage created not by our genes but by our having participated in overlapping conversations over half a century, and we have a common responsibility to preserve and develop this heritage. In the American Society for Cybernetics there is a significant continuing reservoir of shared understanding of a way of being in the world, acting on the wisdom of recognition, which is badly needed.

Reprinted from/first published in Cybernetics & Human Knowing, Vol 8, No. 4, pp. 87–90.
Copyright © Imprint Academic, Exeter, UK.

(continued)

This essay provides a window into the world of cyber-systemic understanding and practice as it can be understood today. Manifest in the essay are tensions between understandings of systems as ontologies and systems as epistemologies (as depicted in Fig. 2.3)[15] as well as tensions between *being systemic* and *doing Systems* i.e., the tension over the extent that one commits to action in the world. What I particularly value about Bateson's piece is the reflexive and systemic nature of her concerns. Gerard Fairtlough similarly arrived at an understanding, through structured inquiry and reflection, that innovation is epitomised by the reflective practitioner, especially one who has been involved in business, scientific research, education, government and NGOs. Geoffrey Vickers is another example of the reflective, systems practitioner. Yet Vickers came to Systems as a means to reflect on his life's work whilst Fairtlough drew on it to act in the world through the setting up and managing of a business.[16] Both pathways seem meaningful.

It seems to me that being systemic is particularly important for understanding situations, for making judgments, and entering into and fostering particular types of conversations. But in a sense being systemic is not enough – it is, after all, only one of the balls to juggle. The phrase 'doing Systems' has come to exemplify for me the transition from *being* to *doing*, and can be seen as a rubric or descriptor which encompasses all of the balls juggled by the aware systems practitioner.

13.4 Doing Systems

13.4.1 Committing to Action (Praxis)

My concern with the transition from *being* to *doing* is theoretically informed and practically and politically important. I share the perspective of Will McWhinney [27] who articulates his concern as the 'uncritical acceptance of an unbounded holism' [p. 93]. He describes a *New Yorker* cartoon of a man reading a small hand-bill posted on a brick wall of an old building, the title for which is: 'The cosmos, a detail'. The point he makes is that this situation applies to us all – our view is always but a detail. McWhinney argues that systems practitioners need to take responsibility for selecting 'what we can best put to use' on the basis that 'finding alternatives is the work of intelligence and selecting what we can manage is a creative act.' In other words systems practice has to involve purposeful action for situation improvement – it is learned and enacted through doing.

Through the isophor of the juggler I have presented and explicated an 'ideal type' model for an aware systems practitioner. Importantly the juggler is an isophor

---

[15] This is evident in the use by Bateson of the phrase 'whole systems' which within the logic of the arguments I have mounted in Chapter 2 is not a particularly helpful or meaningful term.

[16] See also Scholes [37] and Haynes [15] whose systems practice is part of a lineage of SSM thinking and practice.

for embodied systems practice – for doing Systems. The juggler is not a blueprint but a device for understanding and learning about your own systems practice or the systems practice of others. In addition I have purposefully set out to create the circumstances for different forms of systems practice to be appreciated and valued. In all, nine readings appear in the book written by eight different systems practitioners. Each account reveals aspects of value that has emerged from their practice and raises questions in relation to effectiveness.

13.4.2   An Evaluation Framework for Doing Systems

In Chapter 10 I referred to the PersSyst project at The Open University (UK) which set out to effect systemically desirable and culturally feasible changes within the organisation through building capability amongst staff for systems thinking and practice. Naturally in undertaking this systemic inquiry we had to address both effectiveness and evaluation. Armson [1] described the situation faced in the following way:

> It can be argued that evaluation requests are often based on a model in which the results of a development intervention, such as a workshop or training programme, are traceable through a linear chain of causation (intervention leads to improved performance that leads to organizational benefit). Those involved (trainers, participants and others) will understand that the effects of an intervention are much more diverse and that the chain of causes and effects is neither linear nor fully traceable….. those engaged in the action know that success or otherwise is not reducible to simple metrics.

As a part of the ongoing systemic inquiry that underpinned PersSyst, Armson [1] developed a conceptual model of a systems practitioner working in an organisation understood as part of a system of influences (Fig. 13.1). The model exemplifies how a person's perceived performance in their role can be understood as an emergent property of a system of influences, viz.:

- The systems practitioner with their history
- The choice about the framing of a situation
- The approaches the practitioner uses
- The stakeholders, including their boss and their colleagues

Armson's conceptual model can be used as means to understand and evaluate systemically an HR intervention such as a training event. In the case of PersSyst, training events were conducted to build systems practice capability. Historically, evaluation of a training workshop was focused on the individual out of context, not the network of influences shown in Fig. 13.1.[17] However the model provides a much

---

[17] In my experience many change management activities in complex organisations believe naively that enhanced performance can be achieved through staff development forgetting that the capacity to respond – to use new skills or understandings – may be constrained by a range of factors such as structure, processes, relationships (this is a similar set of phenomena as described in Chapter 9 for systems practice).

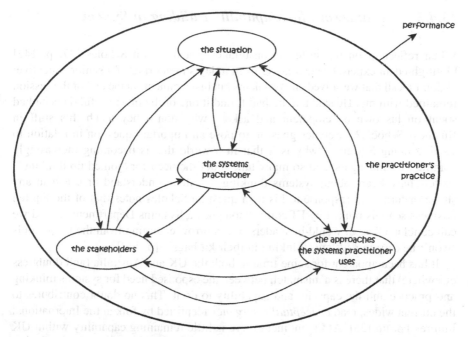

An influence diagram of systems practice

Fig. 13.1 An influence diagram of the factors that give rise to systems practice as a performance which can also act as (i) a design for capability building and/or (ii) a framework for evaluation (Armson [1])

richer and thus satisfactory framework for designing and evaluating HR activity. With elaboration the model can be used to demonstrate how a typical HR intervention (e.g. a one off workshop) in itself has no direct influence on the employee's performance. Armson also makes the point that the model 'carries the clear implication that individual employees are the only stakeholders who can attribute a connection between the intervention and improved performance. They are the only stakeholders engaged at the systemic level from which performance emerges.' This is because the system of influences *is* the 'practice' of the individual shown in Fig. 13.1.

This model is also a framework that can be used to evaluate (whether through self, or 'external' evaluation) the efficacy and effectiveness of an initiative such as PersSyst. The effective introduction of Systems into an organisation requires working with the full set of influences if one is to effect changes to both individual and organisational performance. However the model says nothing about the particular capabilities that might be of concern when educating or enabling enhanced performance or evaluating a systems practitioner.[18]

---

[18]I use capabilities in the sense of being able to act, to create an effective performance. Thus capability is associated with doing or enactment.

13.4.3   A Framework for Capability Building in Systems

When reflecting on his systems consultancy practice, Jim Scholes [37, p. 142] highlighted an experience that arose from a day's mentoring of a senior executive: "I don't recall that we solved any problems but his comments at the end of the session remained with me. He said that he had found it one of the most useful days he had spent on his own development and asked 'why don't they teach this stuff in Business Schools?'" Scholes goes on to pose an important question in relation to valuing doing Systems: 'why is it that in a world that is becoming increasingly complex for managers, and so many talk about the need for "joined up thinking", so few have heard about systems thinking, let alone understand or use it in any shape or form?' In response to his own question Scholes notes that of the top ten business schools ranked by FT.com[19] "none has a Systems Department ....and we can conclude that all would-be leaders of tomorrow emerging from the top schools 'won't be carrying systems thinking in their kit bags'" [p. 143].

It has been apparent for some time in both the UK and Australia (and doubtless elsewhere) that there is a mismatch between the espoused need for systems thinking and practice and the capacity and capability to do it. This no doubt contributes to the current widespread *conceptual emergency* identified by folk at the International Futures Forum [25]. At the moment there is little remaining capability within UK Higher Education to deliver systems-based education and the institutional arrangements for the conservation and flourishing of what capacity there is are not good. From my perspective this appears to be a growing strategic weakness. Let me exemplify the case with one line of argument.

At The Open University we have been in conversation for some time with staff of the National School of Government (NSG)[20] about running a number of events for civil servants to develop systems thinking and practice capability as part of overall capability building in the area of 'sustainability'. The concept note to which we were responding was prepared by the NSG. It was written to describe what they wanted the Higher Education sector to provide and was entitled "*Skills and principles for cross-government working – systems thinking*". In this concept note they argue that:

> Cross-government working remains a high ideal but very hard to make happen in practice. It has been identified as a high priority in the recent capability reviews... At the National School we think that 'systems thinking' and 'systems practice' skills will become increasingly important knowledge and skill areas for civil servants, especially as we expect to encounter problems that are increasingly opaque, complex, uncertain and include conflicting evidence or priorities... The underlying purpose of systems thinking is to sharpen the ability of civil and public servants to anticipate unwanted outcomes and side-effects long before they 'go critical' and to work from a belief that coherent and complementary policy making across government is possible and achievable, even in very challenging areas.

---

[19] This is the url for the Financial Times online – see for example, http://rankings.ft.com/businessschoolrankings/global-mba-rankings (Accessed 10th October 2009)

[20] The NSG is the business school for government operating from its heart [Whitehall] and dedicated to the public sector – (see http://www.nationalschool.gov.uk/about_us/index.asp)

These needs are not limited to the UK; they pertain to other regions as well. For example, in Australian research (but relevant internationally) published as *'Shifting towards sustainability. Six insights into successful organisational change for sustainability'* it was found that the keys to successful change towards sustainability involved an:

> ongoing learning process which actively involves multiple stakeholders in change to achieve sustainability... Learning for Sustainability or Education for Sustainable Development, they found ...involves five key components: Visioning (imagining a better future); Critical thinking and reflection; Participation in decision making; Partnerships; and Systemic thinking.[21]

As mentioned in Chapter 6, some governments have become aware of consistent policy failure around intractable and 'wicked' situations' but at the same time they seem unaware of how to respond or to even know what capability needs to be built. Table 13.1 is a heuristic adapted from a 'systems education matrix' developed by a group of systems practitioners (educators) who met in a 'Fuschl Conversation' in 2008.[22] Their concern was the nature of systems education [20].

The heuristic operates with two learning system models – one called "domain specific" and one "generic" (the right hand columns in Table 13.1). These models apply to the type of work modes that can be pursued by any systems practitioner – to apply or integrate systems thinking and practice in specific domains (e.g. health; IT; local, state or national governance; manufacturing, business etc.) or into other disciplines (e.g., systems geographer; systems engineer; systems ecologist etc.) or to develop systems practice as a generic capability able to be applied anywhere.

Within each learning system model I have identified four levels – the first three, sense making, practical mastery and theoretical mastery emerged from the Fuschl conversation. I have taken the liberty of adding an extra dimension, praxis mastery, which I distinguish from the others. In my use of the heuristic I would consider the set of capabilities under 'generic' to include (in an additive sense) those within the same row (i.e., those listed under 'domain specific').

The heuristic can be further understood as the reification, at this moment in time, of the valuing of different systems practice capabilities – a valuing that has emerged from the conversation and engagement of an expert group.[23]

What is not made readily apparent in Table 13.1 is how effectiveness might be judged or valued in relation to awareness in systems practice, i.e., in effectively

---

[21]Unfortunately this research suggests that the five key components are additive – my own perspective, based on the arguments of this book is that investing in systemic thinking and practice would deliver all of the other components. (See: Hunting SA & Tilbury D. 2006. 'Shifting towards sustainability: Six insights into successful organisational change for sustainability'. Australian Research Institute in Education for Sustainability (ARIES) for the Australian Government Department of the Environment, Water, Heritage and the Arts, Sydney: ARIES).

[22]A 'Fuschl Conversation', initiated in 1980 in the Austrian village of Fuschl, is based on the process of dialogue, of 'meaning flowing through' (see [19, 21]). It can be understood as a collectively guided disciplined inquiry comprising (i) an exploration of issues of social/societal significance, (ii) engaged by scholarly practitioners in self-organised teams, (iii) who select a theme for their conversation, (iv) which is initiated in the course of a preparation phase that leads to an intensive learning phase (see http://www.ifsr.org/node/33 Accessed 9th October 2009).

[23]It is not my intention to expand upon what is known about educating the systems practitioner. There is a significant literature to which I and colleagues have contributed. For those interested see [10, 17, 18, 23, 36] as a start.

Table 13.1 An heuristic for considering the design of 'learning systems' for capability building for systems practice. Two learning systems are described, domain specific and generic. Each have different design considerations and learning outcomes (Adapted from [20])

	Learning system model	
Main learning outcome	Domain specific	Generic
	Summary: Having the ability to integrate systems approaches into one or more disciplines or situations	Summary: Having the ability to understand, apply and relate systems concepts in multiple contexts and/or to add to the Systems knowledge base.
Sense – making	Having the ability to use basic systems concepts to make sense of phenomena, objects and processes in the world	Systems student; Systemically aware citizen
Practical mastery	Having the ability to competently apply Systems concepts for research or practice or the ability to facilitate the learning of Systems by others	Systems practitioner
Theoretical mastery	In a position to add competently to the body of Systems knowledge (viz., philosophy, theory, methodology and praxis) as well as areas of practical application in specific contexts.	Systems facilitator; Creator of circumstances for systemic ways of knowing
Praxis mastery	Able to braid theory and practice in their situation of concern and to manage for emergence	Systemic designer; Aware systems practitioner (able to be reflexively responsible) and to operate systemically and systematically

juggling all of the balls that were explicated in Chapters 5–8. Table 13.2 is an assessment matrix for projects undertaken by students in The Open University course '*Managing complexity: a systems approach*' that is largely constructed around the isophor of the juggler [33]. Although not presented here, the schema has been further elaborated against each of the four balls, Being, Engaging, Contextualising and Managing. It should be possible for anyone concerned with context specific design of leaning systems or engaging in systemic evaluation to draw upon, and adapt, these two frameworks (Tables 13.1 and 13.2).

Unfortunately, awareness within the extant systems community of both need and how to go about capability building seems difficult to translate into mainstream curricula, national capability frameworks and the like. But I think this situation can be changed. It will require the building of a discourse for systems practice as a key enabler for living and adapting in a climate change world. To succeed such a discourse needs to create and foster practices built on an emotion of hope. Hope, after all is a manner of living in the present; it has to do with how we do what we do now, rather than simply imagining a possible future.

Table 13.2 A marking scheme for projects completed by students undertaking an Open University course designed around the systems practitioner as juggler[24]

Project mark	Broad guidelines	Being systemically aware
85–100	Demonstrates ownership in the use of concepts/skills and applies ideas in a logical way, reflectively varying approach with context. Adapts the systems approach in creative ways. The tutor learns from the student as well as vice versa. Clear evidence demonstrated by the student of realistic, astute, practical judgment and perception as an action researcher.	Demonstrates self awareness, awareness of others and an ethical focus. In particular, Project Log and other reports use second-order language (essentially using 'I' and 'we')
70–85	Demonstrates a solid grasp of the material and can apply it over a wide range of contexts. Lacks the completely imaginative reflection of the 85–100 answer. Evidence that report and conclusions are well reasoned. Gives evidence of potential as an action researcher but needs to develop reflective capability.	Demonstrates self-awareness and an ethical focus, as shown by frequent use of second-order language.
55–69	Demonstrates an understanding of the course and the ability to manage an inquiry. Understanding of arguments appears to be incomplete. Demonstration of engaging reflectively is limited.Demonstrates some evidence of being a systems practitioner.	Demonstrates some self-awareness and ethical focus, but not consistently. Language used is mainly first-order.
40–54	Demonstrates that the course material has been read and paid attention to. An instrumentalist approach to the course is adopted. Demonstrates difficulties in contextualising approaches to changing circumstances. Little evidence of engaging reflectively.	Demonstrates some awareness of self and of own effects as a practitioner. Gives little evidence of ethical focus, as corroborated by use of only first-order language.

13.5 Acting in a Climate of Hope in a Climate-Changing World

13.5.1 Valuing in a Context of Hope

At a recent ISSS (International Society for Systems Science) meeting Steve Maharey, now Vice-Chancellor of Massey University, New Zealand (and a former minister in the NZ government) made some convincing arguments about how modern

---

[24] The fail mark in undergraduate courses at The Open University (UK) is set at 40% – hence the cut-off point in this table.

environmentalism had failed because it has not cultivated a climate of hope.[25] Whether you agree with this claim or not does not matter here. The insight I gained was to reflect on what the underlying emotional dynamics were in the various discourses about change (or not) in a climate changing world. What seems clear is that we will not succeed in transforming our situation when the underlying emotions are fear or fundamentalism.

Graham Leicester [24, p. 4] claims that there are three choices we can make in the face of uncertainty and complexity in challenging times: (i) defensive – deny our confusion, reinforce our certainty, stick ever more doggedly to what we know – become fundamentalists; (ii) become destructive by throwing up our hands in despair and admitting that it is all too confusing (thus deciding to remain lost) and (iii) choose growth and transformation. These are essentially personal choices. As choices they can be linked to the claims he and colleagues at the International Futures Forum make about being in a conceptual emergency [25].

For them a conceptual emergency is a personal condition, not an abstract one. They make the very valid points that 'you cannot respond to a conceptual emergency until you know you are in one' and that 'conceptual emergencies are acknowledged by many organizations but often are a product of the fact no single organization has a budget line or sense of responsibility for them'. They also claim that 'it is better to respond by taking action, at any scale, rather than planning to take action – because action triggers learning' [p. 41]. I am not convinced that this last point is automatically the case as I have experienced a lot of action without much learning!

But what of scales beyond the personal? Reflecting on his 40 years of doing systems within the context of the history of ideas Peter Checkland recently observed that this was "but the blink of an eye".[26] His reflection was in the context of concerns for the uptake of systems thinking and practice and particularly the paradigm shift from 'hard' (systematic) to 'soft' (systemic) that he has done much to elucidate. But are these time frames acceptable in a climate changing world? And if the answer is no, what is to be done to change the situation?

Perhaps an answer is to collaborate in building a social climate (discourse) of imagination and hope? Leicester [24, pp. 15–16] argues that in the face of complexity and uncertainty the 'first horizon is failing, the second horizon is innovating, but if there is no vision of a desirable third to which an innovation is heading, change is merely opportunistic.' From this perspective a third metaphor for 'adaptation' can be added to 'adaptation to' and 'adaptation with' that I introduced in Chapter 1.

---

[25]*'People First: How to make environmental sustainability something we can all live with'* 53rd Meeting of the International Society for Systems Sciences (ISSS), University of Queensland, Brisbane, Australia, July 12–17 2009. See http://www.massey.ac.nz/massey/fms//About%20 Massey/Vice-Chancellors-office/Speeches/2009/People_First_July_2009.pdf (Accessed 12th October 2009).

[26]Personal communication, 3rd October 2009, Bourn, UK.

Illustration 13.3

This third metaphor is 'adaptation for'. Individually, nationally and globally we need a discourse and praxis response to this, for as Leister [24] notes: 'without a third horizon vision pulling us forward there can be no such distinction [between innovation that props up the old way and a new, sustainable way] and all innovation will inevitably draw us backward towards the past' [p. 16].

Earlier I asked the question what was it that Checkland, I and other colleagues were valuing in relation to 'the penny dropping' experience of which he spoke. Why is this so important to us? My response is that through the systems practice that we are concerned with the 'penny dropping' phenomenon gives rise to an expansion of what it is to be human. It does so by providing new and different ways of knowing. Choosing to engage with situations as if they were complex and uncertain and engaging in systems practice built on learning and inquiry always results in a wider, more systemic view, thus opening up more choices in a given situation. New horizons of knowing, and thus doing seem essential if one is to build a discourse and praxis of hope.

13.5.2 Opportunities to be Cultivated

I want to finish the book by laying out an agenda for action around some opportunities for building systems practice within a discourse of hope in a climate changing world. This is not an exhaustive listing, but one that might be taken up by policymakers concerned with enhancing horizontal accountability in cabinet-based governance, or developing systemic and adaptive governance arrangements more generally. In concluding with a list I am conscious that this is not a particularly systemic way to finish a book on systems practice (it is however systematic). Hence my final invitation is for you to consider these points in the light of your own circumstances. If they seem germane then incorporate them into your own systemic inquiry into opportunities for building systems practice. By so doing you will have taken the 'design turn' I spoke about in Chapter 10. The points I make based on my experience are:

- Identify opportunities to create demand for systems practice – this is a generic point that needs to be pursued in context specific ways (several of the points below can be seen as 'hows' for this 'what')[27]
- Create systems practitioner posts to company boards, cabinets, major committees, senior teams, large projects[28] (e.g. like the systemically failing £12 billion National Programme for IT, known as NfIT, in the UK NHS) etc. charged with emulating the role of the joker or jester of mediaeval courts[29]
- Build systems practice capability into public sector senior executive recruitment
- Recognise systems practice as part of sustainability capability and build this into recruitment and promotion criteria
- Make the case to national skills councils, as exist in the UK, to invest in systems practice capability[30]
- Introduce systems practice into secondary curricula – the model used in the International Baccalaureate of teaching 'theories of knowledge' is one that could be followed
- Commission and experiment with systemic inquiry in all situations that might better be framed as uncertain, complex, 'wicked' etc.
- Create opportunities to do more systemic action research and at the same time improve quality and conceptual rigour
- See staff induction as a form of systemic inquiry into an organisation – as an opportunity for learning and for building relational capital based on praxis amongst new staff (see [2])
- Experiment with breaking out of hierarchy in organisations through building capability for systems practice. Fairtlough [13] conceptualised the changes he felt were needed to break out of hierarchy as 'creative compartments' which were openness, interaction, smallness, focus and innovation
- Cultivate and find ways to speak about 'emotional intelligence' as part of everyday organisational life
- Appreciate relational thinking and act with awareness that humans are engaged in a co-evolutionary dynamic with the earth. Take responsibility for the futures we can 'design' (i.e., embrace a design turn)

---

[27] This can be described as a "demand pull" strategy compared to a "supply push" strategy.

[28] I base this suggestion on experience of a person who once played this role within the Dupont company. He operated at board level answerable only to the CEO of the day. Given this history in Dupont I have found the current CEO's responses to the global financial crisis of interest (see http://knowledge.wharton.upenn.edu/article.cfm?articleid=2273# Accessed 12th October 2009).

[29] This suggestion needs to be understood within an appreciation of the historical and political significance of a court jester, i.e., "in societies where freedom of speech was not recognized as a right, the court jester – precisely because anything he said was by definition 'a jest' and 'the uttering of a fool' – could speak frankly on controversial issues in a way in which anyone else would have been severely punished for" (see http://en.wikipedia.org/wiki/Jester Accessed 12th October 2009).

[30] e.g. see the UK Commission for Employment and Skills at http://www.ukces.org.uk/(Accessed 10th October 2009).

- Contextualise and add to this list as you see fit
- ... and act on any of these items wherever they are relevant and you find an opening to do so

References

1. Armson, R. (2007) How do we know it's working? Addressing the evaluation conundrum. In Systemic Development: Local Solutions in a Global Environment, Ed. James Sheffield, pp. 253–262, ISCE Publishing USA.
2. Armson, R., Ison, R.L., Short, L., Ramage, M. & Reynolds, M. (2001) Rapid institutional appraisal (RIA): a systemic approach to staff development. Systems Practice & Action Research 14, 763–777.
3. Bateson, G. (1999/1972) Steps to an Ecology of Mind. University Press: Chicago.
4. Bateson, G. (1991/1968) The moral and aesthetic structure of human adaptation, in Sacred Unity, edited by Rodney Donaldson. HarperCollins: New York.
5. Bateson, G., and Bateson, M.C. (1979) Angels Fear: Towards an Epistemology of the Sacred. Macmillan: New York.
6. Bateson, M.C. (1972) Our Own Metaphor. Knopf: New York.
7. Bateson, M.C. (2001) The wisdom of recognition, Cybernetics & Human Knowing 8(4) 87–90.
8. Bateson, M.C. (2004) Willing to Learn: Passages of Personal Discovery. Steerforth Press: Hanover, NH.
9. Bertalanffy, L. von (1968) General Systems Theory. Braziller: New York.
10. Blackmore, C. (Ed.). (2010) Social Learning Systems and Communities of Practice. Springer: London.
11. Christie, C.A., Montrosse, B.E., Klein, B.M. (2005) Emergent design evaluation: A case study, Evaluation & Program Planning 28, 271–277.
12. Fairtlough, G. (2007) The Three Ways of Getting Things Done. Hierarchy, Heterarchy and Autonomy in Organizations. Triarchy Press: Axminster.
13. Fairtlough, G. (2008) No Secrets. Innovation Through Openness. Triarchy Press: Axminster.
14. Giddens, A. (2009) The Politics of Climate Change. Polity: London.
15. Haynes, M. (1995) Soft systems methodology. Modes of practice. In K. Ellis et al. (eds.) Critical Issues in Systems Theory and Practice. Plenum Press: New York.
16. Hodges, J. (2009) Foundations, fallacies, and assumptions of science for livestock in development: use of transgenic animals. UN IAEA-FAO International Symposium on Sustainable Improvement of Animal Production and Health, Vienna 8–11 June.
17. Ison, R.L. (1999) Guest Editorial: Applying systems thinking to higher education. Special Edition Systems Research & Behavioural Science 16, 107–112.
18. Ison, R.L., Blackmore, C.P., Collins, K.B. & Furniss, P. (2007) Systemic environmental decision making: designing learning systems. Kybernetes 36 (9/10) 1340–1361
19. Isaacs, W.N. (1996) The process and potential of dialogue in social change, Educational Technology, January/February, 20–30.
20. Jones, J., Bosch, O., Drack, M., Horiuchi, Y. and Ramage, M. (2009) On the Design of Systems-Oriented University Curricula, The Research Reports of Shibaura Institute of Technology (Social Sciences & Humanities) 43(1), 121–130.
21. Kersten, S. (2000) From debate to dialogue about vegetation management in Western New South Wales Australia. In LEARN Group Eds [M. Cerf, D. Gibbon, B. Hubert, R. Ison, J. Jiggins, M. Paine, J. Proost, N. Röling] (2000) Cow up a Tree. Knowing and Learning for Change in Agriculture. Case Studies from Industrial Countries. INRA (Institut National de la Recherche Agronomique) Editions, Paris. pp. 191–204.

22. Krippendorff, K. (2009) Conversation. Possibilities for its repair and descent into discourse and computation, Constructivist Foundations 4(3), 138–150.
23. Lane, A.B. and Morris, R.M. (2001) Teaching diagramming at a distance: seeing the human wood through the technological trees, Systemic Practice & Action Research 14(6) 715–734.
24. Leicester, G. (2009) Beyond Survival. A short course in pioneering in response to the present crisis. International Futures Forum, Fife.
25. Leicester, G. and O'Hara, M. (2009) Ten Things to Do in a Conceptual Emergency. International Futures Forum, Fife.
26. McCallister, D.M. (1980) Evaluation in Environmental Planning. Assessing Environmental, Social, Economic and Political Trade-offs, The MIT Press: Cambridge, MA.
27. McWhinney, W. (2001) Enabling embodiment in education, Cybernetics & Human Knowing 8(4) 91–93.
28. Mead, Margaret (2000/1942) And Keep Your Powder Dry. Berghahn Books: New York.
29. Mead, Margaret (1999/1964) Continuities in Cultural Evolution. Transaction Press: New Brunswick, NJ.
30. Miller, T.R., Baird, T.D., Littlefield, C.M., Kofnas, G., Chapin, F.S. and Redman, C.L. (2008) Epistemological pluralism: reorganizing interdisciplinary research. Ecology & Society 13(2):46.
31. Nuzzo, Angelica (2001) "Constructing the World of Experience—the 'Enigma' of Beauty and Life." (draft).
32. OECD (Organization for Economic Co-operation and Development) (1986) Methods and Procedures in AID Evaluation, OECD, Paris.
33. Open University (2007) Managing Complexity: A Systems Approach (T306) Technology: Level 3 course, Project Guide. Prepared on behalf of the Course Team by Graham Paton, Bob Saunders, Eddie Hurst, Bob Zimmer and Sue Holwell. Open University: Milton Keynes.
34. O'Malley, M.A., and Dupré, J. (2005). Fundamental issues in systems biology. BioEssays 27, 1270–1276.
35. Patton, M.Q. (1990) Qualitative Evaluation and Research Methods, 2nd edn, Sage Publications: Newbury Park, CA.
36. Salner, M. (1986) Adult cognitive and epistemological development in systems education. Systems Research 3, 225–232.
37. Scholes, J. (2005) Reflections on the relevance of systems thinking to management practice. Systemist 27(2) 132–145.
38. Schotter, J. (1984) Social Accountability and Selfhood. Basil Blackwell: Oxford.
39. Schotter, J. (1993) Conversational Realities. Constructing a Life Through Language. Sage: Newbury Park CA.
40. SLIM (2004) The role of conducive policies for fostering social learning for integrated management of water. SLIM: Open University: Milton Keynes (see http://slim.open.ac.uk).
41. Straton, A. (2006) A complex systems approach to the value of ecological resources, Ecological Economics 56 (2006) 402– 411.
42. Talbott, S. (2009) On making the genome whole. Part 1: Twilight of the Double Helix. NETFUTURE. Science, Technology, and Human Responsibility Issue # 175 (March 12) – http://www.netfuture.org/2009/Mar1209_175.html; (Accessed 2nd October 2010).
43. Wadsworth, Y. (1991) Everyday Evaluation on the Run. Action Research Issues Association (Inc.), Melbourne.
44. Williams, Bob (2009) Systems and Evaluation – see http://users.actrix.co.nz/bobwill/ (Accessed 10th October, 2009)
45. Williams, Bob and Imam, Iraj (Eds) (2007) Systems Concepts in Evaluation: An Expert Anthology. EdgePress/American Evaluation Association: California.

Index

A

Abandoning certainty, 19, 237, 246, 256, 304
Abrogate responsibility, 219
Abstractions, 19, 50, 60, 99, 126, 128, 237, 313
Accommodation(s), 105, 158, 248, 255, 292, 307
Accountability, 70, 124, 225, 268, 311–312, 325
Ackoff, R., 24, 25, 33, 125, 126, 130, 133, 219, 291
Action inquiry
    learning, 246
    research, 246
Activity model, 176, 247, 248, 255
Actor network theory, 48–49
Adaptation
    as a good pair of shoes, 12
    as co-evolution, 13, 260
    as fitting into, 12
    for, 325
    as jigsaw, 12
    with, 324
Adaptive Methodology for Ecosystem Sustainability and Health (AMESH), 139, 140, 148
Adaptive practice, 222
Aesthetic judgment, 315
Aesthetics, 314–316
Agency boundaries, 124
Agricultural policy, 136, 170, 181, 270
Agro-ecosystem assessment, 147
Aha-moment, 17
Air pollution, 65
Alcoholics, 314
American Society of Cybernetics, 313
Amplifiers, 276
Analogies, 22, 25, 224, 238, 314
Anthropologists, 101, 314, 315
Antidotes, 218

Apartheid of the emotions, 218, 235–240, 243, 272
Apology/apologies, 30, 312
Applied systems, 133, 245, 277
Appreciative inquiry, 139, 189
Apprenticeship, 278
Armson, R., 185, 218, 257, 278, 318, 319
Artefacts, 7, 109, 111, 261
As if, 60, 92
Attributing purpose, 156, 157
Audit review evaluation, 310
Australian Murray-Darling Basin Authority, 259
Australian Public Service Commissioner, 118, 124, 194
Australian Seed Industry Study, 267
Authentic, authenticity, 20, 105, 236, 311–313
Authentic conversation, 311–312
Awareness, 5, 7, 17, 25, 27, 28, 33, 47, 50, 51, 58, 62, 81, 87, 92, 97, 98, 104, 105, 107, 111, 112, 114, 118, 121, 123, 134, 135, 139, 148, 153, 157, 159, 174, 176, 181, 186, 188, 191–194, 196, 198, 219, 227, 228, 237, 240, 249, 250, 262, 272, 274–279, 285, 306, 309, 316, 321–323, 326

B

Bali, 314
Bateson, M.C., 96, 229, 240, 274, 313–317
Bawden, R.J., 6, 217, 245
B-ball, 58, 79, 85–114, 135, 181, 185, 313
Bed allocation, 207
Beer, S., 157, 209
Behaviour, 7, 20, 21, 31, 38, 43, 76, 86, 88, 93, 95, 111, 118, 119, 124, 137, 145, 154, 156–164, 170, 173, 174, 181, 196, 199, 221, 235, 239, 249, 262, 273, 283, 299, 308, 310, 312